Contents

Preface *vii*

Chapter 1 **REAL NUMBERS, INTRODUCTION TO ANALYTIC GEOMETRY AND FUNCTIONS** *1*

1.1 Sets, Real Numbers, and Inequalities *1*
1.2 Absolute Value *6*
1.3 The Number Plane and Graphs of Equations *9*
1.4 Distance Formula and Midpoint Formula *12*
1.5 Equations of a Line *15*
1.6 The Circle *19*
1.7 Functions and Their Graphs *23*
1.8 Function Notation, Operations on Functions, and Types of Functions *26*
Review Exercises *32*

Chapter 2 **LIMITS AND CONTINUITY** *36*

2.1 The Limit of a Function *36*
2.2 Theorems on Limits of Functions *44*
2.3 One-Sided Limits *48*
2.4 Infinite Limits *50*
2.5 Continuity of a Function at a Number *54*
2.6 Theorems on Continuity *57*
Review Exercises *61*

Chapter 3 **THE DERIVATIVE** *66*

3.1 The Tangent Line *66*
3.2 Instantaneous Velocity in Rectilinear Motion *70*
3.3 The Derivative of a Function *74*

3.4 Differentiability and Continuity 77
3.5 Some Theorems on Differentiation of Algebraic Functions 81
3.6 The Derivative of a Composite Function 86
3.7 The Derivative of the Power Function for Rational Exponents 89
3.8 Implicit Differentiation 92
3.9 The Derivative as a Rate of Change 97
3.10 Related Rates 100
3.11 Derivatives of Higher Order 103
Review Exercises 107

Chapter 4 TOPICS ON LIMITS, CONTINUITY, AND THE DERIVATIVE 114
4.1 Limits at Infinity 114
4.2 Horizontal and Vertical Asymptotes 119
4.3 Additional Theorems on Limits of Functions 124
4.4 Continuity on an Interval 126
4.5 Maximum and Minimum Values of a Function 131
4.6 Applications Involving an Absolute Extremum on a Closed Interval 135
4.7 Rolle's Theorem and the Mean-Value Theorem 138
Review Exercises 142

Chapter 5 ADDITIONAL APPLICATIONS OF THE DERIVATIVE 146
5.1 Increasing and Decreasing Functions and the First Derivative Test 146
5.2 The Second Derivative Test for Relative Extrema 152
5.3 Additional Problems Involving Absolute Extrema 154
5.4 Concavity and Points of Inflection 159
5.5 Applications to Drawing a Sketch of the Graph of a Function 163
5.6 An Application of the Derivative in Economics 167
Review Exercises 171

Chapter 6 THE DIFFERENTIAL AND ANTIDIFFERENTIATION 178
6.1 The Differential 178
6.2 Differential Formulas 182
6.3 The Inverse of Differentiation 185
6.4 Differential Equations with Variables Separable 189
6.5 Antidifferentiation and Rectilinear Motion 194
6.6 Applications of Antidifferentiation in Economics 198
Review Exercises 201

Chapter 7 THE DEFINITE INTEGRAL 206
7.1 The Sigma Notation 206
7.2 Area 210
7.3 The Definite Integral 216
7.4 Properties of the Definite Integral 225
7.5 The Mean-Value Theorem for Integrals 228
7.6 The Fundamental Theorem of the Calculus 230
Review Exercises 235

Chapter 8 APPLICATIONS OF THE DEFINITE INTEGRAL 242
8.1 Area of a Region in a Plane 242
8.2 Volume of a Solid of Revolution: Circular-Disk and Circular-Ring Methods 250
8.3 Volume of a Solid of Revolution: Cylindrical-Shell Method 256
8.4 Volume of a Solid Having Known Parallel Plane Sections 260
8.5 Work 262
8.6 Liquid Pressure 266
8.7 Center of Mass of a Rod 270
8.8 Center of Mass of a Plane Region 272
8.9 Center of Mass of a Solid of Revolution 281
8.10 Length of Arc of a Plane Curve 286
Review Exercises 289

Appendix 297
Chapter Tests 297
Solutions for Chapter Tests 304

AN OUTLINE FOR THE STUDY OF CALCULUS Volume I

2. $2x^3y + 3xy^3 = 5$ find D_xy

AN OUTLINE FOR THE STUDY OF CALCULUS
Volume I

by

John H. Minnick
DE ANZA COLLEGE

Edited by

Louis Leithold

HARPER & ROW, PUBLISHERS
New York   Hagerstown   San Francisco   London

Sponsoring Editor: George J. Telecki
Designer: Rita Naughton
Production Supervisor: Will C. Jomarrón
Compositor: T. McNabney Composition Service
Printer: The Murray Printing Company
Binder: The Murray Printing Company
Art Studio: J & R Technical Services Inc.

AN OUTLINE FOR THE STUDY OF CALCULUS Volume I

Library of Congress Cataloging in Publication Data

Minnick, John Harper, Date -
   An outline for the study of calculus.

   "The definitions, theorems, and exercises are taken
from The calculus with analytic geometry, third edition,
by Louis Leithold."
   1. Calculus--Outlines, syllabi, etc. I. Title.
QA303.M685   515'.02'02   76-8190
ISBN 0-06-043946-7

Preface

Each section of the outline includes all of the most important definitions and theorems that are usually found in a course in calculus and analytic geometry. Often these are followed by a discussion that elaborates the concepts and presents a summary of problem solving techniques. A selection of exercises with complete and detailed solutions, including all graphs, is given for each section. At the end of each chapter there is a set of review exercises, also with complete solutions. In the Appendix there is a test for each chapter with a time limit indicated, followed by solutions for the test.

For those exercises that are more easily solved by using a computer, general flow charts that show how to apply the computer are given. Each flow chart is followed by a sample program, written in BASIC, that illustrates the solution of a particular exercise. The computer solutions are found in Chapters 7, 16, and 21.

The outline may be used for self study or to supplement any standard three semester course in calculus. Volume I contains Chapters 1-8, Volume II contains Chapters 9-16, and Volume III contains Chapters 17-21. The definitions, theorems, and exercises are taken from *The Calculus with Analytic Geometry, third edition,* by Louis Leithold. The chapter and section numbers and the exercise numbers agree with those used in Leithold. However, the chapter tests found in the Appendix are compiled from test questions that I have used with my own students at De Anza College.

J.H.M

1

Real numbers, introduction to analytic geometry, and functions

1.1 SETS, REAL NUMBERS, AND INEQUALITIES

Inequalities are used to define the limit of a function—a fundamental concept of calculus. The technique for finding the solution set of a first degree inequality is similar to that for solving a first degree equation, except that you must remember to reverse the sense of the inequality when you multiply or divide each member of the inequality by a negative number. However, to solve an inequality that is not first degree requires special techniques that are quite different from those used to solve equations. Frequently, the steps involve finding the union or the intersection of sets. Following is a list of those definitions and theorems that are most frequently used when solving inequalities.

1.1.4 Definition   Let A and B be two sets. The *union* of A and B, denoted by $A \cup B$ and read "A union B," is the set of all elements that are in A or in B or in both A and B.

$$A \cup B = \{x \mid x \in A \text{ or } x \in B\}$$

1.1.5 Definition   Let A and B be two sets. The *intersection* of A and B, denoted by $A \cap B$ and read "A intersection B," is the set of all elements that are in both A and B.

$$A \cap B = \{x \mid x \in A \text{ and } x \in B\}$$

1.1.22 Theorem   If $a < b$, then $a + c < b + c$, and $a - c < b - c$ if c is any real number.

1.1.24 Theorem   If $a < b$ and c is any positive number, then $ac < bc$.

1.1.25 Theorem   If $a < b$ and c is any negative number, then $ac > bc$.

1.1.33 Definition   The *open interval* from a to b denoted by (a, b), is defined by

$$(a, b) = \{x \mid a < x < b\}$$

1.1.34 Definition   The *closed interval* from a to b, denoted by $[a, b]$, is defined by

$$[a, b] = \{x \mid a \leqslant x \leqslant b\}$$

1.1.35 Definition   The *interval half-open on the left*, denoted by $(a, b]$, is defined by

$$(a, b] = \{x \mid a < x \leqslant b\}$$

1.1.36 Definition   The *interval half-open on the right*, denoted by $[a, b)$, is defined by

$$[a, b) = \{x \mid a \leqslant x < b\}$$

1.1.37 Definition
(i) $(a, +\infty) = \{x \mid x > a\}$
(ii) $(-\infty, b) = \{x \mid x < b\}$
(iii) $[a, +\infty) = \{x \mid x \geqslant a\}$
(iv) $(-\infty, b] = \{x \mid x \leqslant b\}$
(v) $(-\infty, +\infty) = R^1$

Exercises 1.1

10. If $A = \{0, 2, 4, 6, 8\}$, $B = \{1, 2, 4, 8\}$, $C = \{1, 3, 5, 7, 9,\}$, and $D = \{0, 3, 6, 9\}$, then represent $(A \cup B) \cap (C \cup D)$ by listing its members within braces.

SOLUTION

$$A \cup B = \{0, 1, 2, 4, 6, 8\}$$
$$C \cup D = \{0, 1, 3, 5, 6, 7, 9\}$$
$$(A \cup B) \cap (C \cup D) = \{0, 1, 6\}$$

In Exercises 14-30, find the solution set of the given inequality and illustrate the set on the real number line.

14. $3x - 5 < \dfrac{3}{4}x + \dfrac{1 - x}{3}$

SOLUTION: First, we "eliminate" the fractions.

$$12(3x - 5) < 12\left(\frac{3}{4}x + \frac{1 - x}{3}\right)$$

$$36x - 60 < 3 \cdot 3x + 4(1 - x)$$
$$36x - 60 < 9x + 4 - 4x$$
$$31x < 64$$
$$x < \tfrac{64}{31}$$

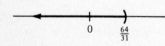

Figure 1.1.14

Hence, the solution set is $(-\infty, \tfrac{64}{31})$, as illustrated in Fig. 1.1.14.

16. $2 \leqslant 5 - 3x < 11$

SOLUTION: We reduce the middle expression to "x" by first adding -5 to each of the three expressions.

$$-3 \leqslant -3x < 6$$

Next, we divide each expression by -3 and reverse the sense of the inequality.

$$1 \geqslant x > -2$$
$$-2 < x \leqslant 1$$

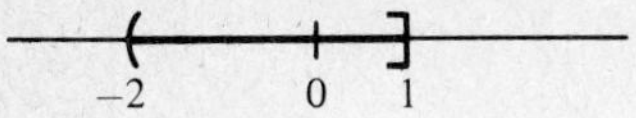

Thus, the solution set is $(-2, 1]$, as illustrated in Fig. 1.1.16.

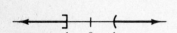

Figure 1.1.16

20. $\dfrac{2}{1-x} \leqslant 1$

SOLUTION: Since the multiplier needed to eliminate the fraction is $1-x$, which may be either positive or negative, we must consider two cases.

Case 1: $1 - x > 0$, or, equivalently, $x < 1$
  Multiplying both sides of the given inequality by $1 - x$, we get

$$2 \leqslant 1 - x$$
$$x \leqslant -1$$

Thus, the solution set for Case 1 is $\{x \mid x < 1 \text{ and } x \leqslant -1\}$, or, equivalently, $\{x \mid x < 1\} \cap \{x \mid x \leqslant -1\} = \{x \mid x \leqslant -1\} = (-\infty, -1]$.

Case 2: $1 - x < 0$, or, equivalently, $x > 1$.
  Multiplying both sides of the given inequality by $1 - x$ and reversing the sense of the inequality, we have

$$2 \geqslant 1 - x$$
$$x \geqslant -1$$

Thus, the solution set for Case 2 is $\{x \mid x > 1\} \cap \{x \mid x \geqslant -1\} = \{x \mid x > 1\} = (1, +\infty)$.
  Finally, the solution set for the given inequality is the union of the solution sets for Case 1 and Case 2, namely $(-\infty, -1] \cup (1, +\infty)$, as shown in Fig. 1.1.20.

Figure 1.1.20

ALTERNATE SOLUTION: We do not eliminate the fraction, but rather we write the given inequality in zero form; that is, with zero on one side, and then simplify.

$$\frac{2}{1-x} - 1 \leqslant 0$$

$$\frac{2 - (1-x)}{1-x} \leqslant 0$$

$$\frac{x+1}{1-x} \leqslant 0$$

$$\frac{x+1}{x-1} \geqslant 0 \tag{1}$$

Note that we reverse the sense of the inequality on the last step, because the multiplier is -1. Next, we consider the factors $x + 1$ and $x - 1$ separately. The factor $x + 1$ has a positive value if $x + 1 > 0$, or, equivalently, if $x > -1$, and $x + 1$ has a negative value if $x < -1$. Similarly, $x - 1$ is positive if $x > 1$ and negative if $x < 1$. Table 20 summarizes this discussion about the factors $x + 1$ and $x - 1$ and also indicates for each interval whether the fraction $(x + 1)/(x - 1)$ is positive or negative. The signs for this fraction are found by using the fact that the quotient of two positive or of two negative numbers is positive, whereas the quotient of one positive and one negative number is negative. Since strict inequality (1) is satisfied whenever the fraction has a positive value, and equation (1) is satisfied if $x = -1$, the solution set for (1) is $\{x \mid x \leqslant -1 \text{ or } x > 1\} = (-\infty, -1] \cup (1, +\infty)$.

Table 20

	$x < -1$	$x = -1$	$-1 < x < 1$	$x = 1$	$x > 1$
$x + 1$	−	0	+	+	+
$x - 1$	−	−	−	0	+
$\dfrac{x + 1}{x - 1}$	+	0	−	does not exist	+

24. $x^2 - 3x + 2 > 0$

SOLUTION: First we factor.

$$(x - 2)(x - 1) > 0 \qquad (2)$$

As in the alternate solution for Exercise 20, we consider the factors separately. Table 24 indicates that $x - 2$ has a positive value if $x > 2$ and a negative value if $x < 2$. Similarly, $x - 1$ is positive if $x > 1$ and negative if $x < 1$. The product $(x - 2)(x - 1)$ is positive whenever both factors are positive or both factors are negative. Thus, the solution set for (2) is $\{x \mid x > 2 \text{ or } x < 1\} = (-\infty, 1) \cup (2, +\infty)$, as illustrated in Fig. 1.1.24.

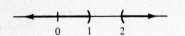

Figure 1.1.24

Table 24

	$x < 1$	$x = 1$	$1 < x < 2$	$x = 2$	$x > 2$
$x - 2$	−	−	−	0	+
$x - 1$	−	0	+	+	+
$(x - 2)(x - 1)$	+	0	−	0	+

28. $2x^2 - 6x + 3 < 0$

SOLUTION: First, we divide by 2.

$$x^2 - 3x + \frac{3}{2} < 0$$

To "complete the square" we add and subtract $(-3/2)^2$.

$$x^2 - 3x + \left(\frac{-3}{2}\right)^2 - \left(\frac{-3}{2}\right)^2 + \frac{3}{2} < 0$$

$$\left(x - \frac{3}{2}\right)^2 - \frac{3}{4} < 0$$

Now we factor this "difference of squares."

$$\left[\left(x - \frac{3}{2}\right) + \frac{\sqrt{3}}{2}\right]\left[\left(x - \frac{3}{2}\right) - \frac{\sqrt{3}}{2}\right] < 0$$

$$\left(x - \frac{3 - \sqrt{3}}{2}\right)\left(x - \frac{3 + \sqrt{3}}{2}\right) < 0 \qquad (3)$$

We consider the factors separately, as in the alternate solution for Exercise 20. As Table 28 indicates, the factor $x - (3 - \sqrt{3})/2$ "changes sign" when $x = (3 - \sqrt{3})/2$, whereas the factor $x - (3 + \sqrt{3})/2$ changes sign when $x = (3 + \sqrt{3})/2$. Since inequality (3) is satisfied whenever the product of the two factors is negative, the solution set is

$$\left\{x \mid \frac{3-\sqrt{3}}{2} < x < \frac{3+\sqrt{3}}{2}\right\} = \left(\frac{3-\sqrt{3}}{2}, \frac{3+\sqrt{3}}{2}\right)$$

as illustrated in Fig. 1.1.28.

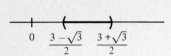

Figure 1.1.28

Table 28

	$-\infty$	$\dfrac{3-\sqrt{3}}{2}$	$\dfrac{3+\sqrt{3}}{2}$	$+\infty$
$x - \dfrac{3-\sqrt{3}}{2}$		$-$	$+$	$+$
$x - \dfrac{3+\sqrt{3}}{2}$		$-$	$-$	$+$
$\left(x - \dfrac{3-\sqrt{3}}{2}\right)\left(x - \dfrac{3+\sqrt{3}}{2}\right)$		$+$	$-$	$+$

We note that the end points in the solution set are the solutions of the equation $2x^2 - 6x + 3 = 0$, which is obtained from the given inequality. Thus, we may find these end points by solving the equation by the quadratic formula.

$$x = \frac{-b \pm \sqrt{b^2 - 4ac}}{2a} = \frac{-(-6) \pm \sqrt{(-6)^2 - 4 \cdot 2 \cdot 3}}{2 \cdot 2} = \frac{3 \pm \sqrt{3}}{2}$$

30. $\dfrac{x+1}{2-x} < \dfrac{x}{3+x}$

SOLUTION: First, we write the inequality in zero form and then simplify.

$$\frac{x+1}{2-x} - \frac{x}{3+x} < 0$$

$$\frac{2x^2 + 2x + 3}{(2-x)(3+x)} < 0$$

Because the coefficient of x in the factor $2 - x$ is negative, we multiply by -1 and reverse the sense of the inequality.

$$\frac{2x^2 + 2x + 3}{(x-2)(x+3)} > 0 \tag{4}$$

We consider each factor separately. First, we complete the square on the expression that appears in the numerator.

$$2x^2 + 2x + 3 = 2\left(x^2 + x + \frac{1}{4}\right) - 2 \cdot \frac{1}{4} + 3$$

$$= 2\left(x + \frac{1}{2}\right)^2 + \frac{5}{2}$$

Because the square of any real number is nonnegative, we see that $2x^2 + 2x + 3$ is always positive. However, $x - 2$ changes sign at $x = 2$, and $x + 3$ changes sign at $x = -3$. Table 30 shows that the solution set for inequality (4) is $(-\infty, -3) \cup (2, +\infty)$, as illustrated in Fig. 1.1.30.

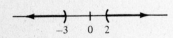

Figure 1.1.30

Table 30

	$x < -3$	$-3 < x < 2$	$x > 2$
$2x^2 + 2x + 3$	$+$	$+$	$+$
$x - 2$	$-$	$-$	$+$
$x + 3$	$-$	$+$	$+$
$\dfrac{2x^2 + 2x + 3}{(x - 2)(x + 3)}$	$+$	$-$	$+$

40. Prove: If $x < y$, then $x < \frac{1}{2}(x + y) < y$

SOLUTION: We prove the inequalities in the conclusion one at a time.

STATEMENT	REASON
$x < y$	Hypothesis
$2x < x + y$	Theorem 1.1.22
$x < \frac{1}{2}(x + y)$	Theorem 1.1.24

Similarly, if $x < y$, then $x + y < 2y$, and $\frac{1}{2}(x + y) < y$. Thus, if $x < y$, then $x < \frac{1}{2}(x + y) < y$.

1.2 ABSOLUTE VALUE

The most important definitions and theorems in this section are the following ones.

1.2.1 Definition

$$|x| = x \quad \text{if} \quad x \geqslant 0$$
$$|x| = -x \quad \text{if} \quad x < 0$$

1.2.2 Theorem If $a > 0$, then $|x| < a$ if and only if $-a < x < a$.

1.2.4 Theorem If $a > 0$, then $|x| > a$ if and only if $x > a$ or $x < -a$.

1.2.6 Theorem $|ab| = |a| \cdot |b|$

1.2.7 Theorem If $b \neq 0$, then $\left|\dfrac{a}{b}\right| = \dfrac{|a|}{|b|}$

1.2.8 Theorem (Triangle Inequality) $|a + b| \leqslant |a| + |b|$

In addition to the above, we often use the following theorem, which is a corollary to exercises 36 and 37 in Exercises 1.1.

Theorem If $a \geqslant 0$, then

    (i) $a < b$ if and only if $a^2 < b^2$
    (ii) $a < b$ if and only if $\sqrt{a} < \sqrt{b}$

Exercises 1.2

---

8. Find the solution set.

$$2x + 3 = |4x + 5| \tag{1}$$

SOLUTION: We consider two cases.

Case 1: $4x + 5 \geqslant 0$

By Definition 1.2.1, $|4x + 5| = 4x + 5$. Thus, Eq. (1) is equivalent to

$$2x + 3 = 4x + 5$$
$$x = -1$$

Since -1 satisfies the Case 1 assumption, $4x + 5 \geqslant 0$, we conclude that -1 also satisfies Eq. (1).

Case 2: $4x + 5 < 0$

By Definition 1.2.1, $|4x + 5| = -(4x + 5)$. Thus, Eq. (1) is equivalent to

$$2x + 3 = -(4x + 5)$$

$$x = -\frac{4}{3}$$

Since $-4/3$ satisfies the Case 2 assumption, $4x + 5 < 0$, we conclude that $-4/3$ also satisfies Eq. (1).

Hence, the solution set for Eq. (1) is

$$\left\{-1\right\} \cup \left\{-\frac{4}{3}\right\} = \left\{-1, -\frac{4}{3}\right\}$$

14. Find the set of all replacements of x for which $\sqrt{x^2 + 2x - 1}$ is real.

SOLUTION: Since the square root of a negative number is not real, x must satisfy the inequality

$$x^2 + 2x - 1 \geqslant 0$$

We complete the square and take the principal square root of each member.

$$x^2 + 2x + 1 \geqslant 2$$
$$(x + 1)^2 \geqslant 2$$
$$|x + 1| \geqslant \sqrt{2} \qquad\qquad (2)$$

Note that we must use absolute value bars to represent the principal square root of $(x + 1)^2$. Now, by Theorem 1.2.4, inequality (2) is satisfied if

$$x + 1 \geqslant \sqrt{2} \quad \text{or} \quad x + 1 \leqslant -\sqrt{2}$$

That is, if

$$x \geqslant -1 + \sqrt{2} \quad \text{or} \quad x \leqslant -1 - \sqrt{2}$$

Hence, the set we seek is $[-1 + \sqrt{2}, +\infty) \cup (-\infty, -1 - \sqrt{2}\,]$.

In Exercises 15-28 find the solution set of the given inequality and illustrate the solution set on the real number line.

18. $|6 - 2x| \geqslant 7$

SOLUTION: By Theorem 1.2.4, the given inequality is satisfied if either $6 - 2x \geqslant 7$ or $6 - 2x \leqslant -7$. We solve each inequality.

$$6 - 2x \geqslant 7 \qquad\qquad\qquad 6 - 2x \leqslant -7$$
$$-2x \geqslant 1 \qquad\qquad\qquad\quad -2x \leqslant -13$$

$$x \leqslant -\frac{1}{2} \qquad\qquad\qquad\quad x \geqslant \frac{13}{2}$$

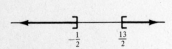

Figure 1.2.18

The solution set is $(-\infty, -\frac{1}{2}] \cup [\frac{13}{2}, +\infty)$ and is illustrated in Fig. 1.2.18.

20. $|3 + 2x| < |4 - x|$

SOLUTION: First, we square each member.

$$|3 + 2x|^2 < |4 - x|^2$$
$$9 + 12x + 4x^2 < 16 - 8x + x^2$$
$$3x^2 + 20x - 7 < 0$$
$$(x + 7)(3x - 1) < 0 \qquad (3)$$

As Table 20 indicates, the factor $x + 7$ changes sign at $x = -7$, whereas $3x - 1$ changes sign at $x = \frac{1}{3}$. Since inequality (3) is satisfied whenever $(x + 7)(3x - 1)$ is negative, the table indicates that the solution set is $(-7, \frac{1}{3})$, and is illustrated in Fig. 1.2.20.

Figure 1.2.20

Table 20

	$-\infty$	-7	$\frac{1}{3}$	$+\infty$
$x + 7$		$-$	$+$	$+$
$3x - 1$		$-$	$-$	$+$
$(x + 7)(3x - 1)$		$+$	$-$	$+$

24. $\left| \dfrac{6 - 5x}{3 + x} \right| \leqslant \dfrac{1}{2}$

SOLUTION: If $3 + x \neq 0$, we may multiply on both sides of the inequality by the positive number $2|3 + x|$.

$$2|3 + x| \cdot \left| \frac{6 - 5x}{3 + x} \right| \leqslant 2|3 + x| \cdot \frac{1}{2}$$

$$2|6 - 5x| \leqslant |3 + x|$$
$$[2|6 - 5x|]^2 \leqslant |3 + x|^2$$
$$4(36 - 60x + 25x^2) \leqslant 9 + 6x + x^2$$
$$100x^2 - 240x + 144 \leqslant x^2 + 6x + 9$$
$$99x^2 - 246x + 135 \leqslant 0$$
$$33x^2 - 82x + 45 \leqslant 0$$
$$(11x - 9)(3x - 5) \leqslant 0 \qquad (4)$$

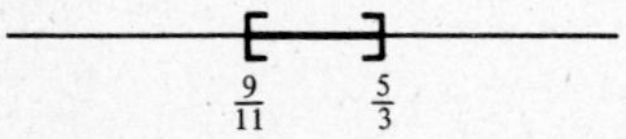

As Table 24 indicates, the factor $11x - 9$ changes sign at $x = \frac{9}{11}$, and the factor $3x - 5$ changes sign at $x = \frac{5}{3}$. Since the strict inequality (4) is satisfied whenever $(11x - 9)(3x - 5)$ is negative, and the equation (4) is satisfied either when $x = \frac{9}{11}$ or when $x = \frac{5}{3}$, by Table 24 we see that the solution set for (4) is $[\frac{9}{11}, \frac{5}{3}]$, and this set is illustrated in Fig. 1.2.24. Note that our assumption $3 + x \neq 0$ is satisfied by every element in $[\frac{9}{11}, \frac{5}{3}]$.

Figure 1.2.24

Table 24

	$x < \frac{9}{11}$	$x = \frac{9}{11}$	$\frac{9}{11} < x < \frac{5}{3}$	$x = \frac{5}{3}$	$x > \frac{5}{3}$
$11x - 9$	$-$	0	$+$	$+$	$+$
$3x - 5$	$-$	$-$	$-$	0	$+$
$(11x - 9)(3x - 5)$	$+$	0	$-$	0	$+$

32. Solve for x and use absolute value bars to write the answer.

$$\frac{x + 5}{x + 3} < \frac{x + 1}{x - 1}$$

SOLUTION:

$$\frac{x+5}{x+3} - \frac{x+1}{x-1} < 0$$

$$\frac{(x+5)(x-1) - (x+1)(x+3)}{(x+3)(x-1)} < 0$$

$$\frac{-8}{(x+3)(x-1)} < 0$$

$$\frac{8}{(x+3)(x-1)} > 0 \tag{5}$$

Since inequality (5) is satisfied if and only if $(x+3)(x-1)$ is positive, we see that (5) is equivalent to

$$(x+3)(x-1) > 0$$
$$x^2 + 2x - 3 > 0$$
$$x^2 + 2x + 1 > 4$$
$$(x+1)^2 > 4$$
$$\sqrt{(x+1)^2} > \sqrt{4}$$
$$|x+1| > 2$$

34. Prove: $|a| - |b| \leqslant |a - b|$

SOLUTION: We have the following equation

$$|a| = |(a-b) + b| \tag{6}$$

From Theorem 1.2.8 we have

$$|(a-b) + b| \leqslant |a-b| + |b| \tag{7}$$

Substituting from Eq. (6) into inequality (7), we have

$$|a| \leqslant |a-b| + |b|$$
$$|a| - |b| \leqslant |a-b|$$

1.3 THE NUMBER PLANE AND GRAPHS OF EQUATIONS

To draw a sketch of the graph of an equation by plotting points is slow and inaccurate unless a computer is used to make the calculations and many points are found. As we proceed through the course, we will discover theorems that will enable us to draw a sketch of a graph by plotting only a few points. The first of these theorems is the test for symmetry.

1.3.6 Theorem

The graph of an equation in x and y is

(i) symmetric with respect to the x-axis if and only if an equivalent equation is obtained when y is replaced by $-y$ in the equation.

(ii) symmetric with respect to the y-axis if and only if an equivalent equation is obtained when x is replaced by $-x$ in the equation.

(iii) symmetric with respect to the origin if and only if an equivalent equation is obtained when x is replaced by $-x$ and y is replaced by $-y$ in the equation.

It is important to remember that if $x > 0$, then $\sqrt{x}$ represents only the *positive* square root of x. Thus, $\sqrt{25} \neq \pm 5$. Rather, $\sqrt{25} = 5$ and $-\sqrt{25} = -5$.

Exercises 1.3

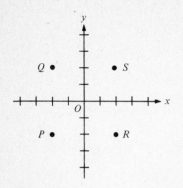

Figure 1.3.4

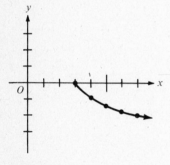

Figure 1.3.10

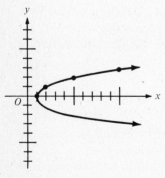

Figure 1.3.14

4. Let $P = (-2, -2)$. Plot P and points $Q, R,$ and S so that P and Q are symmetric with respect to the x-axis, P and R are symmetric with respect to the y-axis, and P and S are symmetric with respect to the origin.

SOLUTION: By Theorem 1.3.6, to find Q we replace y by $-y$ in P. Hence, $Q = (-2, 2)$. Similarly, to find R we replace x by $-x$ and obtain $R = (2, -2)$, and to find S we replace x by $-x$ and y by $-y$ to obtain $S = (2, 2)$. The points are plotted in Fig. 1.3.4.

In Exercises 7-28 draw a sketch of the graph of the equation.

10. $y = -\sqrt{x - 3}$

SOLUTION: Since the square root of a negative number is not real, we must choose as replacements for x only those numbers that satisfy $x - 3 \geqslant 0$, or, equivalently, $x \geqslant 3$. For each such replacement we use the given equation to calculate y. Table 10 gives the results of such calculations. We use a decimal approximation for y whenever y is irrational. A sketch of the graph is obtained by plotting the points from Table 10 and drawing a smooth curve that contains these points, as shown in Fig. 1.3.10.

Table 10

x	3	4	5	6	7
y	0	-1	-1.4	-1.7	-2

14. $x = y^2 + 1$

SOLUTION: Replacing y by $-y$ in the equation gives

$$x = (-y)^2 + 1$$

which is equivalent to

$$x = y^2 + 1$$

By Theorem 1.3.6, the graph is symmetric with respect to the x-axis. We plot several points for which $y \geqslant 0$, draw a smooth curve that contains them, and use symmetry to complete the sketch. For Table 14 we choose y first and then calculate x from the given equation. The sketch of the graph is shown in Figure 1.3.14.

Table 14

x	1	2	5	10
y	0	1	2	3

18. $y = -|x| + 2$

SOLUTION: Replacing x by $-x$ in the equation gives

$$y = -|-x| + 2 \qquad \text{(1)}$$

and since $|-x| = |x|$, Eq. (1) is equivalent to

$$y = -|x| + 2$$

By Theorem 1.3.6, the graph is symmetric with respect to the y-axis. We plot several points for which $x \geqslant 0$, draw a smooth curve that contains them, and use symmetry

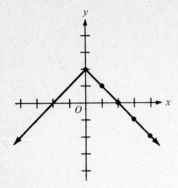

Figure 1.3.18

to complete the sketch. Because $|x| = x$ if $x \geq 0$, the given equation is equivalent to

$$y = -x + 2 \quad \text{if } x \geq 0 \tag{2}$$

We use Eq. (2) to calculate y for several replacements of x and show the result in Table 18. The graph is shown in Fig. 1.3.18.

Table 18

x	0	1	2	3	4
y	2	1	0	−1	−2

20. $4x^2 - 9y^2 = 36$

SOLUTION: All three tests for symmetry are satisfied. We plot points for which $x \geq 0$, $y \geq 0$. Then we use symmetry to complete the sketch. First, we solve the equation for y, taking only the principal square root.

$$y^2 = \frac{4x^2 - 36}{9}$$

$$y^2 = \frac{4}{9}(x^2 - 9)$$

$$y = \frac{2}{3}\sqrt{x^2 - 9} \tag{3}$$

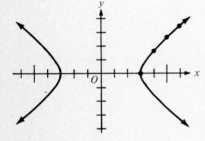

Figure 1.3.20

Because $x^2 - 9 \geq 0$ and $x \geq 0$, then $x \geq 3$. The values of y in Table 20 are computed from Eq. (3). We use decimal approximations whenever y is irrational. The sketch of the graph is shown in Fig. 1.3.20.

Table 20

x	3	4	5	6
y	0	1.8	2.7	3.5

22. $y^2 = 4x^3$

SOLUTION: The graph is symmetric with respect to the x-axis. We plot points for which $y \geq 0$ and use symmetry to complete the sketch. The values of y in Table 22 are calculated from the equation

$$y = 2\sqrt{x^3}$$

where we choose $x \geq 0$ so that $x^3 \geq 0$ and y is real. The sketch of the graph is shown in Fig. 1.3.22.

Table 22

x	0	1	2	3
y	0	2	5.7	10.4

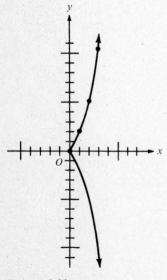

Figure 1.3.22

24. $3x^2 - 13xy - 10y^2 = 0$

SOLUTION: The given equation is equivalent to

$$(3x + 2y)(x - 5y) = 0 \tag{4}$$

Because $ab = 0$ if and only if $a = 0$ or $b = 0$, we have

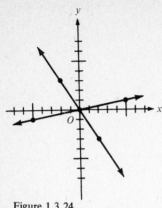

Figure 1.3.24

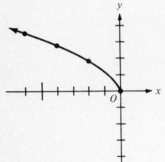

Figure 1.3.30(a)

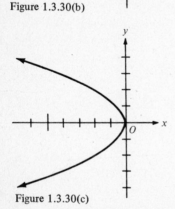

Figure 1.3.30(b)

$$3x + 2y = 0 \tag{5}$$

or

$$x - 5y = 0 \tag{6}$$

Thus, the graph of Eq. (4) is the union of the graphs of Eqs. (5) and (6). From Eq. (5) we have

$$y = -\frac{3x}{2}$$

and Table 24a. For Eq. (6) we have

$$y = \frac{x}{5}$$

and Table 24b. The sketch of the graph of the given equation is shown in Fig. 1.3.24.

Table 24a

x	-2	0	2
y	3	0	-3

Table 24b

x	-5	0	5
y	-1	0	1

30. Draw a sketch of the graph of each of the following equations:

(a) $y = \sqrt{-2x}$ (b) $y = -\sqrt{-2x}$ (c) $y^2 = -2x$

SOLUTION:

(a) Because $-2x \geqslant 0$, $x \leqslant 0$. The values of y in Table 30a are calculated from the equation $y = \sqrt{-2x}$. A sketch of the graph is shown in Fig. 1.3.30a.

Table 30a

x	0	-2	-4	-6
y	0	2	2.8	3.5

(b) The values of y in Table 30b are calculated from the equation $y = -\sqrt{-2x}$. A sketch of the graph is shown in Fig. 1.3.30b.

Table 30b

x	0	-2	-4	-6
y	0	-2	-2.8	-3.5

(c) If

$$y^2 = -2x$$

then $y = \pm\sqrt{-2x}$. Hence, the graph of $y^2 = -2x$ is the union of the graphs in parts (a) and (b). See Fig. 1.3.30c.

Figure 1.3.30(c)

1.4 DISTANCE FORMULA AND MIDPOINT FORMULA

The distance formula is given by the following theorem.

1.4.1 Theorem The undirected distance between the two points $P_1(x_1, y_1)$ and $P_2(x_2, y_2)$ is given by

$$|\overline{P_1P_2}| = \sqrt{(x_2 - x_1)^2 + (y_2 - y_1)^2}$$

Theorem The midpoint of the segment with end points $P_1(x_1, y_1)$ and $P_2(x_2, y_2)$ is $M(\bar{x}, \bar{y})$ with

$$\bar{x} = \frac{x_1 + x_2}{2} \qquad \bar{y} = \frac{y_1 + y_2}{2}$$

Exercises 1.4

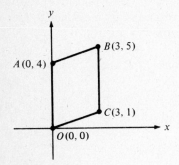

Figure 1.4.6

6. Find the midpoints of the diagonals of the quadrilateral whose vertices are $(0, 0), (0, 4), (3, 5),$ and $(3, 1)$.

SOLUTION: Let $O = (0, 0)$, $A = (0, 4)$, $B = (3, 5)$, and $C = (3, 1)$ as illustrated in Fig. 1.4.6. Let $M_1(\bar{x}_1, \bar{y}_1)$ be the midpoint of OB. Then

$$\bar{x}_1 = \frac{0 + 3}{2} = \frac{3}{2} \qquad \bar{y}_1 = \frac{0 + 5}{2} = \frac{5}{2}$$

Hence, $M_1 = (\frac{3}{2}, \frac{5}{2})$.
Let $M_2(\bar{x}_2, \bar{y}_2)$ be the midpoint of AC. Then

$$\bar{x}_2 = \frac{0 + 3}{2} = \frac{3}{2} \qquad \bar{y}_2 = \frac{4 + 1}{2} = \frac{5}{2}$$

Hence, $M_2 = (\frac{3}{2}, \frac{5}{2})$.

12. Determine whether or not the points $(14, 7), (2, 2)$ and $(-4, -1)$ lie on a line.

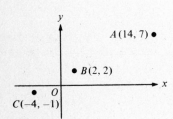

Figure 1.4.12

SOLUTION: Let $A = (14, 7)$, $B = (2, 2)$ and $C = (-4, -1)$. (See Fig. 1.4.12.) The points $A, B,$ and C lie on a line if and only if

$$|\overline{CB}| + |\overline{BA}| = |\overline{CA}|$$

$$|\overline{CB}| = \sqrt{(2 + 4)^2 + (2 + 1)^2} = \sqrt{45} = 3\sqrt{5}$$
$$|\overline{BA}| = \sqrt{(14 - 2)^2 + (7 - 2)^2} = \sqrt{169} = 13$$
$$|\overline{CA}| = \sqrt{(14 + 4)^2 + (7 + 1)^2} = \sqrt{388} = 2\sqrt{97}$$

Because $3\sqrt{5} + 13 \neq 2\sqrt{97}$, we conclude that the points are not collinear.

16. Given the two points $A(-3, 4)$ and $B(2, 5)$, find the coordinates of P on the line through A and B such that P is: (a) twice as far from A as from B, and (b) twice as far from B as from A. (Point P is not between A and B.)

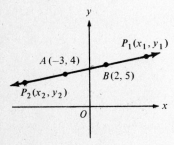

Figure 1.4.16

SOLUTION:
(a) Refer to Fig. 1.4.16. Let $P_1(x_1, y_1) = P$. Since $A, B,$ and P_1 lie on a line,

$$|\overline{AB}| + |\overline{BP_1}| = |\overline{AP_1}| \tag{1}$$

Furthermore,

$$|\overline{AP_1}| = 2 \cdot |\overline{BP_1}| \tag{2}$$

Substituting from Eq. (2) into Eq. (1) we have

$$|\overline{AB}| + |\overline{BP_1}| = 2 \cdot |\overline{BP_1}|$$
$$|\overline{AB}| = |\overline{BP_1}|$$

Hence, B is the midpoint of AP_1. We use the midpoint formula.

$$2 = \frac{x_1 + (-3)}{2} \qquad 5 = \frac{y_1 + 4}{2}$$

$$x_1 = 7 \qquad\qquad y_1 = 6$$

Thus, $P_1 = (7, 6)$.

(b) Let $P_2(x_2, y_2) = P$. By a method similar to that used in part (a) we find that A is the midpoint of BP_2. Thus,

$$-3 = \frac{x_2 + 2}{2} \qquad 4 = \frac{y_2 + 5}{2}$$

$$x_2 = -8 \qquad y_2 = 3$$

Therefore, $P_2 = (-8, 3)$.

20. Find the coordinates of P, which is on the line segment P_1P_2 and which is three times as far from P_1 as it is from P_2, if $P_1 = (1, 3)$ and $P_2 = (6, 2)$.

SOLUTION: Refer to Fig. 1.4.20. Let $P = (x, y)$. Since $P_1, P,$ and P_2 are collinear.

$$|\overline{P_1P}| + |\overline{PP_2}| = |\overline{P_1P_2}| \tag{3}$$

We are given that

$$|\overline{PP_2}| = \frac{1}{3} \cdot |\overline{P_1P}| \tag{4}$$

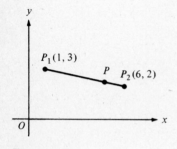

$P_1(1, 3)$

P $P_2(6, 2)$

Figure 1.4.20

Substituting from Eq. (4) into Eq. (3) we have

$$|\overline{P_1P}| + \frac{1}{3}|\overline{P_1P}| = |\overline{P_1P_2}|$$

$$\frac{4}{3} \cdot |\overline{P_1P}| = |\overline{P_1P_2}|$$

$$\frac{|\overline{P_1P}|}{|\overline{P_1P_2}|} = \frac{3}{4}$$

We use the formulas of Exercise 18 with $r_1 = 3$ and $r_2 = 4$.

$$x = \frac{(r_2 - r_1)x_1 + r_1x_2}{r_2} = \frac{(4 - 3) \cdot 1 + 3 \cdot 6}{4} = \frac{19}{4}$$

$$y = \frac{(r_2 - r_1)y_1 + r_1y_2}{r_2} = \frac{(4 - 3) \cdot 3 + 3 \cdot 2}{4} = \frac{9}{4}$$

Thus, $P = (\frac{19}{4}, \frac{9}{4})$.

24. Find an equation whose graph is the circle that is the set of all points that are at a distance of 4 units from the point $(1, 3)$.

SOLUTION: Let C be the point $(1, 3)$ and let $P(x, y)$ be any point on the circle. Then $|\overline{CP}| = 4$, and by the distance formula,

$$\sqrt{(x - 1)^2 + (y - 3)^2} = 4$$
$$(x - 1)^2 + (y - 3)^2 = 16$$
$$x^2 + y^2 - 2x - 6y - 6 = 0$$

28. Prove analytically that the midpoint of the hypoteneuse of any right triangle is equidistant from each of the three vertices.

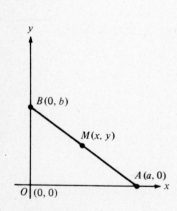

$B(0, b)$

$M(x, y)$

$A(a, 0)$

$O(0, 0)$

Figure 1.4.28

SOLUTION: Refer to Fig. 1.4.28. The hypothesis and conclusion are as follows.

HYPOTHESIS: Triangle AOB is a right triangle, and M is the midpoint of AB.

CONCLUSION: $|\overline{AM}| = |\overline{BM}| = |\overline{OM}|$

PROOF: Since $M(x, y)$ is the midpoint of AB with $A = (a, 0)$ and $B = (0, b)$, by the midpoint formula

$$x = \frac{a+0}{2} = \frac{a}{2} \qquad y = \frac{0+b}{2} = \frac{b}{2} \tag{5}$$

Then by the distance formula and Eqs. (5)

$$|\overline{AM}| = \sqrt{(x-a)^2 + (y-0)^2} = \sqrt{\left(\frac{a}{2} - a\right)^2 + \left(\frac{b}{2}\right)^2} = \frac{\sqrt{a^2+b^2}}{2}$$

$$|\overline{BM}| = \sqrt{(x-0)^2 + (y-b)^2} = \sqrt{\left(\frac{a}{2}\right)^2 + \left(\frac{b}{2} - b\right)^2} = \frac{\sqrt{a^2+b^2}}{2}$$

$$|\overline{OM}| = \sqrt{(x-0)^2 + (y-0)^2} = \sqrt{\left(\frac{a}{2}\right)^2 + \left(\frac{b}{2}\right)^2} = \frac{\sqrt{a^2+b^2}}{2}$$

Therefore,

$$|\overline{AM}| = |\overline{BM}| = |\overline{OM}|$$

1.5 EQUATIONS OF A LINE

The graph of every first degree equation in x and y is a line. To draw a sketch of a line, we need either two points on the line or one point and the slope of the line. Conversely, if we are given either two points on a line or one point and the slope of the line, we can find an equation of the line, and the equation will be of first degree in x and y.

1.5.1 Definition If $P_1(x_1, y_1)$ and $P_2(x_2, y_2)$ are two points on line ℓ, which is not parallel to the y-axis, then the *slope of* ℓ, denoted by m, is given by

$$m = \frac{y_2 - y_1}{x_2 - x_1}$$

1.5.2 Theorem If A and B are not both zero, the graph of $Ax + By + C = 0$ is a straight line.

1.5.3 Theorem If ℓ_1 and ℓ_2 are two distinct nonvertical lines having slopes m_1 and m_2, respectively, then ℓ_1 and ℓ_2 are parallel if and only if $m_1 = m_2$.

1.5.4 Theorem If neither line ℓ_1 nor line ℓ_2 is vertical, and if m_1 is the slope of ℓ_1 and m_2 is the slope of ℓ_2, then ℓ_1 and ℓ_2 are perpendicular if and only if $m_1 m_2 = -1$.

The following standard forms may be used to find an equation of a line. It is understood that (x_1, y_1) and (x_2, y_2) are two particular points on the line, m is the slope of the line, and a and b are, respectively, the x and y intercepts of the line.

STANDARD FORMS FOR EQUATION OF A LINE

Two-Point Form $\qquad y - y_1 = \dfrac{y_2 - y_1}{x_2 - x_1}(x - x_1)$

Point-Slope Form $\qquad y - y_1 = m(x - x_1)$

Slope-Intercept Form $\qquad y = mx + b$

Intercept Form $\qquad \dfrac{x}{a} + \dfrac{y}{b} = 1$

Exercises 1.5

4. Find the slope of the line through the points $(-2.1, 0.3)$ and $(2.3, 1.4)$.

SOLUTION:

$$m = \frac{y_2 - y_1}{x_2 - x_1} = \frac{1.4 - 0.3}{2.3 - (-2.1)} = \frac{1.1}{4.4} = \frac{1}{4}$$

10. Find an equation of the line through $(1, 4)$ and parallel to the line $2x - 5y + 7 = 0$.

SOLUTION: Let ℓ be the required line and let ℓ_1 be the line whose equation is

$$2x - 5y + 7 = 0 \qquad (1)$$

We shall find the slope of ℓ and then use the point-slope form to find an equation of ℓ. Because ℓ and ℓ_1 are parallel, it follows from Theorem 1.5.3 that the two lines have the same slope. To find the slope of ℓ_1 we solve Eq. (1) for y. We obtain

$$y = \frac{2}{5}x + \frac{7}{5} \qquad (2)$$

Comparison of Eq. (2) with the slope-intercept form shows that the slope of ℓ_1 is $\frac{2}{5}$. Thus, the slope of ℓ is $\frac{2}{5}$. Now we may obtain an equation of ℓ by using the point-slope form with $(x_1, y_1) = (1, 4)$ and $m = \frac{2}{5}$. We have

$$y - 4 = \frac{2}{5}(x - 1)$$
$$2x - 5y + 18 = 0$$

14. Find an equation of the line for which the slope is -2 and the x intercept is 4.

SOLUTION: We cannot use the slope-intercept form, because this form involves the y intercept, not the x intercept. However, since the x intercept is 4, then $(4, 0)$ is a point on the line. Thus, using the point-slope form, we have

$$y - 0 = -2(x - 4)$$
$$2x + y - 8 = 0$$

18. Three consecutive vertices of a parallelogram are $(-4, 1), (2, 3)$ and $(8, 9)$. Find the coordinates of the fourth vertex.

SOLUTION: Let $A(-4, 1), B(2, 3)$ and $C(8, 9)$ be the three given vertices, and let $D(x, y)$ be the fourth vertex. See Fig. 1.5.18. We shall find D by finding the simultaneous solution for equations of lines DC and AD. Since DC is parallel to AB, the slope of DC equals the slope of AB. Moreover,

$$\text{The slope of } AB = \frac{3 - 1}{2 - (-4)} = \frac{1}{3}$$

By the point-slope form with $m = \frac{1}{3}$ and $(x_1, y_1) = (8, 9)$, an equation for DC is

$$y - 9 = \frac{1}{3}(x - 8)$$

$$y = \frac{1}{3}x + \frac{19}{3} \qquad (3)$$

The slope of AD equals the slope of BC.

$$\text{The slope of } BC = \frac{9 - 3}{8 - 2} = 1$$

By the point-slope form with $m = 1$ and $(x_1, y_1) = (-4, 1)$, we obtain an equation for AD. It is

$$y - 1 = 1 \cdot (x + 4)$$
$$y = x + 5 \qquad (4)$$

Hence, from Eqs. (3) and (4) we have

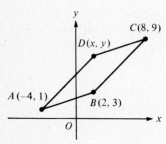

Figure 1.5.18

$$\frac{1}{3}x + \frac{19}{3} = x + 5$$

$$x = 2 \tag{5}$$

By substitution from Eq. (5) into Eq. (4), we have $y = 7$. Therefore, the fourth vertex is $(2, 7)$.

24. Find an equation of the line which has equal intercepts and which passes through the point $(8, -6)$.

SOLUTION: The intercept form of an equation of a line is

$$\frac{x}{a} + \frac{y}{b} = 1$$

Because we are given that the intercepts are equal, $a = b$. Thus, the required line has an equation of the form

$$\frac{x}{a} + \frac{y}{a} = 1 \tag{6}$$

Because the line contains $(8, -6)$, this pair must satisfy Eq. (6). Therefore,

$$\frac{8}{a} + \frac{-6}{a} = 1$$

$$a = 2$$

It follows from Eq. (6) that an equation for the required line is

$$\frac{x}{2} + \frac{y}{2} = 1$$

$$x + y - 2 = 0$$

26. Find equations of the perpendicular bisectors of the sides of the triangle $A(-1, -3)$, $B(5, -3)$ and $C(5, 5)$, and prove that they meet in a point.

SOLUTION: Refer to Fig. 1.5.26. Let ℓ_1 be the perpendicular bisector of AB. Since AB is parallel to the x-axis, then ℓ_1 is parallel to the y-axis, and an equation for ℓ_1 is of the form $x = x_1$. Since the midpoint of AB is $(2, -3)$, ℓ_1 contains $(2, -3)$. Thus, an equation of ℓ_1 is

$$x = 2 \tag{7}$$

Let ℓ_2 be the perpendicular bisector of BC. Since BC is parallel to the y-axis, ℓ_2 is parallel to the x-axis and an equation of ℓ_2 is of the form $y = y_1$. Since the midpoint of BC is $(5, 1)$, then ℓ_2 contains $(5, 1)$. An equation for ℓ_2 is

$$y = 1 \tag{8}$$

Let ℓ_3 be the perpendicular bisector of AC. If m_3 is the slope of AC, then

$$m_3 = \frac{5 - (-3)}{5 - (-1)} = \frac{4}{3}$$

By Theorem 1.5.4, the slope of ℓ_3 is $-\frac{3}{4}$. Since the midpoint of AC is $(2, 1)$, then ℓ_3 contains $(2, 1)$. Thus, the point-slope form of the equation for ℓ_3 is

$$y - 1 = -\frac{3}{4}(x - 2)$$

or

$$3x + 4y = 10 \tag{9}$$

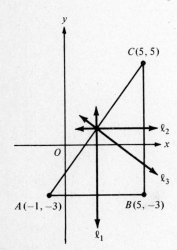

Figure 1.5.26

By Eqs. (7) and (8) we see that lines ℓ_1 and ℓ_2 intersect at point $(2, 1)$. Because $(2, 1)$ also satisfies Eq. (9), the point lies on line ℓ_3. Thus $\ell_1, \ell_2,$ and ℓ_3 meet at the point $(2, 1)$.

28. Let ℓ_1 be the line $A_1x + B_1y + C_1 = 0$, and let ℓ_2 be the line $A_2x + B_2y + C_2 = 0$. If ℓ_1 is not parallel to ℓ_2 and if k is any constant, the equation $A_1x + B_1y + C_1 + k(A_2x + B_2y + C_2) = 0$ represents an unlimited number of lines. Prove that each of these lines contains the point of intersection of ℓ_1 and ℓ_2.

PROOF: Let $P_0(x_0, y_0)$ be the point of intersection of ℓ_1 and ℓ_2. Since P_0 lies on lines ℓ_1 and $\ell_2, (x_0, y_0)$ must satisfy the equations for both ℓ_1 and ℓ_2. That is

$$A_1x_0 + B_1y_0 + C_1 = 0$$

and

$$A_2x_0 + B_2y_0 + C_2 = 0$$

If we add the first equation plus k times the second, we have

$$A_1x_0 + B_1y_0 + C_1 + k(A_2x_0 + B_2y_0 + C_2) = 0$$

Thus, the equation $A_1x + B_1y + C_1 + k(A_2x + B_2y + C_2) = 0$ is satisfied by the point $P_0(x_0, y_0)$, and hence any line represented by this equation must contain P_0.

32. Prove analytically that the line segments joining consecutive midpoints of the sides of any quadrilateral form a parallelogram.

SOLUTION: Refer to Fig. 1.5.32.

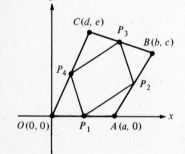

Figure 1.5.32

HYPOTHESIS:

$OABC$ is a quadrilateral.
P_1 is the midpoint of OA.
P_2 is the midpoint of AB.
P_3 is the midpoint of BC.
P_4 is the midpoint of CO.

CONCLUSION: $P_1P_2P_3P_4$ is a parallelogram.

PROOF: Since $O = (0, 0)$, $A = (a, 0)$, and P_1 is the midpoint of OA, then by the midpoint formula

$$P_1 = \left(\frac{a}{2}, 0\right)$$

Since $B = (b, c)$ and P_2 is the midpoint of AB, then

$$P_2 = \left(\frac{a + b}{2}, \frac{c}{2}\right)$$

Since $C = (d, e)$ and P_3 is the midpoint of BC, then

$$P_3 = \left(\frac{b + d}{2}, \frac{c + e}{2}\right)$$

Since P_4 is the midpoint of CO, then

$$P_4 = \left(\frac{d}{2}, \frac{e}{2}\right)$$

We show that P_1P_2 is parallel to P_4P_3 and P_2P_3 is parallel to P_1P_4. Let

$$m_1 = \text{the slope of } P_1P_2 = \frac{\dfrac{c}{2} - 0}{\dfrac{a+b}{2} - \dfrac{a}{2}} = \frac{c}{b}$$

$$m_2 = \text{the slope of } P_4P_3 = \frac{\dfrac{e}{2} - \dfrac{c+e}{2}}{\dfrac{d}{2} - \dfrac{b+d}{2}} = \frac{c}{b}$$

Since $m_1 = m_2$ by Theorem 1.5.3, P_1P_2 is parallel to P_4P_3.
Let

$$m_3 = \text{the slope of } P_2P_3 = \frac{\dfrac{c+e}{2} - \dfrac{c}{2}}{\dfrac{b+d}{2} - \dfrac{a+b}{2}} = \frac{e}{d-a}$$

$$m_4 = \text{the slope of } P_1P_4 = \frac{\dfrac{e}{2} - 0}{\dfrac{d}{2} - \dfrac{a}{2}} = \frac{e}{d-a}$$

Since $m_3 = m_4$, then P_2P_3 is parallel to P_1P_4.

Because the opposite sides of quadrilateral $P_1P_2P_3P_4$ are parallel, the quadrilateral is a parallelogram.

1.6 THE CIRCLE

If the center and radius of a circle are known, we can find an equation of the circle of the form $x^2 + y^2 + Dx + Ey + F = 0$. However, not every equation of this form has a circle for its graph. We complete the square of the terms in x and y to determine whether the graph is in fact a circle, a point circle, or the empty set. From algebra we know that to complete the square for an expression of the form $x^2 + bx$ we must add $(b/2)^2$, and

$$x^2 + bx + \left(\frac{b}{2}\right)^2 = \left(x + \frac{b}{2}\right)^2$$

1.6.1 Definition   A *circle* is the set of all points in a plane equidistant from a fixed point in the plane. The fixed point is called the *center* of the circle, and the measure of the constant equal distance is called the *radius* of the circle.

1.6.2 Theorem   The circle with center at the point $C(h, k)$ and radius r has *center radius* form.

$$(x - h)^2 + (y - k)^2 = r^2$$

The *general* form of an equation of a circle is

$$x^2 + y^2 + Dx + Ey + F = 0$$

1.6.3 Theorem   The graph of any equation of the form

$$x^2 + y^2 + Dx + Ey + F = 0$$

is either a circle, a point circle, or it is the empty set.

Exercises 1.6

4.   Write an equation in both center-radius form and general form for the circle with center $C(-1, 1)$ and radius $r = 2$.

SOLUTION: In the formula of Theorem 1.6.2 let $h = -1$, $k = 1$ and $r = 2$. Then the center-radius form is

$$(x + 1)^2 + (y - 1)^2 = 4 \tag{1}$$

To find the general form we write Eq. (1) as

$$x^2 + 2x + 1 + y^2 - 2y + 1 = 4$$
$$x^2 + y^2 + 2x - 2y - 2 = 0$$

10. Find an equation for each circle that is tangent to the line $3x + 4y - 16 = 0$ at $(4, 1)$ and with $r = 5$.

SOLUTION: Let ℓ be the given line and let $C(h, k)$ be the center of the required circle. Figure 1.6.10 suggests that there are two such circles, and we must find the center of each. Since $T(4, 1)$ is the point of tangency of the circle and line ℓ, the line CT is perpendicular to line ℓ. The point-slope form of the equation of line ℓ is found by solving the equation for y. Thus,

$$y = -\frac{3}{4}x + 4$$

Hence, the slope of line ℓ is $-\frac{3}{4}$. Therefore, the slope of CT is $\frac{4}{3}$. Using the slope formula with points C and T, we obtain

$$\frac{k - 1}{h - 4} = \frac{4}{3}$$

$$k - 1 = \frac{4}{3}(h - 4) \tag{2}$$

Moreover, since $|\overline{CT}|$ is the radius 5, the distance formula gives

$$\sqrt{(h - 4)^2 + (k - 1)^2} = 5 \tag{3}$$

If we substitute from Eq. (2) into Eq. (3), we have

$$\sqrt{(h - 4)^2 + \frac{16}{9}(h - 4)^2} = 5$$

$$\sqrt{\frac{25}{9}(h - 4)^2} = 5$$

$$\frac{5}{3}|h - 4| = 5$$

$$|h - 4| = 3$$

There are two solutions to this equation.

$$h - 4 = 3 \quad \text{and} \quad h - 4 = -3$$
$$h = 7 \qquad\qquad h = 1$$

Substituting $h = 7$ into Eq. (2), we obtain $k = 5$. Substituting $h = 1$ into Eq. (2), we obtain $k = -3$. Hence, the centers of the two possible circles are $(7, 5)$ and $(1, -3)$. The center-radius form for the equations of the circles are

$$(x - 7)^2 + (y - 5)^2 = 25$$

and

$$(x - 1)^2 + (y + 3)^2 = 25$$

14. Find the center and radius of the circle $x^2 + y^2 - 10x - 10y + 25 = 0$ and draw a sketch of the graph.

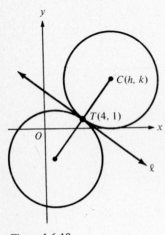

Figure 1.6.10

SOLUTION: We complete the square.

$$(x^2 - 10x) + (y^2 - 10y) \quad\quad = -25$$
$$[x^2 - 10x + (-5)^2] + [y^2 - 10y + (-5)^2] = -25 + (-5)^2 + (-5)^2$$
$$(x - 5)^2 + (y - 5)^2 \quad\quad = 5^2$$

By Theorem 1.6.2, the center is at $(5, 5)$ and the radius is 5. A sketch of the graph is shown in Fig. 1.6.14.

In Exercises 15-20 determine whether the graph is a circle, a point-circle, or the empty set.

16. $4x^2 + 4y^2 + 24x - 4y + 1 = 0$

SOLUTION: We divide by 4 and then complete the square.

$$(x^2 + 6x) + (y^2 - y) \quad\quad = -\tfrac{1}{4}$$
$$[x^2 + 6x + 3^2] + [y^2 - y + (-\tfrac{1}{2})^2] = -\tfrac{1}{4} + 3^2 + (-\tfrac{1}{2})^2$$
$$(x + 3)^2 + (y - \tfrac{1}{2})^2 \quad\quad = 3^2$$

This is an equation of a circle with center at $(-3, \tfrac{1}{2})$ and radius 3.

18. $x^2 + y^2 + 2x - 4y + 5 = 0$

SOLUTION:

$$(x^2 + 2x) + (y^2 - 4y) = -5$$
$$(x^2 + 2x + 1) + (y^2 - 4y + 4) = -5 + 1 + 4$$
$$(x + 1)^2 + (y - 2)^2 = 0$$

Since the only solution for this equation occurs when each square has value zero, the only solution is $(-1, 2)$. Therefore, the graph is a point-circle.

20. $9x^2 + 9y^2 + 6x - 6y + 5 = 0$

SOLUTION:

$$\left(x^2 + \frac{2}{3}x\right) + \left(y^2 - \frac{2}{3}y\right) = -\frac{5}{9}$$
$$\left[x^2 + \frac{2}{3}x + \left(\frac{1}{3}\right)^2\right] + \left[y^2 - \frac{2}{3}y + \left(-\frac{1}{3}\right)^2\right] = -\frac{5}{9} + \left(\frac{1}{3}\right)^2 + \left(-\frac{1}{3}\right)^2$$
$$\left(x + \frac{1}{3}\right)^2 + \left(y - \frac{1}{3}\right)^2 = -\frac{1}{3} \tag{4}$$

The sum of two squares cannot be negative. There is no solution to Eq. (4). Therefore, the graph is the empty set.

24. Find an equation of each of the two lines having slope $-\dfrac{4}{3}$ that are tangent to the circle $x^2 + y^2 + 2x - 8y - 8 = 0$.

SOLUTION: By completing the square, an equation of the circle is

$$(x + 1)^2 + (y - 4)^2 = 25 \tag{5}$$

Therefore, $C(-1, 4)$ is the center of the circle, and 5 is its radius. Let ℓ_1 and ℓ_2 be the required lines and let $T(x_1, y_1)$ be the point at which ℓ_1 is tangent to the circle, as illustrated in Fig. 1.6.24. Since CT is perpendicular to ℓ_1 and since the slope of ℓ_1 is $-\tfrac{4}{3}$, the slope of CT is $\tfrac{3}{4}$. Using the slope formula for line CT with points C and T we obtain

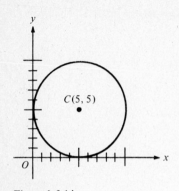

Figure 1.6.14

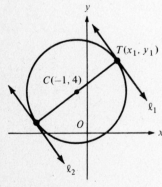

Figure 1.6.24

$$\frac{y_1 - 4}{x_1 + 1} = \frac{3}{4}$$

$$y_1 - 4 = \frac{3}{4}(x_1 + 1) \tag{6}$$

Because T is on the circle, the coordinates of T must satisfy Eq. (5). Therefore,

$$(x_1 + 1)^2 + (y_1 - 4)^2 = 25 \tag{7}$$

Substituting from Eq. (6) and into Eq. (7), we obtain

$$(x_1 + 1)^2 + \frac{9}{16}(x_1 + 1)^2 = 25$$

$$\frac{25}{16}(x_1 + 1)^2 = 25$$

$$(x_1 + 1)^2 = 16$$

The two solutions to this equation are

$$x_1 + 1 = 4 \quad \text{and} \quad x_1 + 1 = -4$$
$$x_1 = 3 \quad \text{and} \quad x_1 = -5$$

If $x_1 = 3$ is substituted into Eq. (6) we obtain $y_1 = 7$. Therefore $T = (3, 7)$ and the point slope form for line ℓ_1 is

$$y - 7 = -\frac{4}{3}(x - 3)$$

If $x_1 = -5$ is substituted into Eq. (6) we obtain $y_1 = 1$. Therefore, the point-slope form for ℓ_2 is

$$y - 1 = -\frac{4}{3}(x + 5)$$

26. Prove analytically that a line from the center of a circle bisecting any chord is perpendicular to the chord.

SOLUTION:

HYPOTHESIS: The circle in Fig. 1.6.26 has its center at the origin, and points $P(r, 0)$ and $Q(a, b)$ are the endpoints of a chord. Furthermore, M is the midpoint of line segment PQ.

CONCLUSION: OM is perpendicular to PQ.

PROOF: Since M is the midpoint of PQ, then by the midpoint formula

$$M = \left(\frac{a + r}{2}, \frac{b}{2}\right)$$

Let

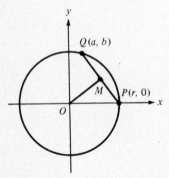

Figure 1.6.26

$$m_1 = \text{the slope of } OM = \frac{\frac{b}{2} - 0}{\frac{a + r}{2} - 0} = \frac{b}{a + r}$$

$$m_2 = \text{the slope of } PQ = \frac{b - 0}{a - r} = \frac{b}{a - r}$$

Then

$$m_1 \cdot m_2 = \frac{b^2}{a^2 - r^2} \tag{8}$$

Because the given circle has center at O and radius r, an equation for the circle is

$$x^2 + y^2 = r^2 \tag{9}$$

Because the point $Q(a, b)$ lies on the circle, (a, b) must satisfy Eq. (9). Thus,

$$\begin{aligned} a^2 + b^2 &= r^2 \\ a^2 - r^2 &= -b^2 \end{aligned} \tag{10}$$

Substitution from Eq. (10) into Eq. (8) gives

$$m_1 \cdot m_2 = \frac{b^2}{-b^2} = -1$$

Therefore, OM is perpendicular to PQ.

1.7 FUNCTIONS AND THEIR GRAPHS

If for each replacement of the independent variable x there corresponds only one value of y, then y *is a function of* x. The set that contains each real pair (x, y) in the correspondence is called a *function,* and the set of all points in R^2 that correspond to this set is the *graph* of the function. Any line parallel to the y-axis intersects the graph of a function in at most one point. Furthermore, any graph that intersects every line parallel to the y-axis in at most one point is the graph of a function. The set of all real replacements for x that result in a real value for y is called the *domain* of the function. And the set of all those values of y is called the *range* of the function.

We can often determine the domain of a function that is defined by an equation in x and y by first solving the equation for y. Then, if addition, subtraction, and multiplication are the only operations indicated in the equation, the domain is $(-\infty, +\infty)$. However, if there is division indicated, we must exclude from the domain those replacements for x that result in division by zero. And if the square root sign appears, we must exclude from the domain those replacements for x that result in a negative value under the radical.

Thus, if

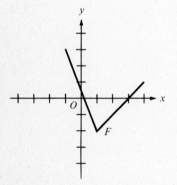

Figure 1.7(a)

$$f = \{(x, y) \mid y = 3x^5 - x^2 + x - 3\}$$

the domain of f is $(-\infty, +\infty)$, since the only operations are addition, subtraction, and multiplication. If

$$g = \left\{(x, y) \mid y = \frac{x - 3}{x^2 - 4}\right\}$$

the domain of g is $\{x \mid x \neq \pm 2\}$, since ± 2 are the only replacements for x that result in division by zero. If

$$h = \{(x, y) \mid y = \sqrt{3 - x}\}$$

the domain of h is $\{x \mid 3 - x \geq 0\} = \{x \mid x \leq 3\} = (-\infty, 3]$, because the expression under the radical sign must be nonnegative.

If the graph of a function is known, both the domain and range of the function can be determined by examining the graph. Let F be the function whose graph consists of the two line segments shown in Fig. 1.7(a). To find the domain of F we draw a line segment from every point on the graph of F perpendicular to the x-axis. Then the domain of F is $[-1, 4]$, as illustrated in Fig. 1.7(b). To find the range of F we

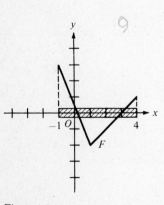

Figure 1.7(b)

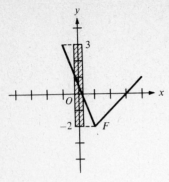

Figure 1.7(c)

draw a line segment from every point on the graph of F perpendicular to the y-axis. The range of F is $[-2, 3]$, as illustrated in Fig. 1.7(c).

Exercises 1.7

In Exercises 1-10 find the domain and range of the given function and draw a sketch of the graph.

4. $G = \{(x, y) \mid y = \sqrt{x + 1}\}$

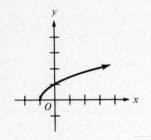

Figure 1.7.4

SOLUTION: Since the square root of any nonnegative number is real and $x + 1 \geqslant 0$ when $x \geqslant -1$, the domain of G is $[-1, +\infty)$. We choose x from this domain, calculate y, and plot the point corresponding to each pair (x, y). A sketch of the graph of G is shown in Fig. 1.7.4. From the graph we see that the range is $[0, +\infty)$.

10. $\left\{(x, y) \mid y = \dfrac{4x^2 - 1}{2x + 1}\right\}$

SOLUTION:

$$y = \frac{(2x + 1)(2x - 1)}{2x + 1} \tag{1}$$

The domain of F is $\{x \mid 2x + 1 \neq 0\} = \{x \mid x \neq -\frac{1}{2}\}$. If $x \neq -\frac{1}{2}$, we may divide the numerator and denominator of the fraction in Eq. (1) by $2x + 1$ and obtain

$$y = 2x - 1 \quad \text{if } x \neq -\tfrac{1}{2} \tag{2}$$

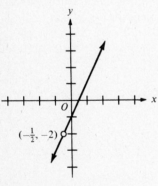

Figure 1.7.10

Then the graph of F is the line $y = 2x - 1$ with the point $(-\frac{1}{2}, -2)$ deleted, as illustrated in Fig. 1.7.10. The y-coordinate of the point $(-\frac{1}{2}, -2)$ is found by substituting $x = -\frac{1}{2}$ in Eq. (2). From the graph we see that the range is $\{y \mid y \neq -2\}$.

In Exercises 11-34 the function is the set of all ordered pairs (x, y) satisfying the given equation. Find the domain and range of the function, and draw a sketch of the graph of the function.

14. $\phi: y = \begin{cases} x + 5 & \text{if } x < -5 \\ \sqrt{25 - x^2} & \text{if } -5 \leqslant x \leqslant 5 \\ x - 5 & \text{if } 5 < x \end{cases}$

SOLUTION: The domain of ϕ is given and is $(-\infty, +\infty)$. The function ϕ is the union of three functions, which we consider separately. Let

$$\phi_1: y = x + 5, \ x < -5$$

The graph of ϕ_1 is that part of the line $y = x + 5$ which lies to the left of the point $(-5, 0)$. Let

$$\phi_2: y = \sqrt{25 - x^2}, \ -5 \leqslant x \leqslant 5 \tag{3}$$

Squaring on both sides of Eq. (3) we obtain

$$x^2 + y^2 = 25$$

which is an equation of the circle with center at the origin and radius 5. However, $y \geqslant 0$ since the radical sign in Eq. (3) indicates the nonnegative square root. Hence, the graph of ϕ_2 is the semicircle that does not lie below the x-axis. Let

$$\phi_3: \ y = x - 5, \ x > 5$$

The graph of ϕ_3 is that part of the line $y = x - 5$ which lies to the right of $(5, 0)$. The graph of ϕ is the union of the graphs of $\phi_1, \phi_2,$ and $\phi_3,$ and is shown in Fig. 1.7.14. From the graph we see that the range of ϕ is $(-\infty, +\infty)$.

18. $G: \ y = \dfrac{(x^2 + 3x - 4)(x^2 - 5x + 6)}{(x^2 - 3x + 2)(x - 3)}$

SOLUTION:

$$y = \frac{(x + 4)(x - 1)(x - 2)(x - 3)}{(x - 1)(x - 2)(x - 3)} \tag{4}$$

The denominator of this fraction is zero if $x = 1,$ $x = 2,$ or $x = 3$. Hence, the domain of G is $\{x \,|\, x \neq 1, x \neq 2, x \neq 3\}$. If x is any number in this domain, we may divide the numerator and denominator of the fraction in Eq. (4) by the common factors and obtain

$$y = x + 4 \quad \text{if} \ \ x \neq 1, x \neq 2, x \neq 3$$

The graph of G is the line $y = x + 4$ with the points $(1, 5), (2, 6), (3, 7)$ deleted. See Fig. 1.7.18. From the graph we see that the range is $\{y \,|\, y \neq 5, y \neq 6, y \neq 7\}$.

22. $f: \ y = \dfrac{x^3 + 3x^2 + x + 3}{x + 3}$

SOLUTION: The domain of f is $\{x \,|\, x + 3 \neq 0\} = \{x \,|\, x \neq -3\}$. If $x \neq -3$ we may use long division to reduce the fraction.

$$\begin{array}{r}
x^2 + 1 \\
x + 3 \overline{)x^3 + 3x^2 + x + 3} \\
\underline{x^3 + 3x^2 } \\
x + 3 \\
\underline{x + 3}
\end{array}$$

Thus, $y = x^2 + 1$ if $x \neq -3$.

We plot some points and obtain the graph of f which is shown in Fig. 1.7.22. Note that $(-3, 10)$ is deleted, because $x \neq -3$. From the graph we see that the range is $[1, +\infty)$.

26. $g: \ y = |x| \cdot |x - 1|$

SOLUTION: Recall that $|a| = a$ if $a \geqslant 0$, and $|a| = -a$ if $a < 0$. Therefore, y is real for every real replacement of x, and the domain is $(-\infty, +\infty)$. Furthermore if $x \geqslant 1$, then $|x - 1| = x - 1$ and $|x| = x$, and

$$|x| \cdot |x - 1| = x^2 - x \quad \text{if} \ \ x \geqslant 1$$

If $0 \leqslant x < 1$, then $|x - 1| = -(x - 1) = -x + 1$ and $|x| = x$, and

$$|x| \cdot |x - 1| = -x^2 + x \quad \text{if} \ \ 0 \leqslant x < 1$$

If $x < 0$, then $|x - 1| = -(x - 1) = -x + 1$ and $|x| = -x$, and

$$|x| \cdot |x - 1| = x^2 - x \quad \text{if} \ \ x < 0$$

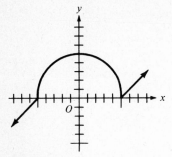

Figure 1.7.14

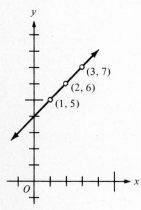

Figure 1.7.18

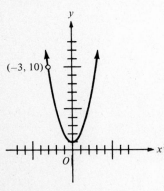

Figure 1.7.22

Therefore,

$$y = \begin{cases} x^2 - x & \text{if } x < 0 \\ -x^2 + x & \text{if } 0 \leqslant x < 1 \\ x^2 - x & \text{if } x \geqslant 1 \end{cases}$$

By plotting some points we obtain a sketch of the graph shown in Fig. 1.7.26. The graph shows that the range is $[0, +\infty)$.

30. $H: y = |x| + [\![x]\!]$

SOLUTION: The domain is $(-\infty, +\infty)$.
If $x \geqslant 0$, $|x| = x$. Furthermore,

If $0 \leqslant x < 1$, $[\![x]\!] = 0$; so $y = x$
If $1 \leqslant x < 2$, $[\![x]\!] = 1$; so $y = x + 1$
If $2 \leqslant x < 3$, $[\![x]\!] = 2$; so $y = x + 2$

and so on.

If $x < 0$, $|x| = -x$. Furthermore:

If $-1 \leqslant x < 0$, $[\![x]\!] = -1$; so $y = -x - 1$
If $-2 \leqslant x < -1$, $[\![x]\!] = -2$; so $y = -x - 2$

and so on.
In Fig. 1.7.30 we have a sketch of the graph of H. The range is the union of $(-1, 1)$ and each interval of the form $[2n, 2n + 1)$, where n is a positive integer.

34. $h: y = \dfrac{|x|}{[\![x]\!]}$

SOLUTION: If $0 \leqslant x < 1$, $[\![x]\!] = 0$, and y is not defined. Thus, the domain of h is $(-\infty, 0) \cup [1, +\infty)$. If $x \geqslant 1$, $|x| = x$. Furthermore,

If $1 \leqslant x < 2$, $[\![x]\!] = 1$; so $y = x$

If $2 \leqslant x < 3$, $[\![x]\!] = 2$; so $y = \dfrac{x}{2}$

If $3 \leqslant x < 4$, $[\![x]\!] = 3$; so $y = \dfrac{x}{3}$

and so on.
If $x < 0$, $|x| = -x$. Furthermore,

If $-1 \leqslant x < 0$, $[\![x]\!] = -1$; so $y = \dfrac{-x}{-1} = x$

If $-2 \leqslant x < -1$, $[\![x]\!] = -2$; so $y = \dfrac{-x}{-2} = \dfrac{x}{2}$

If $-3 \leqslant x < -2$, $[\![x]\!] = -3$; so $y = \dfrac{-x}{-3} = \dfrac{x}{3}$

and so on.
A sketch of the graph is shown in Fig. 1.7.34. The range is $[-1, 0) \cup [1, 2)$.

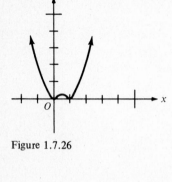

Figure 1.7.26

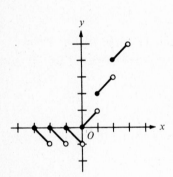

Figure 1.7.30

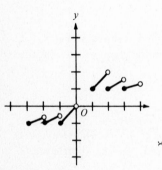

Figure 1.7.34

1.8 FUNCTION NOTATION, OPERATIONS ON FUNCTIONS, AND TYPES OF FUNCTIONS

If $f(x)$ is an expression whose only variable is x, then the equation $y = f(x)$ defines a function, f, where $f = \{(x, y) | y = f(x)\}$. If a is any constant that is an element of the domain of f, then $f(a)$ represents the number in the range of f that corresponds to a. This number may be found by replacing x by a in the expression $f(x)$. Moreover, if a is any expression containing a variable, then $f(a)$ represents the expression that results from replacing x by a in the expression $f(x)$.

1.8.1 Definition Given two functions f and g:

(i) their *sum*, denoted by $(f + g)$, is the function defined by $(f + g)(x) = f(x) + g(x)$;

(ii) their *difference*, denoted by $(f - g)$, is the function defined by $(f - g)(x) = f(x) - g(x)$;

(iii) their *product*, denoted by $(f \cdot g)$, is the function defined by $(f \cdot g)(x) = f(x) \cdot g(x)$;

(iv) their *quotient*, denoted by (f/g), is the function defined by $(f/g)(x) = f(x)/g(x)$.

For (i), (ii), and (iii) the domain is the intersection of the domains for f and g, and for (iv) the domain is the set of all elements in the intersection of the domains for f and g for which $g(x) \neq 0$.

1.8.2 Definition Given two functions f and g, the *composite function*, denoted by $f \circ g$, is defined by $(f \circ g)(x) = f(g(x))$, and the domain of $f \circ g$ is the set of all numbers x in the domain of g for which $g(x)$ is in the domain of f.

1.8.3 Definition

(i) A function f is said to be an *even* function if for every x in the domain of f, $f(-x) = f(x)$.

(ii) A function f is said to be an *odd* function if for every x in the domain of f, $f(-x) = -f(x)$.

Exercises 1.8

2. Given: $g(x) = 3x^2 - 4$, find

(a) $g(-4)$　　(b) $g(\frac{1}{2})$　　(c) $g(x^2)$　　(d) $g(3x^2 - 4)$　　(e) $g(x - h)$

(f) $g(x) - g(h)$　　(g) $\dfrac{g(x + h) - g(x)}{h}$, $h \neq 0$

SOLUTION:

(a) $g(-4) = 3(-4)^2 - 4 = 44$

(b) $g(\frac{1}{2}) = 3(\frac{1}{2})^2 - 4 = \frac{-13}{4}$

(c) $g(x^2) = 3(x^2)^2 - 4 = 3x^4 - 4$

(d) $g(3x^2 - 4) = 3(3x^2 - 4)^2 - 4 = 3(9x^4 - 24x^2 + 16) - 4$
$\qquad\qquad = 27x^4 - 72x^2 + 44$

(e) $g(x - h) = 3(x - h)^2 - 4 = 3x^2 - 6xh + 3h^2 - 4$

(f) $g(x) - g(h) = (3x^2 - 4) - (3h^2 - 4) = 3x^2 - 3h^2$

(g) $\dfrac{g(x + h) - g(x)}{h} = \dfrac{[3(x + h)^2 - 4] - [3x^2 - 4]}{h}$

$\qquad = \dfrac{3x^2 + 6xh + 3h^2 - 4 - 3x^2 + 4}{h}$

$\qquad = \dfrac{6xh + 3h^2}{h}$

$\qquad = \dfrac{h(6x + 3h)}{h}$

$\qquad = 6x + 3h$, if $h \neq 0$

6. Given $f(t) = (|3 + t| - |t| - 3)/t$, express $f(t)$ without absolute value bars if:
(a) $t > 0$,　　(b) $-3 \leqslant t < 0$;　　(c) $t < -3$.

SOLUTION:

(a) Since $t > 0$, then $|3 + t| = 3 + t$ and $|t| = t$. Hence,

$$f(t) = \frac{(3 + t) - t - 3}{t} = 0$$

(b) Because $t < 0$, then $|t| = -t$. Since $t \geq -3$, then $3 + t \geq 0$; thus $|3 + t| = 3 + t$. Hence,

$$f(t) = \frac{(3 + t) - (-t) - 3}{t} = \frac{2t}{t} = 2$$

(c) Since $t < -3$, then $t < 0$ and $3 + t < 0$; therefore, $|t| = -t$ and $|3 + t| = -(3 + t)$. Hence,

$$f(t) = \frac{-(3 + t) - (-t) - 3}{t} = \frac{-6}{t}$$

10. Let $f(x) = \sqrt{x - 2}$ and $g(x) = 1/x$. Define each of the following functions and determine its domain.

(a) $f + g$ (b) $f - g$ (c) $f \cdot g$ (d) f/g (e) g/f (f) $f \circ g$
(g) $g \circ f$

SOLUTION: First, we note that the domain of f is $\{x \,|\, x - 2 \geq 0\} = [2, +\infty)$. And the domain of g is $\{x \,|\, x \neq 0\}$.

(a) $(f + g)(x) = f(x) + g(x) = \sqrt{x - 2} + \dfrac{1}{x}$

The domain is the intersection of the domains for f and g, namely $[2, +\infty)$.

(b) $(f - g)(x) = f(x) - g(x) = \sqrt{x - 2} - \dfrac{1}{x}$

The domain is $[2, +\infty)$.

(c) $(f \cdot g)(x) = f(x) \cdot g(x) = \sqrt{x - 2} \cdot \dfrac{1}{x} = \dfrac{\sqrt{x - 2}}{x}$

The domain is $[2, +\infty)$.

(d) $(f/g)(x) = \dfrac{f(x)}{g(x)} = \dfrac{\sqrt{x - 2}}{1/x} = x\sqrt{x - 2}$

The domain is $[2, +\infty)$.

(e) $(g/f)(x) = \dfrac{g(x)}{f(x)} = \dfrac{1/x}{\sqrt{x - 2}} = \dfrac{1}{x\sqrt{x - 2}}$

The domain is $(2, +\infty)$. Note that 2 is not in the domain of g/f because $f(2) = 0$.

(f) $(f \circ g)(x) = f(g(x)) = f\left(\dfrac{1}{x}\right) = \sqrt{\dfrac{1}{x} - 2}$

Because the domain of f is $\{x \,|\, x \geq 2\}$ and the domain of g is $\{x \,|\, x \neq 0\}$, the domain of $f \circ g$ is $\{x \,|\, x \neq 0 \text{ and } g(x) \geq 2\}$. If $g(x) \geq 2$, then

$$\frac{1}{x} \geq 2$$

$$\frac{1}{x} - 2 \geq 0$$

$$\frac{1 - 2x}{x} \geq 0$$

$$\frac{2x-1}{x} \leqslant 0 \tag{1}$$

The factor $2x - 1$ changes sign at $x = \frac{1}{2}$ and the factor x changes sign at $x = 0$, as shown in Table 10. The strict inequality (1) is satisfied whenever the fraction has a negative value, and Eq. (1) is satisfied if $x = \frac{1}{2}$. Thus, Table 10 shows that the domain of $f \circ g$ is $(0, \frac{1}{2}]$.

Table 10

	$x < 0$	$x = 0$	$0 < x < \frac{1}{2}$	$x = \frac{1}{2}$	$x > \frac{1}{2}$
$2x - 1$	$-$	$-$	$-$	0	$+$
x	$-$	0	$+$	$+$	$+$
$\dfrac{2x-1}{x}$	$+$	does not exist	$-$	0	$+$

(g) $(g \circ f)(x) = g(f(x)) = g(\sqrt{x-2}) = \dfrac{1}{\sqrt{x-2}}$

Because the domain of f is $\{x \mid x \geqslant 2\}$ and the domain of g is $\{x \mid x \neq 0\}$, the domain of $g \circ f$ is $\{x \mid x \geqslant 2$ and $f(x) \neq 0\}$. If $f(x) \neq 0$, then $\sqrt{x-2} \neq 0$, or, equivalently, $x \neq 2$. Therefore, the domain of $g \circ f$ is $\{x \mid x \geqslant 2$ and $x \neq 2\} = (2, +\infty)$.

18. What is the function that is both even and odd?

SOLUTION: Let F be a function that is both even and odd. Because F is even

$$F(-x) = F(x) \tag{2}$$

Because F is odd

$$F(-x) = -F(x) \tag{3}$$

Substituting from Eq. (2) into Eq. (3), we obtain

$$F(x) = -F(x)$$
$$2 \cdot F(x) = 0$$
$$F(x) = 0$$

In Exercises 21-34 draw a sketch of the graph and state the domain and range of the given function.

22. $\operatorname{sgn} x = \begin{cases} -1 & \text{if } x < 0 \\ 0 & \text{if } x = 0 \\ 1 & \text{if } x > 0 \end{cases}$

SOLUTION: In Fig. 1.8.22 we have a sketch of the graph of $\operatorname{sgn}(x)$. The domain of the function is $(-\infty, +\infty)$, and the range of the function is $\{-1, 0, 1\}$.

26. $f(x) = \operatorname{sgn} x^2 - \operatorname{sgn} x$

SOLUTION: Because

$$\operatorname{sgn} x = \begin{cases} -1 & \text{if } x < 0 \\ 0 & \text{if } x = 0 \\ 1 & \text{if } x > 0 \end{cases}$$

then

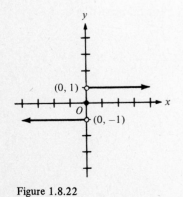

Figure 1.8.22

$$\text{sgn } x^2 = \begin{cases} -1 & \text{if } x^2 < 0 \\ 0 & \text{if } x^2 = 0 \\ 1 & \text{if } x^2 > 0 \end{cases}$$

Because $x^2 > 0$ if $x > 0$, and $x^2 > 0$ if $x < 0$, we have

$$\text{sgn } x^2 = \begin{cases} 1 & \text{if } x < 0 \\ 0 & \text{if } x = 0 \\ 1 & \text{if } x > 0 \end{cases}$$

Thus,

$$\text{sgn } x^2 - \text{sgn } x = \begin{cases} 1 - (-1) & \text{if } x < 0 \\ 0 - 0 & \text{if } x = 0 \\ 1 - 1 & \text{if } x > 0 \end{cases}$$

or, equivalently,

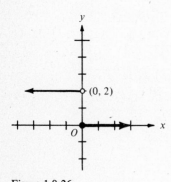

Figure 1.8.26

$$f(x) = \begin{cases} 2 & \text{if } x < 0 \\ 0 & \text{if } x = 0 \\ 0 & \text{if } x > 0 \end{cases}$$

A sketch of the graph of f is shown in Fig. 1.8.26. The domain of f is $(-\infty, +\infty)$, and the range of f is $\{0, 2\}$.

28. $F(x) = (x + 1) \cdot U(x + 1)$

SOLUTION: Because

$$U(x) = \begin{cases} 0 & \text{if } x < 0 \\ 1 & \text{if } x \geq 0 \end{cases}$$

then

$$U(x + 1) = \begin{cases} 0 & \text{if } x + 1 < 0 \\ 1 & \text{if } x + 1 \geq 0 \end{cases}$$

or, equivalently,

$$U(x + 1) = \begin{cases} 0 & \text{if } x < -1 \\ 1 & \text{if } x \geq -1 \end{cases}$$

Hence,

$$(x + 1) \cdot U(x + 1) = \begin{cases} (x + 1) \cdot 0 & \text{if } x < -1 \\ (x + 1) \cdot 1 & \text{if } x \geq -1 \end{cases}$$

or, equivalently,

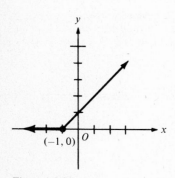

Figure 1.8.28

$$F(x) = \begin{cases} 0 & \text{if } x < -1 \\ x + 1 & \text{if } x \geq -1 \end{cases}$$

A sketch of the graph of F is shown in Fig. 1.8.28. The domain of F is $(-\infty, +\infty)$, and the range of F is $[0, +\infty)$.

32. $f(x) = \text{sgn } x \cdot U(x + 1)$

SOLUTION: From the definition of $\text{sgn } x$ we have

$$\text{sgn } x = \begin{cases} -1 & \text{if } x < 0 \\ 0 & \text{if } x = 0 \\ 1 & \text{if } x > 0 \end{cases}$$

From Exercise 28 we have

$$U(x+1) = \begin{cases} 0 & \text{if } x < -1 \\ 1 & \text{if } x \geq -1 \end{cases}$$

Thus, we must consider four cases: $x < -1$, $-1 \leq x < 0$, $x = 0$, and $x > 0$. Then

$$\text{sgn}(x) \cdot U(x+1) = \begin{cases} -1 \cdot 0 & \text{if } x < -1 \\ -1 \cdot 1 & \text{if } -1 \leq x < 0 \\ 0 \cdot 1 & \text{if } x = 0 \\ 1 \cdot 1 & \text{if } x > 0 \end{cases}$$

or, equivalently,

$$f(x) = \begin{cases} 0 & \text{if } x < -1 \\ -1 & \text{if } -1 \leq x < 0 \\ 0 & \text{if } x = 0 \\ 1 & \text{if } x > 0 \end{cases}$$

A sketch of the graph of f is shown in Fig. 1.8.32. The domain of f is $(-\infty, +\infty)$, and the range of f is $\{-1, 0, 1\}$.

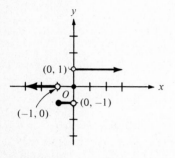

Figure 1.8.32

36. Let $f(x) = \begin{cases} 0 & \text{if } x < 0 \\ 2x & \text{if } 0 \leq x \leq 1 \\ 0 & \text{if } x > 1 \end{cases}$

and

$$g(x) = \begin{cases} 1 & \text{if } x < 0 \\ \frac{1}{2}x & \text{if } 0 \leq x \leq 1 \\ 1 & \text{if } x > 1 \end{cases}$$

Find the domain and range of $g \circ f$ and formulas for $(g \circ f)(x)$.

SOLUTION:

$$(g \circ f)(x) = g(f(x)) = \begin{cases} g(0) & \text{if } x < 0 \\ g(2x) & \text{if } 0 \leq x \leq 1 \\ g(0) & \text{if } x > 1 \end{cases} \tag{4}$$

and

$$g(0) = \frac{1}{2} \cdot 0 = 0 \tag{5}$$

$$g(2x) = \begin{cases} 1 & \text{if } 2x < 0 \\ \frac{1}{2}(2x) & \text{if } 0 \leq 2x \leq 1 \\ 1 & \text{if } 2x > 1 \end{cases}$$

or, equivalently,

$$g(2x) = \begin{cases} 1 & \text{if } x < 0 \\ x & \text{if } 0 \leq x \leq \frac{1}{2} \\ 1 & \text{if } x > \frac{1}{2} \end{cases} \tag{6}$$

By substituting Eqs. (5) and (6) into (4), we obtain

$$(g \circ f)(x) = \begin{cases} 0 & \text{if } x < 0 \\ x & \text{if } 0 \leq x \leq \frac{1}{2} \\ 1 & \text{if } \frac{1}{2} < x \leq 1 \\ 0 & \text{if } x > 1 \end{cases}$$

The domain of $g \circ f$ is $(-\infty, +\infty)$ and the range of $g \circ f$ is $[0, \frac{1}{2}] \cup \{1\}$.

Review Exercises

4. Find the solution set of the given inequality and illustrate the solution on the real number line.

$$\frac{3}{x+4} < \frac{2}{x-5}$$

SOLUTION:

$$\frac{3}{x+4} - \frac{2}{x-5} < 0$$

$$\frac{3(x-5) - 2(x+4)}{(x+4)(x-5)} < 0$$

$$\frac{x-23}{(x+4)(x-5)} < 0$$

As Table 4 indicates, $x - 23$ changes sign at $x = 23$; $x + 4$ changes sign at $x = -4$; and $x - 5$ changes sign at $x = 5$. Since inequality (1) is satisfied whenever the fraction has a negative value, the solution set is $(-\infty, -4) \cup (5, 23)$. It is illustrated in Fig. 1.4R.

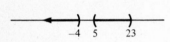

Figure 1.4R

Table 4

	$-\infty$	-4	5	23	$+\infty$
$x - 23$	$-$	$-$	$-$	$+$	
$x + 4$	$-$	$+$	$+$	$+$	
$x - 5$	$-$	$-$	$+$	$+$	
$\dfrac{x - 23}{(x+4)(x-5)}$	$-$	$+$	$-$	$+$	
	$x < -4$	$-4 < x < 5$	$5 < x < 23$	$x > 23$	

8. Prove that the points $(1, -1), (3, 2)$, and $(7, 8)$ are collinear in two ways: (a) by using the distance formula; (b) by using slopes.

SOLUTION:

(a) Let $A = (1, -1)$, $B = (3, 2)$, and $C = (7, 8)$. Then

$$|\overline{AB}| = \sqrt{(3-1)^2 + (2+1)^2} = \sqrt{13}$$
$$|\overline{BC}| = \sqrt{(7-3)^2 + (8-2)^2} = \sqrt{52} = 2\sqrt{13}$$
$$|\overline{AC}| = \sqrt{(7-1)^2 + (8+1)^2} = \sqrt{117} = 3\sqrt{13}$$

Since $|\overline{AB}| + |\overline{BC}| = |\overline{AC}|$, then A, B, and C are collinear.

(b) Slope of $AB = \dfrac{2+1}{3-1} = \dfrac{3}{2}$

Slope of $BC = \dfrac{8-2}{7-3} = \dfrac{3}{2}$

Since the slope of AB equals the slope of BC, then A, B, and C are collinear.

12. Show that the triangle with vertices at $(-8, 1), (-1, -6)$ and $(2, 4)$ is isosceles and find its area.

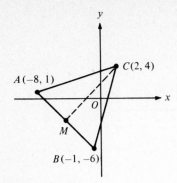

Figure 1.12R

SOLUTION: Refer to Fig. 1.12R. Let $A(-8, 1)$, $B(-1, -6)$ and $C(2, 4)$ be the given vertices. Then

$$|\overline{AB}| = \sqrt{(-1 + 8)^2 + (-6 - 1)^2} = \sqrt{98} = 7\sqrt{2}$$
$$|\overline{BC}| = \sqrt{(2 + 1)^2 + (4 + 6)^2} = \sqrt{109}$$
$$|\overline{AC}| = \sqrt{(2 + 8)^2 + (4 - 1)^2} = \sqrt{109}$$

Because $|\overline{BC}| = |\overline{AC}|$, triangle ABC is isosceles. Its base is line segment AB, and its vertex is point C. To find its area we use the formula

$$A = \tfrac{1}{2}bh \qquad (2)$$

Because $|\overline{AB}| = 7\sqrt{2}$, we have $b = 7\sqrt{2}$. The altitude of an isosceles triangle is the distance between its vertex and the midpoint of its base. Let M be the midpoint of the base, AB. Then $M = (-\tfrac{9}{2}, -\tfrac{5}{2})$, and

$$h = |CM| = \sqrt{\left(-\frac{9}{2} - 2\right)^2 + \left(-\frac{5}{2} - 4\right)^2} = \frac{13\sqrt{2}}{2}$$

By substitution in formula (2) we have

$$A = \frac{1}{2}(7\sqrt{2})\left(\frac{13\sqrt{2}}{2}\right)$$

$$A = \frac{91}{2}$$

14. Two opposite vertices of a square are $(3, -4)$ and $(9, -4)$. Find the other vertices.

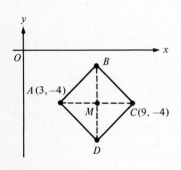

Figure 1.14R

SOLUTION: Let $ABCD$ be the square with $A = (3, -4)$ and $C = (9, -4)$ as shown in Fig. 1.14R. Let M be the midpoint of diagonal AC. Then $M = (6, -4)$. Since $ABCD$ is a square, then M is also the midpoint of diagonal BD. Moreover, BD is parallel to the y-axis because AC is parallel to the x-axis and the diagonals of a square are perpendicular. Then, since $|\overline{AM}| = 3$, we have $|\overline{BM}| = 3$; so $B = (6, -1)$. Similarly, $D = (6, -7)$.

18. Find an equation of the circle having as its diameter the common chord of the two circles $x^2 + y^2 + 2x - 2y - 14 = 0$ and $x^2 + y^2 - 4x + 4y - 2 = 0$.

SOLUTION: We first find the two points of intersection of the circles. If we subtract the given equations, we have

$$x^2 + y^2 + 2x - 2y - 14 = 0 \qquad (3)$$
$$\underline{x^2 + y^2 - 4x + 4y - 2 = 0} \qquad (4)$$
$$6x - 6y - 12 = 0$$

$$y = x - 2 \qquad (5)$$

By substitution from Eq. (5) into Eq. (3), we have

$$x^2 + (x - 2)^2 + 2x - 2(x - 2) - 14 = 0$$
$$x^2 - 2x - 3 = 0$$
$$(x - 3)(x + 1) = 0$$

$$x = 3 \qquad x = -1$$
$$y = 1 \qquad y = -3$$

Hence, $A(3, 1)$ and $B(-1, -3)$ are the end points of the common chord. The midpoint of AB is $C(1, -1)$, which is the center of the required circle. Moreover, the radius is $|\overline{AC}| = \sqrt{8}$; so an equation of the circle is

$$(x - 1)^2 + (y + 1)^2 = 8$$

28. Let $f(x) = \sqrt{x}$ and $g(x) = 1/x^2$. Define each of the following functions and determine its domain:

(a) $f + g$ (b) $f - g$ (c) $f \cdot g$ (d) f/g (e) g/f (f) $f \circ g$
(g) $g \circ f$

SOLUTION: First we note that the domain of f is $\{x \mid x \geqslant 0\}$ and the domain of g is $\{x \mid x \neq 0\}$.

(a) $(f + g)(x) = \sqrt{x} + 1/x^2$
The domain of $f + g$ is the intersection of the domains of f and g, namely $(0, +\infty)$.

(b) $(f - g)(x) = \sqrt{x} - 1/x^2$
The domain of $f - g$ is $(0, +\infty)$.

(c) $(f \cdot g)(x) = \sqrt{x} \cdot 1/x^2 = \dfrac{\sqrt{x}}{x^2}$

The domain of $f \cdot g$ is $(0, +\infty)$.

(d) $(f/g)(x) = \dfrac{\sqrt{x}}{\dfrac{1}{x^2}} = x^2 \cdot \sqrt{x}$

The domain of f/g is the set of all elements in the intersection of the domains of f and g for which $g(x) \neq 0$. Thus, the domain of f/g is $(0, +\infty)$.

(e) $(g/f)(x) = \dfrac{\dfrac{1}{x^2}}{\sqrt{x}} = \dfrac{1}{x^2\sqrt{x}}$

The domain of g/f is $(0, +\infty)$.

(f) $(f \circ g)(x) = f(g(x)) = f(1/x^2) = \sqrt{1/x^2} = \dfrac{1}{|x|}$

The domain of $f \circ g$ is the set of all x in the domain of g such that $g(x)$ is in the domain of f. That is, $\{x \mid x \neq 0$ and $g(x) \geqslant 0\} = \{x \mid x \neq 0$ and $1/x^2 \geqslant 0\} = \{x \mid x \neq 0\}$ since $1/x^2 > 0$ if $x \neq 0$.

(g) $(g \circ f)(x) = g(f(x)) = g(\sqrt{x}) = 1/(\sqrt{x})^2 = \dfrac{1}{x}$

The domain of $g \circ f$ is the set of all x in the domain of f such that $f(x)$ is in the domain of g. That is, $\{x \mid x \geqslant 0$ and $f(x) \neq 0\} = \{x \mid x \geqslant 0$ and $\sqrt{x} \neq 0\} = \{x \mid x > 0\} = (0, +\infty)$.

30. Prove that the two lines $A_1x + B_1y + C_1 = 0$ and $A_2x + B_2y + C_2 = 0$ are parallel if and only if $A_1B_2 - A_2B_1 = 0$.

SOLUTION: Let ℓ_1 be the line $A_1x + B_1y + C_1 = 0$ and let ℓ_2 be the line $A_2x + B_2y + C_2 = 0$. If neither line is vertical, then $B_1 \neq 0$ and $B_2 \neq 0$, and ℓ_1 and ℓ_2 are parallel if and only if their slopes are equal. Solving the equation of ℓ_1 for y, we obtain

$$y = -\frac{A_1}{B_1}x - \frac{C_1}{B_1}$$

Thus, the slope of ℓ_1 is $-A_1/B_1$. Solving the equation of ℓ_2 for y, we obtain

$$y = -\frac{A_2}{B_2}x - \frac{C_2}{B_2}$$

Thus, the slope of ℓ_2 is $-A_2/B_2$. Hence, the lines are parallel if and only if

$$-\frac{A_1}{B_1} = -\frac{A_2}{B_2}$$

$$-A_1 B_2 = -A_2 B_1$$
$$A_1 B_2 - A_2 B_1 = 0$$

If ℓ_1 is vertical, then $B_1 = 0$, and we must show that ℓ_1 and ℓ_2 are parallel if and only if $A_1 B_2 = 0$. Suppose that $A_1 B_2 = 0$. Since $B_1 = 0$, then $A_1 \neq 0$, because ℓ_1 would not exist if both $B_1 = 0$ and $A_1 = 0$. Therefore, $B_2 = 0$, ℓ_2 is vertical, and hence ℓ_2 is parallel to ℓ_1. To prove the "only if" part, suppose that ℓ_2 is parallel to ℓ_1. Then ℓ_2 is vertical. Hence $B_2 = 0$, and thus $A_1 B_2 = 0$.

34. Prove analytically that the set of points equidistant from two given points is the perpendicular bisector of the line segment joining the two points.

SOLUTION: Choose the coordinate axes so that the origin is at one of the given points and the other given point is on the x-axis. (See Fig. 1.34R.)

HYPOTHESIS: $P(x, y)$ is any point equidistant from O and $A(a, 0)$.

CONCLUSION: P lies on the perpendicular bisector of line segment OA.

PROOF:

$$|\overline{OP}| = \sqrt{x^2 + y^2}$$
$$|\overline{AP}| = \sqrt{(x - a)^2 + y^2}$$

Because P is equidistant from O and A, then $|\overline{AP}| = |\overline{OP}|$. Thus,

$$\sqrt{(x - a)^2 + y^2} = \sqrt{x^2 + y^2}$$
$$x^2 - 2ax + a^2 + y^2 = x^2 + y^2$$
$$2ax = a^2$$

$$x = \frac{a}{2}$$

Note that $a \neq 0$ because O and A are distinct points.

Because $x = a/2$, it follows that P lies on the line having the equation $x = a/2$. This line is the perpendicular bisector of the line segment OA.

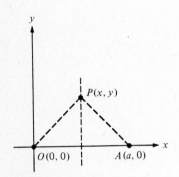

Figure 1.34R

2
Limits and continuity

2.1 THE LIMIT OF A FUNCTION   The definition for the limit of a function is as follows.

2.1.1 Definition   Let f be a function that is defined at every number in some open interval containing a, except possibly at the number a itself. The *limit of* f(x) *as* x *approaches* a *is* L, written as

$$\lim_{x \to a} f(x) = L$$

if for every $\epsilon > 0$, however small, there exists a $\delta > 0$, such that

$$|f(x) - L| < \epsilon \quad \text{whenever} \quad 0 < |x - a| < \delta$$

Note that $0 < |x - a|$ implies that $x \neq a$. This hypothesis must be used whenever $f(a)$ is not defined, as in Exercise 8.

To find δ we usually factor the expression $|f(x) - L|$ into the form $|x - a| \cdot |g(x)|$. If $f(x)$ is a first degree polynomial, then $|g(x)|$ is a constant, and we find δ as in Exercise 10. However, if $f(x)$ is not a first degree polynomial, then $|g(x)|$ is not a constant, and we must find an upper bound for $|g(x)|$, as in Exercise 14. We may often find this upper bound if we assume that $\delta \leqslant 1$. However, as illustrated in Exercise 18, we must sometimes make a different assumption.

Finally, we note that Definition 2.1.1 does not tell us how to find L. In Section 2.2 we have theorems that can be used to find L.

Exercises 2.1

In Exercises 1–8, we are given $f(x)$, a, and L, as well as $\lim_{x \to a} f(x) = L$. Determine a number δ for the given ϵ such that $|f(x) - L| < \epsilon$ whenever $0 < |x - a| < \delta$.

4. $\lim\limits_{x \to -2} (2 + 5x) = -8; \epsilon = 0.002$

SOLUTION: We are given that $f(x) = 2 + 5x$, $a = -2$ and $L = -8$. We want to find δ such that

$$|(2 + 5x) - (-8)| < 0.002 \quad \text{whenever} \quad 0 < |x - (-2)| < \delta$$

Because

$$|(2 + 5x) - (-8)| = |5x + 10|$$
$$= 5|x + 2|$$

and

$$|x - (-2)| = |x + 2|$$

we must find δ such that

$$5|x + 2| < 0.002 \quad \text{whenever} \quad 0 < |x + 2| < \delta$$

or, equivalently,

$$|x + 2| < 0.0004 \quad \text{whenever} \quad 0 < |x + 2| < \delta$$

Therefore, we take $\delta = 0.0004$

8. $\lim\limits_{x \to \frac{1}{3}} \dfrac{9x^2 - 1}{3x - 1} = 2; \epsilon = 0.01$

SOLUTION: We are given that $f(x) = \dfrac{9x^2 - 1}{3x - 1}$, $a = \frac{1}{3}$, and $L = 2$. We want to find δ such that

$$\left| \frac{9x^2 - 1}{3x - 1} - 2 \right| < 0.01 \quad \text{whenever} \quad 0 < \left| x - \frac{1}{3} \right| < \delta \tag{1}$$

If $0 < |x - \frac{1}{3}|$, then $0 < |3x - 1|$ and $3x - 1 \neq 0$. Thus,

$$\frac{9x^2 - 1}{3x - 1} - 2 = \frac{(3x + 1)(3x - 1)}{3x - 1} - 2$$

$$= (3x + 1) - 2$$
$$= 3x - 1 \tag{2}$$

By substitution from (2) into (1) we obtain

$$|3x - 1| < 0.01 \quad \text{whenever} \quad 0 < |x - \tfrac{1}{3}| < \delta$$

or, equivalently,

$$|x - \tfrac{1}{3}| < \tfrac{0.01}{3} \quad \text{whenever} \quad 0 < |x - \tfrac{1}{3}| < \delta$$

Therefore, we take $\delta = \frac{0.01}{3} = \frac{1}{300}$.

In Exercises 9-29, establish the limit by using Definition 2.1.1; that is, for any $\epsilon > 0$, find a $\delta > 0$, such that $|f(x) - L| < \epsilon$ whenever $0 < |x - a| < \delta$.

10. $\lim\limits_{x \to -2} (7 - 2x) = 11$

SOLUTION: For any $\epsilon > 0$ we must find a $\delta > 0$ such that

$$|(7 - 2x) - 11| < \epsilon \quad \text{whenever} \quad 0 < |x - (-2)| < \delta \tag{1}$$

because

$$|(7 - 2x) - 11| = |-2x - 4|$$
$$= |-2| \cdot |x + 2|$$
$$= 2|x + 2| \tag{2}$$

and

$$|x - (-2)| = |x + 2| \tag{3}$$

by substitution from (2) and (3) into (1) we obtain

$$2|x + 2| < \epsilon \quad \text{whenever} \quad 0 < |x + 2| < \delta$$

or, equivalently,

$$|x + 2| < \frac{\epsilon}{2} \quad \text{whenever} \quad 0 < |x + 2| < \delta$$

Therefore we must have $\delta \leqslant \frac{1}{2}\epsilon$, and we may prove the limit as follows.

PROOF: For any $\epsilon > 0$ let $\delta = \frac{1}{2}\epsilon$. Then $\delta > 0$ and if $0 < |x - (-2)| < \delta$, it follows that

$$|x + 2| < \delta$$

$$|x + 2| < \frac{\epsilon}{2}$$

$$2|x + 2| < \epsilon \tag{4}$$

By substitution from (2) into (4) we obtain

$$|(7 - 2x) - 11| < \epsilon$$

Therefore, by Definition 2.1.1, $\lim\limits_{x \to -2} (7 - 2x) = 11$

14. $\lim\limits_{x \to -3} x^2 = 9$

SOLUTION: For any $\epsilon > 0$ we must find a $\delta > 0$ such that

$$|x^2 - 9| < \epsilon \quad \text{whenever} \quad 0 < |x - (-3)| < \delta$$

or, equivalently,

$$|x + 3| \cdot |x - 3| < \epsilon \quad \text{whenever} \quad 0 < |x + 3| < \delta \tag{1}$$

We find an upper bound for $|x - 3|$. Suppose that $\delta \leqslant 1$. Then whenever $0 < |x + 3| < \delta$, it follows that

$$|x + 3| < 1$$
$$-1 < x + 3 < 1$$
$$-4 < x < -2$$
$$-7 < x - 3 < -5$$
$$5 < -x + 3 < 7$$

and because $|-x + 3| = |x - 3|$, we have

$$5 < |x - 3| < 7 \tag{2}$$

Because $|x + 3| < \delta$, by (2) and Theorem 1.1.26 (If $0 < a < b$ and $0 < c < d$, then $ac < bd$) we obtain

$$|x + 3| \cdot |x - 3| < 7\delta \tag{3}$$

Therefore, by (1) and (3) we must have $7\delta \leqslant \epsilon$, or $\delta \leqslant \frac{1}{7}\epsilon$. We may now prove the limit.

PROOF: For any $\epsilon > 0$ let $\delta = \min(1, \frac{1}{7}\epsilon)$. Then $\delta > 0, \delta \leqslant 1$ and $\delta \leqslant \frac{1}{7}\epsilon$. Whenever $0 < |x - (-3)| < \delta$, it follows that $|x + 3| < \delta$, and hence,

$$|x + 3| < \frac{\epsilon}{7} \tag{4}$$

Moreover, since $|x + 3| < 1$ inequality (2) follows. By (2), (4), and Theorem 1.1.26 we obtain

$$|x + 3| \cdot |x - 3| < \frac{\epsilon}{7} \cdot 7$$

or, equivalently,

$$|x^2 - 9| < \epsilon$$

Thus, by Definition 2.1.1, $\lim_{x \to -3} x^2 = 9$.

16. $\lim_{x \to 6} \dfrac{x}{x - 3} = 2$

SOLUTION: For any $\epsilon > 0$ we must find a $\delta > 0$ such that

$$\left|\frac{x}{x - 3} - 2\right| < \epsilon \quad \text{whenever} \quad 0 < |x - 6| < \delta \tag{1}$$

Because

$$\left|\frac{x}{x - 3} - 2\right| = \left|\frac{x - 2(x - 3)}{x - 3}\right|$$

$$= \left|\frac{-x + 6}{x - 3}\right|$$

$$= |x - 6| \cdot \frac{1}{|x - 3|} \tag{2}$$

by substitution from (2) into (1), we see that

$$|x - 6| \cdot \frac{1}{|x - 3|} < \epsilon \quad \text{whenever} \quad 0 < |x - 6| < \delta \tag{3}$$

Therefore, we must find an upper bound for $1/|x - 3|$. Suppose $\delta \leqslant 1$. Then whenever $0 < |x - 6| < \delta$, we have

$$|x - 6| < 1$$
$$-1 < x - 6 < 1$$
$$2 < x - 3 < 4$$
$$2 < |x - 3| < 4$$

Taking the reciprocal of each member and reversing the sense of the inequality, we have

$$\frac{1}{4} < \frac{1}{|x - 3|} < \frac{1}{2} \tag{4}$$

Because $|x - 6| < \delta$, by (4) and Theorem 1.1.26, we have

$$|x - 6| \cdot \frac{1}{|x - 3|} < \delta \cdot \frac{1}{2} \tag{5}$$

Thus, by (5) and (3) we must have $\frac{1}{2}\delta \leqslant \epsilon$ or $\delta \leqslant 2\epsilon$. The proof follows.

PROOF: For any $\epsilon > 0$ let $\delta = \min(1, 2\epsilon)$. Then $\delta > 0, \delta \leqslant 1$, and $\delta \leqslant 2\epsilon$. Whenever $0 < |x - 6| < \delta$, it follows that $|x - 6| < \delta$ and hence,

$$|x - 6| < 2\epsilon \tag{6}$$

Furthermore, since $|x - 6| < 1$ inequality (4) follows. By (4), (6), and Theorem 1.1.26, we obtain

$$|x - 6| \cdot \frac{1}{|x - 3|} < 2\epsilon \cdot \frac{1}{2}$$

which, by substitution from (2), is equivalent to

$$\left| \frac{x}{x - 3} - 2 \right| < \epsilon$$

Therefore, by Definition 2.1.1

$$\lim_{x \to 6} \frac{x}{x - 3} = 2$$

18. $\lim_{x \to -4} \dfrac{1}{x + 3} = -1$

SOLUTION: For any $\epsilon > 0$ we must find a $\delta > 0$ such that

$$\left| \frac{1}{x + 3} - (-1) \right| < \epsilon \quad \text{whenever} \quad 0 < |x - (-4)| < \delta \tag{1}$$

Because

$$\left| \frac{1}{x + 3} - (-1) \right| = \left| \frac{x + 4}{x + 3} \right| \tag{2}$$

by substitution from Eq. (2) into (1), we must have

$$|x + 4| \cdot \frac{1}{|x + 3|} < \epsilon \quad \text{whenever} \quad 0 < |x + 4| < \delta \tag{3}$$

We must find an upper bound for $1/|x + 3|$. If $\delta \leqslant \frac{1}{2}$ and $0 < |x + 4| < \delta$, then

$$|x + 4| < \tfrac{1}{2}$$
$$-\tfrac{1}{2} < x + 4 < \tfrac{1}{2}$$
$$-\tfrac{3}{2} < x + 3 < -\tfrac{1}{2}$$
$$\tfrac{1}{2} < -(x + 3) < \tfrac{3}{2}$$
$$\tfrac{1}{2} < |-(x + 3)| < \tfrac{3}{2}$$

Taking reciprocals and replacing $|-(x + 3)|$ with $|x + 3|$, we obtain

$$\frac{2}{3} < \frac{1}{|x + 3|} < 2 \tag{4}$$

Note that we are unable to find an upper bound for $1/|x + 3|$ if we assume that $\delta \leqslant 1$ as in Exercise 16. For then if $0 < |x + 4| < \delta$, it follows that

$$|x + 4| < 1$$
$$-1 < x + 4 < 1$$
$$-2 < x + 3 < 0$$
$$0 < |x + 3| < 2 \tag{5}$$

Because the reciprocal of 0 does not exist, we cannot find an upper bound for $1/|x + 3|$ from inequality (5).

Since $|x + 4| < \delta$, by (4) and Theorem 1.1.26 we have

$$|x + 4| \cdot \frac{1}{|x + 3|} < 2\delta \tag{6}$$

By (3) and (6) we must have $2\delta \leqslant \epsilon$ or $\delta \leqslant \frac{1}{2}\epsilon$. The proof of the limit follows.

PROOF: For any $\epsilon > 0$ let $\delta = \min(\frac{1}{2}, \frac{1}{2}\epsilon)$. Then $\delta > 0, \delta \leqslant \frac{1}{2}$, and $\delta \leqslant \frac{1}{2}\epsilon$. Thus, whenever $0 < |x - (-4)| < \delta$, it follows that $|x + 4| < \delta$. Hence,

$$|x + 4| < \tfrac{1}{2}\epsilon \tag{7}$$

Moreover, since $|x + 4| < \frac{1}{2}$, inequality (4) follows. By (4), (7), and Theorem 1.1.26, we have

$$|x + 4| \cdot \frac{1}{|x + 3|} < \frac{\epsilon}{2} \cdot 2$$

which, by substitution from Eq. (2), is equivalent to

$$\left| \frac{1}{x + 3} - (-1) \right| < \epsilon$$

Therefore, by Definition 2.1.1

$$\lim_{x \to -4} \frac{1}{x + 3} = -1$$

24. $\displaystyle \lim_{x \to 1} \frac{1}{\sqrt{5 - x}} = \frac{1}{2}$

SOLUTION: For any $\epsilon > 0$ we must find a $\delta > 0$ such that

$$\left| \frac{1}{\sqrt{5 - x}} - \frac{1}{2} \right| < \epsilon \quad \text{whenever} \quad 0 < |x - 1| < \delta \tag{1}$$

Because

$$
\begin{aligned}
\left| \frac{1}{\sqrt{5 - x}} - \frac{1}{2} \right| &= \left| \frac{2 - \sqrt{5 - x}}{2\sqrt{5 - x}} \right| \\
&= \left| \frac{2 - \sqrt{5 - x}}{2\sqrt{5 - x}} \cdot \frac{2 + \sqrt{5 - x}}{2 + \sqrt{5 - x}} \right| \\
&= \left| \frac{4 - (5 - x)}{2\sqrt{5 - x}\,(2 + \sqrt{5 - x})} \right| \\
&= \left| \frac{-1 + x}{2\sqrt{5 - x}\,(2 + \sqrt{5 - x})} \right| \\
&= |x - 1| \cdot \frac{1}{2\sqrt{5 - x}\,(2 + \sqrt{5 - x})} \tag{2}
\end{aligned}
$$

By substitution from Eq. (2) into (1) we must have

$$|x - 1| \cdot \frac{1}{2\sqrt{5 - x}\,(2 + \sqrt{5 - x})} < \epsilon \quad \text{whenever} \quad 0 < |x - 1| < \delta \tag{3}$$

We find an upper bound for the fraction in (3). Suppose that $\delta \leqslant 1$. Whenever $0 < |x - 1| < \delta$, we have

$$
\begin{aligned}
|x - 1| &< 1 \\
-1 < x - 1 \quad &< 1 \\
0 < x \quad &< 2 \\
-2 < -x \quad &< 0 \\
3 < 5 - x \quad &< 5 \\
\sqrt{3} < \sqrt{5 - x} \quad &< \sqrt{5} \tag{4}
\end{aligned}
$$

$$2 + \sqrt{3} < 2 + \sqrt{5 - x} < 2 + \sqrt{5} \tag{5}$$

By (4), (5), and Theorem 1.1.26, we have

$$\sqrt{3}\,(2+\sqrt{3}) < \sqrt{5-x}\,(2+\sqrt{5-x}) < \sqrt{5}\,(2+\sqrt{5})$$

$$\frac{1}{\sqrt{5}\,(2+\sqrt{5})} < \frac{1}{\sqrt{5-x}\,(2+\sqrt{5-x})} < \frac{1}{\sqrt{3}\,(2+\sqrt{3})}$$

Hence

$$\frac{1}{2\sqrt{5-x}\,(2+\sqrt{5-x})} < \frac{1}{2\sqrt{3}\,(2+\sqrt{3})} \tag{6}$$

Because $|x-1| < \delta$, by (6) and Theorem 1.1.26 we have

$$|x-1| \cdot \frac{1}{2\sqrt{5-x}\,(2+\sqrt{5-x})} < \frac{\delta}{2\sqrt{3}\,(2+\sqrt{3})} \tag{7}$$

By (3) and (7) we must have

$$\frac{\delta}{2\sqrt{3}\,(2+\sqrt{3})} \leqslant \epsilon$$

$$\delta \leqslant 2\sqrt{3}\,(2+\sqrt{3})\epsilon$$

The proof of the limit follows.

PROOF: For any $\epsilon > 0$ let $\delta = \min(1, 2\sqrt{3}\,(2+\sqrt{3})\epsilon)$. Then $\delta > 0$, $\delta \leqslant 1$, and $\delta \leqslant 2\sqrt{3}\,(2+\sqrt{3})\epsilon$. Whenever $0 < |x-1| < \delta$, it follows that

$$|x-1| < 2\sqrt{3}\,(2+\sqrt{3})\epsilon \tag{8}$$

Moreover, since $|x-1| < 1$, inequality (6) follows. By (6), (8), and Theorem 1.1.26, we obtain

$$|x-1| \cdot \frac{1}{2\sqrt{5-x}\,(2+\sqrt{5-x})} < 2\sqrt{3}\,(2+\sqrt{3})\epsilon \cdot \frac{1}{2\sqrt{3}\,(2+\sqrt{3})} \tag{9}$$

By substitution from Eq. (2) into (9) we have

$$\left| \frac{1}{\sqrt{5-x}} - \frac{1}{2} \right| < \epsilon$$

Therefore, by Definition 2.1.1

$$\lim_{x \to 1} \frac{1}{\sqrt{5-x}} = \frac{1}{2}$$

28. Prove that $\displaystyle\lim_{x \to a} \sqrt[3]{x} = \sqrt[3]{a}$.

[Hint: $a^3 - b^3 = (a-b)(a^2 + ab + b^2)$.]

SOLUTION: First, we assume that $a > 0$.

For any $\epsilon > 0$ we must find a $\delta > 0$ such that

$$|\sqrt[3]{x} - \sqrt[3]{a}| < \epsilon \quad \text{whenever} \quad 0 < |x-a| < \delta \tag{1}$$

Because $a^3 - b^3 = (a-b)(a^2 + ab + b^2)$,

$$x - a = (\sqrt[3]{x} - \sqrt[3]{a})(\sqrt[3]{x^2} + \sqrt[3]{xa} + \sqrt[3]{a^2})$$

and

$$|\sqrt[3]{x} - \sqrt[3]{a}| = |x-a| \cdot \frac{1}{|\sqrt[3]{x^2} + \sqrt[3]{xa} + \sqrt[3]{a^2}|} \tag{2}$$

We find an upper bound for the fraction on the right side of Eq. (2).

Suppose $\delta \leqslant a$. Whenever $0 < |x - a| < \delta$, we have

$$|x - a| < a$$
$$-a < x - a < a$$
$$0 < x < 2a$$

Because $x > 0$ and $a > 0$, then $\sqrt[3]{x^2} > 0$, $\sqrt[3]{xa} > 0$, and hence

$$|\sqrt[3]{x^2} + \sqrt[3]{xa} + \sqrt[3]{a^2}| > \sqrt[3]{a^2}$$

$$\frac{1}{|\sqrt[3]{x^2} + \sqrt[3]{xa} + \sqrt[3]{a^2}|} < \frac{1}{\sqrt[3]{a^2}} \qquad (3)$$

Because $|x - a| < \delta$, by (3) and Theorem 1.1.26 we have

$$|x - a| \cdot \frac{1}{|\sqrt[3]{x^2} + \sqrt[3]{xa} + \sqrt[3]{a^2}|} < \delta \cdot \frac{1}{\sqrt[3]{a^2}} \qquad (4)$$

By (1), (2), and (4) we see that we must have

$$\frac{\delta}{\sqrt[3]{a^2}} \leqslant \epsilon$$

$$\delta \leqslant \sqrt[3]{a^2}\,\epsilon$$

Next, we assume that $a < 0$. Then Eq. (2) remains true, and we can obtain (3) as follows. Suppose $\delta \leqslant -a$. Whenever $0 < |x - a| < \delta$, we have

$$|x - a| < -a$$
$$a < x - a < -a$$
$$2a < x < 0$$

Because $x < 0$ and $a < 0$, then $\sqrt[3]{x^2} > 0$, $\sqrt[3]{xa} > 0$, and hence inequality (3) follows. From (3) we obtain (4), and thus $\delta \leqslant \sqrt[3]{a^2}\,\epsilon$.

The proof of the limit follows.

PROOF: If $a \neq 0$, for any $\epsilon > 0$ let $\delta = \min(|a|, \sqrt[3]{a^2}\,\epsilon)$. Then $\delta > 0$, $\delta \leqslant |a|$ and $\delta \leqslant \sqrt[3]{a^2}\,\epsilon$. Whenever $0 < |x - a| < \delta$, it follows that

$$|x - a| < \sqrt[3]{a^2}\,\epsilon \qquad (5)$$

Moreover, since $\delta \leqslant |a|$, then $\delta \leqslant a$ if $a > 0$ and $\delta \leqslant -a$ if $a < 0$. In either case, inequality (3) follows. By (3), (5), and Theorem 1.1.26, we obtain

$$|x - a| \cdot \frac{1}{|\sqrt[3]{x^2} + \sqrt[3]{xa} + \sqrt[3]{a^2}|} < \sqrt[3]{a^2}\,\epsilon \cdot \frac{1}{\sqrt[3]{a^2}} \qquad (6)$$

By substitution from Eq. (2) into (6) we have

$$|\sqrt[3]{x} - \sqrt[3]{a}| < \epsilon$$

If $a = 0$, for any $\epsilon > 0$ let $\delta = \epsilon^3$. Whenever $0 < |x - a| < \delta$, we have $0 < |x| < \epsilon^3$. Hence, $|\sqrt[3]{x}| < \sqrt[3]{\epsilon^3} = \epsilon$, and since $\sqrt[3]{a} = 0$,

$$|\sqrt[3]{x} - \sqrt[3]{a}| < \epsilon$$

Then, by Definition 2.1.1

$$\lim_{x \to a} \sqrt[3]{x} = \sqrt[3]{a}$$

30. Prove that, if $f(x) = g(x)$ for all values of x except $x = a$, then $\lim\limits_{x \to a} f(x) = \lim\limits_{x \to a} g(x)$ if the limits exist.

SOLUTION: Let $\lim\limits_{x \to a} g(x) = L$. We must show that $\lim\limits_{x \to a} f(x) = L$. Then for any $\epsilon > 0$

we must find a $\delta > 0$ such that $|f(x) - L| < \epsilon$ whenever $0 < |x - a| < \delta$.

Because $\lim\limits_{x \to a} g(x) = L$ by Definition 2.1.1, for any $\epsilon > 0$ there must be some $\delta_1 > 0$ such that

$$|g(x) - L| < \epsilon \quad \text{whenever} \quad 0 < |x - a| < \delta_1 \tag{1}$$

PROOF: For any $\epsilon > 0$, let $\delta = \delta_1$. Whenever $0 < |x - a| < \delta$, it follows that $0 < |x - a| < \delta_1$, and by (1) we have

$$|g(x) - L| < \epsilon \tag{2}$$

Because $0 < |x - a|$, then $x \neq a$, and by hypothesis

$$f(x) = g(x) \tag{3}$$

By substitution from (3) into (2) we obtain

$$|f(x) - L| < \epsilon$$

Hence, by Definition 2.1.1, $\lim\limits_{x \to a} f(x) = L = \lim\limits_{x \to a} g(x)$.

2.2 THEOREMS ON LIMITS OF FUNCTIONS

The following theorems may often be used to find the limit of a function.

2.2.1 Limit Theorem 1 If m and b are any constants,

$$\lim_{x \to a} (mx + b) = ma + b$$

2.2.2 Limit Theorem 2 If c is any constant, then for any number a, $\lim\limits_{x \to a} c = c$

2.2.3 Limit Theorem 3 $\lim\limits_{x \to a} x = a$

2.2.4 Limit Theorem 4 If $\lim\limits_{x \to a} f(x) = L$ and $\lim\limits_{x \to a} g(x) = M$, then $\lim\limits_{x \to a} [f(x) \pm g(x)] = L \pm M$

2.2.5 Limit Theorem 5 If $\lim\limits_{x \to a} f_1(x) = L_1, \lim\limits_{x \to a} f_2(x) = L_2, \ldots,$ and $\lim\limits_{x \to a} f_n(x) = L_n$, then

$$\lim_{x \to a} [f_1(x) \pm f_2(x) \pm \cdots \pm f_n(x)] = L_1 \pm L_2 \pm \cdots \pm L_n$$

2.2.6 Limit Theorem 6 If $\lim\limits_{x \to a} f(x) = L$ and $\lim\limits_{x \to a} g(x) = M$, then $\lim\limits_{x \to a} f(x) \cdot g(x) = L \cdot M$

2.2.7 Limit Theorem 7 If $\lim\limits_{x \to a} f_1(x) = L_1, \lim\limits_{x \to a} f_2(x) = L_2, \cdots,$ and $\lim\limits_{x \to a} f_n(x) = L_n$, then

$$\lim_{x \to a} [f_1(x) f_2(x) \cdots f_n(x)] = L_1 L_2 \cdots L_n$$

2.2.8 Limit Theorem 8 If $\lim\limits_{x \to a} f(x) = L$ and n is any positive integer, then

$$\lim_{x \to a} [f(x)]^n = L^n$$

2.2.9 Limit Theorem 9 If $\lim\limits_{x \to a} f(x) = L$ and $\lim\limits_{x \to a} g(x) = M$, and $M \neq 0$, then

$$\lim_{x \to a} \frac{f(x)}{g(x)} = \frac{L}{M}$$

2.2.10 Limit Theorem 10 If $\lim\limits_{x \to a} f(x) = L$, then

$$\lim_{x \to a} \sqrt[n]{f(x)} = \sqrt[n]{L}$$

if $L > 0$ and n is any positive integer, or if $L \leqslant 0$ and n is a positive odd integer.

Limit Theorems 1-8 imply that the limit as x approaches a of any polynomial function defined by $P(x)$, is $P(a)$ (the value of the polynomial if x is replaced by a), and Limit Theorem 9 implies that the limit as x approaches a of any rational function defined by $R(x) = P(x)/Q(x)$, is $R(a)$ (the value of the rational function if x is replaced by a), provided that $Q(a) \neq 0$.

Note that we cannot use Limit Theorem 9 to find the limit of a fraction if the limit of the denominator is zero. However, when the limit of both the numerator and denominator is zero, we may often find the limit by first factoring and then dividing both the numerator and denominator by the common factor, as illustrated in Exercise 6. In so doing, we make use of the following theorem which follows from Exercise 30 of Exercises 2.1.

Theorem If $\lim\limits_{x \to a} g(x) = L$ and $f(x) = g(x)$ for all $x \neq a$, then $\lim\limits_{x \to a} f(x) = L$.

We cannot use Limit Theorems 1-10 to find the limit of a fraction for which the limit of the denominator is zero and for which the limit of the numerator is not zero. In Section 2.4 we have a theorem for finding the limit of such a fraction.

Exercises 2.2

In Exercises 1-17, find the value of the limit, and when applicable, indicate the limit theorems being used.

2. $\lim\limits_{y \to -1} (y^3 - 2y^2 + 3y - 4)$

SOLUTION:

$$\lim_{y \to -1} (y^3 - 2y^2 + 3y - 4) = \lim_{y \to -1} y^3 - \lim_{y \to -1} 2y^2 + \lim_{y \to -1} (3y - 4) \quad \text{(L.T. 5)}$$

$$= (\lim_{y \to -1} y)^3 - \lim_{y \to -1} 2 \cdot (\lim_{y \to -1} y)^2 + \lim_{y \to -1} (3y - 4)$$
$$\text{(L.T. 8 and L.T. 6)}$$

$$= (-1)^3 - 2 \cdot (-1)^2 + [3(-1) - 4] \quad \text{(L.T. 3, L.T. 2,}$$
$$\text{and L.T. 1)}$$

$$= -10$$

6. $\lim\limits_{s \to 1} \dfrac{s^3 - 1}{s - 1}$

SOLUTION: We cannot use Limit Theorem 9 because the limit of the denominator is 0. However, by factoring we have

$$\lim_{s \to 1} \frac{s^3 - 1}{s - 1} = \lim_{s \to 1} \frac{(s - 1)(s^2 + s + 1)}{s - 1}$$

$$= \lim_{s \to 1} (s^2 + s + 1) \quad \text{(Ex. 30 in Ex. 2.1)}$$

$$= \lim_{s \to 1} s^2 + \lim_{s \to 1} (s + 1) \quad \text{(L.T. 4)}$$

$$= \lim_{s \to 1} s \, \lim_{s \to 1} s + \lim_{s \to 1} (s + 1) \quad \text{(L.T. 6)}$$

$$= 1 \cdot 1 + (1 + 1) \quad \text{(L.T. 3 and L.T. 1)}$$
$$= 3$$

12. $\lim\limits_{t \to \frac{3}{2}} \sqrt{\dfrac{8t^3 - 27}{4t^2 - 9}}$

SOLUTION:

$$\lim_{t \to \frac{3}{2}} \sqrt{\frac{8t^3 - 27}{4t^2 - 9}} = \sqrt{\lim_{t \to \frac{3}{2}} \frac{8t^3 - 27}{4t^2 - 9}} \qquad \text{(L.T. 10)}$$

$$= \sqrt{\lim_{t \to \frac{3}{2}} \frac{(2t - 3)(4t^2 + 6t + 9)}{(2t - 3)(2t + 3)}}$$

$$= \sqrt{\lim_{t \to \frac{3}{2}} \frac{4t^2 + 6t + 9}{2t + 3}} \qquad \text{(Ex. 30 in Ex. 2.1)}$$

$$= \sqrt{\frac{\displaystyle\lim_{t \to \frac{3}{2}} (4t^2 + 6t + 9)}{\displaystyle\lim_{t \to \frac{3}{2}} (2t + 3)}} \qquad \text{(L.T. 9, because } \lim_{t \to \frac{3}{2}} (2t + 3) \neq 0)$$

$$= \sqrt{\frac{\displaystyle\lim_{t \to \frac{3}{2}} 4 \cdot \lim_{t \to \frac{3}{2}} t \cdot \lim_{t \to \frac{3}{2}} t + \lim_{t \to \frac{3}{2}} (6t + 9)}{\displaystyle\lim_{t \to \frac{3}{2}} (2t + 3)}} \qquad \text{(L.T. 4 and L.T. 7)}$$

$$= \sqrt{\frac{4 \cdot \frac{3}{2} \cdot \frac{3}{2} + (6 \cdot \frac{3}{2} + 9)}{2 \cdot \frac{3}{2} + 3}} \qquad \text{(L.T. 1, L.T. 2, and L.T. 3)}$$

$$= \tfrac{3}{2}\sqrt{2}$$

14. $\displaystyle\lim_{t \to 0} \frac{2 - \sqrt{4 - t}}{t}$

SOLUTION: The limit of both the numerator and denominator is zero. We rationalize the numerator.

$$\lim_{t \to 0} \frac{2 - \sqrt{4 - t}}{t} = \lim_{t \to 0} \frac{2 - \sqrt{4 - t}}{t} \cdot \frac{2 + \sqrt{4 - t}}{2 + \sqrt{4 - t}}$$

$$= \lim_{t \to 0} \frac{4 - (4 - t)}{t(2 + \sqrt{4 - t})}$$

$$= \lim_{t \to 0} \frac{t}{t(2 + \sqrt{4 - t})}$$

$$= \lim_{t \to 0} \frac{1}{(2 + \sqrt{4 - t})} \qquad \text{(Ex. 30 in Ex. 2.1)}$$

$$= \frac{\displaystyle\lim_{t \to 0} 1}{\displaystyle\lim_{t \to 0} (2 + \sqrt{4 - t})} \qquad \text{(L.T. 9 because } \lim_{t \to 0} (2 + \sqrt{4 - t}) \neq 0)$$

$$= \frac{\displaystyle\lim_{t \to 0} 1}{\displaystyle\lim_{t \to 0} 2 + \sqrt{\lim_{t \to 0} (4 - t)}} \qquad \text{(L.T. 4 and L.T. 10)}$$

$$= \frac{1}{2 + \sqrt{4}} = \frac{1}{4} \qquad \text{(L.T. 1 and L.T. 2)}$$

22. Given that f is the function defined by

$$f(x) = \begin{cases} 2x - 1 & \text{if } x \neq 2 \\ 1 & \text{if } x = 2 \end{cases}$$

(a) Find $\lim\limits_{x \to 2} f(x)$, and show that $\lim\limits_{x \to 2} f(x) \neq f(2)$

(b) Draw a sketch of the graph of f.

SOLUTION: (a) To find $\lim\limits_{x \to 2} f(x)$ we consider replacements of x that are close to 2 but not equal to 2. Thus,

$$\lim_{x \to 2} f(x) = \lim_{x \to 2} (2x - 1)$$
$$= 3$$

and because $f(2) = 1$, then $\lim\limits_{x \to 2} f(x) \neq f(2)$.

(b) The graph of f is shown in Fig. 2.2.22.

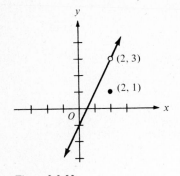

Figure 2.2.22

26. Prove that if $\lim\limits_{x \to a} f(x) = L$, then

$$\lim_{x \to a} (f(x) - L) = 0$$

SOLUTION:

$$\lim_{x \to a} (f(x) - L) = \lim_{x \to a} f(x) - \lim_{x \to a} L \qquad \text{(L.T. 4)}$$
$$= L - L \qquad\qquad \text{(Hypothesis and L.T. 2)}$$
$$= 0$$

27. Using Definition 2.1.1, prove that if

$$\lim_{x \to a} f(x) = L \quad \text{and} \quad \lim_{x \to a} g(x) = 0$$

then

$$\lim_{x \to a} f(x) \cdot g(x) = 0$$

SOLUTION: For any $\epsilon > 0$ we must find a $\delta > 0$ such that

$$|f(x) \cdot g(x)| < \epsilon \quad \text{whenever} \quad 0 < |x - a| < \delta \qquad\qquad \textbf{(1)}$$

Because $\lim\limits_{x \to a} f(x) = L$ for any $\epsilon_1 > 0$, there is a $\delta_1 > 0$ such that

$$|f(x) - L| < \epsilon_1 \quad \text{whenever} \quad 0 < |x - a| < \delta_1$$

In particular, if $\epsilon_1 = 1$, we have

$$|f(x) - L| < 1 \quad \text{whenever} \quad 0 < |x - a| < \delta_1 \qquad\qquad \textbf{(2)}$$

By Exercise 34 in Exercises 1.2 ($|a| - |b| \leq |a - b|$), we have

$$|f(x)| - |L| \leq |f(x) - L| \qquad\qquad \textbf{(3)}$$

From (3) and (2) we obtain

$$|f(x)| - |L| < 1 \quad \text{whenever} \quad 0 < |x - a| < \delta_1$$

or, equivalently,

$$|f(x)| < 1 + |L| \quad \text{whenever} \quad 0 < |x - a| < \delta_1 \qquad\qquad \textbf{(4)}$$

Because $\lim\limits_{x \to a} g(x) = 0$, for any $\epsilon_2 > 0$ there is a $\delta_2 > 0$ such that

$$|g(x)| < \epsilon_2 \quad \text{whenever} \quad 0 < |x - a| < \delta_2 \qquad\qquad \textbf{(5)}$$

For any $\epsilon > 0$, let $\epsilon_2 = \dfrac{\epsilon}{1 + |L|}$ and (5) becomes

$$|g(x)| < \frac{\epsilon}{1 + |L|} \quad \text{whenever} \quad 0 < |x - a| < \delta_2 \tag{6}$$

Choose $\delta = \min(\delta_1, \delta_2)$. Then $\delta \leqslant \delta_1$ and $\delta \leqslant \delta_2$. Whenever $0 < |x - a| < \delta$ it follows that $0 < |x - a| < \delta_1$ and by (4) we obtain

$$|f(x)| < 1 + |L| \tag{7}$$

Moreover, since $0 < |x - a| < \delta_2$, by (6) we obtain

$$|g(x)| < \frac{\epsilon}{1 + |L|} \tag{8}$$

and from (7) and (8)

$$|f(x) \cdot g(x)| = |f(x)| \cdot |g(x)| < (1 + |L|) \cdot \frac{\epsilon}{1 + |L|} = \epsilon$$

whenever $0 < |x - a| < \delta$. Therefore,

$$\lim_{x \to a} f(x) \cdot g(x) = 0$$

28. Prove Limit Theorem 6: If $\lim\limits_{x \to a} f(x) = L$ and $\lim\limits_{x \to a} g(x) = M$, then $\lim\limits_{x \to a} [f(x) \cdot g(x)] = L \cdot M$.

SOLUTION: By algebra

$$f(x) \cdot g(x) = [f(x) - L]g(x) + L[g(x) - M] + L \cdot M \tag{1}$$

Because $\lim\limits_{x \to a} f(x) = L$, by Exercise 26,

$$\lim_{x \to a} [f(x) - L] = 0 \tag{2}$$

Because $\lim\limits_{x \to a} g(x) = M$, by Exercise 27 and (2),

$$\lim_{x \to a} [f(x) - L]g(x) = 0 \tag{3}$$

Moreover, by Exercise 26

$$\lim_{x \to a} [g(x) - M] = 0$$

Hence, by Exercise 27

$$\lim_{x \to a} L[g(x) - M] = 0 \tag{4}$$

Therefore by (1), L.T. 5, (3), (4), and L.T. 2, we have

$$\lim_{x \to a} f(x) \cdot g(x) = \lim_{x \to a} [f(x) - L]g(x) + \lim_{x \to a} L[g(x) - M] + \lim_{x \to a} L \cdot M$$
$$= 0 + 0 + L \cdot M$$
$$= L \cdot M$$

2.3 ONE-SIDED LIMITS   The definitions for one-sided limits are as follows.

2.3.1 Definition   Let f be a function that is defined at every number in some open interval (a, c). Then the *limit of* f(x), *as* x *approaches* a *from the right, is* L, written

$$\lim_{x \to a^+} f(x) = L$$

if for any $\epsilon > 0$, however small, there exists a $\delta > 0$ such that

$$|f(x) - L| < \epsilon \quad \text{whenever} \quad 0 < x - a < \delta$$

2.3.2 Definition   Let f be a function that is defined at every number in some open interval (d, a). Then the *limit of* f(x), *as* x *approaches* a *from the left, is* L, written

$$\lim_{x \to a^-} f(x) = L$$

if for any $\epsilon > 0$, however small, there exists a $\delta > 0$ such that

$$|f(x) - L| < \epsilon \quad \text{whenever} \quad -\delta < x - a < 0$$

We note that the inequality $0 < x - a < \delta$, which appears in Definition 2.3.1, is equivalent to $a < x < a + \delta$. Similarly, $-\delta < x - a < 0$, which appears in Definition 2.3.2, is equivalent to $a - \delta < x < a$.

It can be shown that Limit Theorems 1–10 hold if "$x \to a$" is replaced by "$x \to a^+$" or "$x \to a^-$". Moreover, the following theorem follows from the definitions.

2.3.3 Theorem $\lim_{x \to a} f(x) = L$ if and only if $\lim_{x \to a^+} f(x) = L$ and $\lim_{x \to a^-} f(x) = L$.

Exercises 2.3

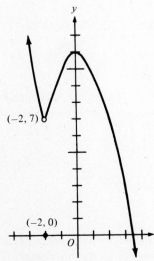

Figure 2.3.2

In Exercises 1–10, draw a sketch of the graph and find the indicated limit if it exists; if the limit does not exist, give the reason.

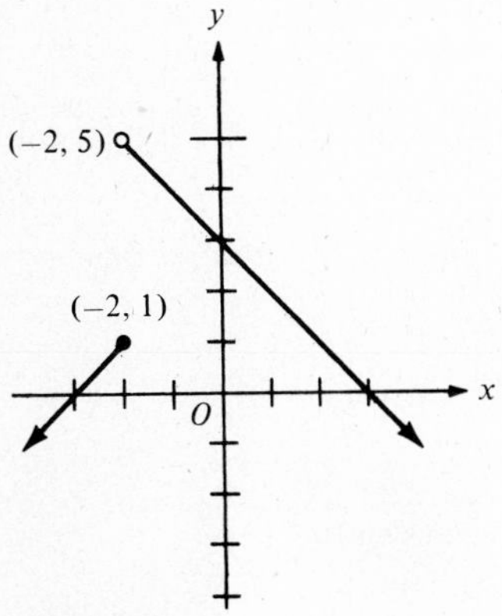

2. $g(s) = \begin{cases} s + 3 & \text{if } s \leq -2 \\ 3 - s & \text{if } -2 < s \end{cases}$

(a) $\lim_{s \to -2^+} g(s)$ (b) $\lim_{s \to -2^-} g(s)$ (c) $\lim_{s \to -2} g(s)$

SOLUTION: A sketch of the graph is shown in Fig. 2.3.2.

(a) Because $g(s) = 3 - s$ if $s > -2$, then

$$\lim_{s \to -2^+} g(s) = \lim_{s \to -2^+} (3 - s) = 5$$

(b) Because $g(s) = s + 3$ if $s < -2$, then

$$\lim_{s \to -2^-} g(s) = \lim_{s \to -2^-} (s + 3) = 1$$

(c) Because $\lim_{s \to -2^+} g(s) \neq \lim_{s \to -2^-} g(s)$, by Theorem 2.3.3 $\lim_{s \to -2} g(s)$ does not exist.

6. $g(t) = \begin{cases} 3 + t^2 & \text{if } t < -2 \\ 0 & \text{if } t = -2 \\ 11 - t^2 & \text{if } -2 < t \end{cases}$

(a) $\lim_{t \to -2^+} g(t)$ (b) $\lim_{t \to -2^-} g(t)$ (c) $\lim_{t \to -2} g(t)$

SOLUTION: A sketch of the graph of g is shown in Fig. 2.3.6.

(a) Because $g(t) = 11 - t^2$ if $t > -2$, then

$$\lim_{t \to -2^+} g(t) = \lim_{t \to -2^+} (11 - t^2) = 7$$

(b) Because $g(t) = 3 + t^2$ if $t < -2$, then

$$\lim_{t \to -2^-} g(t) = \lim_{t \to -2^-} (3 + t^2) = 7$$

(c) Because $\lim_{t \to -2^+} g(t) = \lim_{t \to -2^-} g(t) = 7$, then by Theorem 2.3.3 $\lim_{t \to -2} g(t) = 7$.

Figure 2.3.6

10. The absolute value of the signum function (see Exercise 22 in Exercises 1.8)

(a) $\lim\limits_{x \to 0^+} |\text{sgn } x|$ (b) $\lim\limits_{x \to 0^-} |\text{sgn } x|$ (c) $\lim\limits_{x \to 0} |\text{sgn } x|$

SOLUTION: From Exercise 22 in Exercises 1.8

$$\text{sgn } x = \begin{cases} -1 & \text{if } x < 0 \\ 0 & \text{if } x = 0 \\ 1 & \text{if } x > 0 \end{cases}$$

Therefore

$$|\text{sgn } x| = \begin{cases} 1 & \text{if } x < 0 \\ 0 & \text{if } x = 0 \\ 1 & \text{if } x > 0 \end{cases}$$

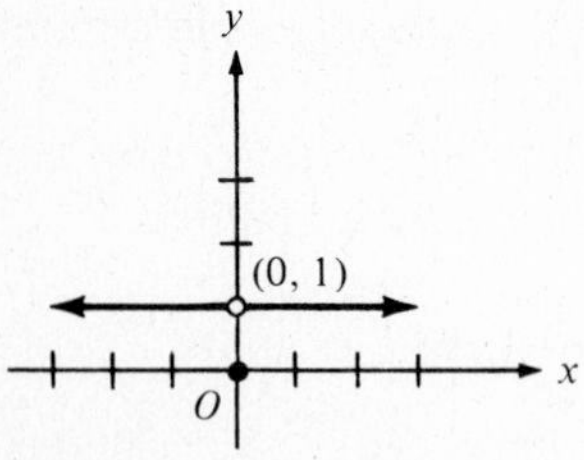

A sketch of the graph of the function is shown in Fig. 2.3.10.

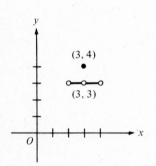

Figure 2.3.10

(a) Because $\text{sgn } x = 1$ if $x > 0$, then $|\text{sgn } x| = 1$ if $x > 0$, and

$$\lim_{x \to 0^+} |\text{sgn } x| = \lim_{x \to 0^+} (1) = 1$$

(b) Because $\text{sgn } x = -1$ if $x < 0$, then $|\text{sgn } x| = 1$ if $x < 0$, and

$$\lim_{x \to 0^-} |\text{sgn } x| = \lim_{x \to 0^-} (1) = 1$$

(c) Because $\lim\limits_{x \to 0^+} |\text{sgn } x| = \lim\limits_{x \to 0^-} |\text{sgn } x| = 1$, by Theorem 2.3.3 $\lim\limits_{x \to 0} |\text{sgn } x| = 1$.

14. Let $G(x) = [\![x]\!] + [\![4 - x]\!]$. Draw a sketch of the graph of G. Find, if they exist:

(a) $\lim\limits_{x \to 3^+} G(x)$ (b) $\lim\limits_{x \to 3^-} G(x)$ (c) $\lim\limits_{x \to 3} G(x)$

SOLUTION:

If $2 < x < 3$, then $[\![x]\!] = 2$ and $[\![4 - x]\!] = 1$; so $G(x) = 3$.
If $x = 3$, then $[\![x]\!] = 3$ and $[\![4 - x]\!] = 1$; so $G(x) = 4$.
If $3 < x < 4$, then $[\![x]\!] = 3$ and $[\![4 - x]\!] = 0$; so $G(x) = 3$.

A sketch of the graph of G for $2 < x < 4$ is shown in Fig. 2.3.14.

(a) $\lim\limits_{x \to 3^+} G(x) = \lim\limits_{x \to 3^+} 3 = 3$

(b) $\lim\limits_{x \to 3^-} G(x) = \lim\limits_{x \to 3^-} 3 = 3$

(c) Because $\lim\limits_{x \to 3^+} G(x) = \lim\limits_{x \to 3^-} G(x) = 3$, by Theorem 2.3.3 $\lim\limits_{x \to 3} G(x) = 3$.

Figure 2.3.14

2.4 INFINITE LIMITS

The definitions for infinite limits are as follows:

2.4.1 Definition

Let f be a function that is defined at every number in some open interval containing a, except possibly at the number a itself. *As* x *approaches* a, f(x) *increases without bound,* which is written

$$\lim_{x \to a} f(x) = +\infty$$

if for any number $N > 0$, there exists a $\delta > 0$ such that

$$f(x) > N \quad \text{whenever} \quad 0 < |x - a| < \delta$$

2.4.2 Definition

Let f be a function that is defined at every number in some open interval containing a, except possibly at the number a itself. *As* x *approaches* a, f(x) *decreases without bound,* which is written

$$\lim_{x \to a} f(x) = -\infty$$

if for any number $N < 0$ there exists a $\delta > 0$ such that

$$f(x) < N \quad \text{whenever} \quad 0 < |x - a| < \delta$$

We may also define one-sided limits that are infinite. The following theorems result from such definitions.

2.4.3 Limit Theorem 11 If r is any positive integer, then

(i) $\displaystyle\lim_{x \to 0^+} \frac{1}{x^r} = +\infty$

(ii) $\displaystyle\lim_{x \to 0^-} \frac{1}{x^r} = \begin{cases} -\infty & \text{if } r \text{ is odd} \\ +\infty & \text{if } r \text{ is even} \end{cases}$

2.4.4 Limit Theorem 12 If a is any real number, and if $\displaystyle\lim_{x \to a} f(x) = 0$ and $\displaystyle\lim_{x \to a} g(x) = c$, where c is a constant not equal to 0, then

(i) if $c > 0$ and if $f(x) \to 0$ through positive values of $f(x)$,

$$\lim_{x \to a} \frac{g(x)}{f(x)} = +\infty$$

(ii) if $c > 0$ and if $f(x) \to 0$ through negative values of $f(x)$,

$$\lim_{x \to a} \frac{g(x)}{f(x)} = -\infty$$

(iii) if $c < 0$ and if $f(x) \to 0$ through positive values of $f(x)$,

$$\lim_{x \to a} \frac{g(x)}{f(x)} = -\infty$$

(iv) if $c < 0$ and if $f(x) \to 0$ through negative values of $f(x)$,

$$\lim_{x \to a} \frac{g(x)}{f(x)} = +\infty$$

We may now find the limit of a fraction if either the numerator or denominator, but not both, has limit zero. If the numerator has limit zero and the denominator has a limit that is not zero, then by Limit Theorem 9 the limit of the fraction is zero. If the denominator has limit zero and the numerator has a limit that is not zero, then by Limit Theorem 12, the limit of the fraction is either $+\infty$ or $-\infty$, depending on which case of L.T. 12 is satisfied.

However, if both the numerator and denominator have limit zero, there are no theorems that tell us what the limit is. We must use a "trick" to find the limit of such a fraction, if the limit exists. We have so far considered two such tricks:

1. Factor the numerator and denominator and divide both by the common factors.
2. Rationalize either the numerator or denominator and proceed as in (1).

We will learn additional "tricks" as we proceed through the book.

Exercises 2.4

In Exercises 1–14, evaluate the limit.

2. $\displaystyle\lim_{x \to 3^+} \frac{4x^2}{9 - x^2}$

SOLUTION: $\lim\limits_{x \to 3^+} (4x^2) = 36$ and $\lim\limits_{x \to 3^+} (9 - x^2) = 0$

Furthermore, since $x \to 3^+$, then $x > 3$ and $9 - x^2 < 0$. Thus, $9 - x^2$ approaches 0 through negative values. By Limit Theorem 12(ii)

$$\lim_{x \to 3^+} \frac{4x^2}{9 - x^2} = -\infty$$

6. $\lim\limits_{x \to 0^+} \dfrac{\sqrt{3 + x^2}}{x}$

SOLUTION: $\lim\limits_{x \to 0^+} \sqrt{3 + x^2} = \sqrt{3}$ and $\lim\limits_{x \to 0^+} x = 0$

Moreover, since $x \to 0^+$, x approaches 0 through positive values. Thus, by Limit Theorem 12(i),

$$\lim_{x \to 0^+} \frac{\sqrt{3 + x^2}}{x} = +\infty$$

10. $\lim\limits_{x \to 4^-} \dfrac{\sqrt{16 - x^2}}{x - 4}$

SOLUTION: Because both the numerator and denominator have limit 0, we cannot use Limit Theorem 12. We must factor. Because $x \to 4^-$, then $x < 4$ or, equivalently, $4 - x > 0$. Thus,

$$x - 4 = -(4 - x) = -\sqrt{(4 - x)^2}$$

Therefore,

$$\begin{aligned}
\lim_{x \to 4^-} \frac{\sqrt{16 - x^2}}{x - 4} &= \lim_{x \to 4^-} \frac{\sqrt{(4 - x)(4 + x)}}{-\sqrt{(4 - x)^2}} \\
&= \lim_{x \to 4^-} \frac{\sqrt{4 - x} \cdot \sqrt{4 + x}}{-\sqrt{4 - x} \cdot \sqrt{4 - x}} \\
&= \lim_{x \to 4^-} \frac{\sqrt{4 + x}}{-\sqrt{4 - x}}
\end{aligned} \tag{1}$$

Because

$$\lim_{x \to 4^-} \sqrt{4 + x} = \sqrt{8}$$

and

$$\lim_{x \to 4^-} -\sqrt{4 - x} = 0$$

and $-\sqrt{4 - x}$ approaches 0 through negative values, by Limit Theorem 12(ii) and Eq. (1) we have

$$\lim_{x \to 4^-} \frac{\sqrt{16 - x^2}}{x - 4} = -\infty$$

12. $\lim\limits_{x \to 1^-} \dfrac{[\![x^2]\!] - 1}{x^2 - 1}$

SOLUTION: Since $x \to 1^-$, then $x < 1$. If $0 < x < 1$, then $[\![x^2]\!] = 0$. Thus,

$$\lim_{x \to 1^-} [\![x^2]\!] - 1 = -1$$

Because $x < 1$ and $x^2 - 1 < 0$ if $-1 < x < 1$, then

$$\lim_{x \to 1^-} (x^2 - 1) = 0$$

and $(x^2 - 1)$ approaches 0 through negative values. Therefore, by Limit Theorem 12(iv),

$$\lim_{x \to 1^-} \frac{[\![x^2]\!] - 1}{x^2 - 1} = +\infty$$

14. $\displaystyle \lim_{s \to 2^-} \left(\frac{1}{s - 2} - \frac{3}{s^2 - 4} \right)$

SOLUTION:

$$\lim_{s \to 2^-} \left(\frac{1}{s - 2} - \frac{3}{s^2 - 4} \right) = \lim_{s \to 2^-} \left(\frac{s + 2}{(s - 2)(s + 2)} - \frac{3}{(s - 2)(s + 2)} \right)$$

$$= \lim_{s \to 2^-} \frac{s - 1}{(s - 2)(s + 2)}$$

Moreover,

$$\lim_{s \to 2^-} (s - 1) = 1 \quad \text{and} \quad \lim_{s \to 2^-} (s - 2)(s + 2) = 0$$

Because $s \to 2^-$, then $s - 2 < 0$; thus $(s - 2)(s + 2)$ approaches 0 through negative values. Therefore, by Limit Theorem 12(ii),

$$\lim_{s \to 2^-} \left(\frac{1}{s - 2} - \frac{3}{s^2 - 4} \right) = -\infty$$

18. Use Definition 2.4.1 to prove that

$$\lim_{x \to -3} \left| \frac{5 - x}{3 + x} \right| = +\infty$$

SOLUTION: For any $N > 0$ we must find a $\delta > 0$ such that

$$\left| \frac{5 - x}{3 + x} \right| > N \quad \text{whenever} \quad 0 < |x + 3| < \delta$$

or, equivalently, by taking reciprocals

$$|x + 3| \cdot \frac{1}{|5 - x|} < \frac{1}{N} \quad \text{whenever} \quad 0 < |x + 3| < \delta \tag{1}$$

We find an upper bound for $1/|5 - x|$. Suppose that $\delta \leq 1$. Whenever $0 < |x + 3| < \delta$, then

$$|x + 3| < 1$$
$$-1 < x + 3 < 1$$
$$-4 < x < -2$$
$$2 < -x < 4$$
$$7 < 5 - x < 9$$
$$7 < |5 - x| < 9$$

$$\frac{1}{9} < \frac{1}{|5 - x|} < \frac{1}{7} \tag{2}$$

Because $|x + 3| < \delta$, by (2) we obtain

$$|x + 3| \cdot \frac{1}{|5 - x|} < \frac{\delta}{7} \tag{3}$$

From (3) and (1) we see that we want

$$\frac{\delta}{7} \leqslant \frac{1}{N}$$

$$\delta \leqslant \frac{7}{N}$$

The proof follows.

PROOF: For any $N > 0$ let $\delta = \min(1, 7/N)$. Then $\delta > 0, \delta \leqslant 1$, and $\delta \leqslant 7/N$. Whenever $0 < |x + 3| < \delta$, then

$$|x + 3| < \frac{7}{N} \tag{4}$$

Moreover, since $|x + 3| < 1$, then inequality (2) holds. From (2) and (4) we obtain

$$|x + 3| \cdot \frac{1}{|5 - x|} < \frac{7}{N} \cdot \frac{1}{7}$$

or, equivalently,

$$\left| \frac{5 - x}{x + 3} \right| > N$$

Therefore, by Definition 2.4.1

$$\lim_{x \to -3} \left| \frac{5 - x}{x + 3} \right| = +\infty$$

2.5 CONTINUITY OF A FUNCTION AT A NUMBER

There are many theorems having a hypothesis that includes the condition that a function be continuous at a number a. Hence, we must be able to determine whether or not a function is continuous or discontinuous at a. If there is a break in the graph of f at the point where $x = a$, then f is discontinuous at a. Often it is not convenient to draw a sketch of the graph to observe whether or not a function is continuous at a number. For such cases we have the following analytic definition of continuity at a number.

2.5.1 Definition

The function f is said to be *continuous* at the number a if and only if the following three conditions are satisfied:

(i) $f(a)$ exists
(ii) $\lim_{x \to a} f(x)$ exists
(iii) $\lim_{x \to a} f(x) = f(a)$.

If one or more of these three conditions fails to hold at a, the function f is said to be *discontinuous* at a.

We note that we may have to consider one-sided limits and use Theorem 2.3.3 to determine whether or not condition (ii) in Definition 2.5.1 is satisfied. This is illustrated in Exercise 10. If a function f is discontinuous at a, but condition (ii) is satisfied at a (that is, if f is discontinuous at a, but $\lim_{x \to a} f(x)$ exists and is L), then the discontinuity at a is removable, and we may remove the discontinuity by defining $f(a) = L$. (L cannot be either $+\infty$ or $-\infty$.) If condition (ii) in Definition 2.5.1 is not satisfied at a, then the function has an essential discontinuity at a, and the discontinuity cannot be removed.

Exercises 2.5

In Exercises 1–22, draw a sketch of the graph of the function; then by observing where there are breaks in the graph, determine the values of the independent variable at which the function is discontinuous and show why Definition 2.5.1 is not satisfied at each discontinuity.

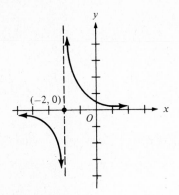

Figure 2.5.2

2. $g(x) = \begin{cases} \dfrac{1}{x+2} & \text{if } x \neq -2 \\ 0 & \text{if } x = -2 \end{cases}$

SOLUTION: A sketch of the graph of g is shown in Fig. 2.5.2. There is a break in the graph at the point where $x = -2$; so we use Definition 2.5.1 at $a = -2$ to show there is a discontinuity.

Because $g(-2) = 0$, condition (i) is satisfied. Because

$$\lim_{x \to -2^+} g(x) = +\infty \quad \text{and} \quad \lim_{x \to -2^-} g(x) = -\infty$$

condition (ii) is not satisfied. Therefore, g is discontinuous at -2.

6. $f(x) = \dfrac{x^3 - 2x^2 - 11x + 12}{x^2 - 5x + 4}$

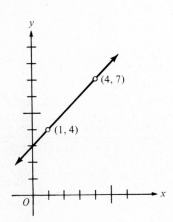

Figure 2.5.6

SOLUTION: $x^2 - 5x + 4 = (x - 4)(x - 1)$. We factor the numerator of $f(x)$ by dividing with the factors $x - 4$ and $x - 1$. Thus,

$$f(x) = \frac{(x - 4)(x - 1)(x + 3)}{(x - 4)(x - 1)}$$

or, equivalently,

$$f(x) = x + 3 \quad \text{if } x \neq 4, \ x \neq 1$$

A sketch of the graph of f is shown in Fig. 2.5.6. There are breaks in the graph at the points where $x = 1$ and $x = 4$.

Because division by zero is not defined, we see that neither $f(1)$ nor $f(4)$ is defined. Thus, in Definition 2.5.1 condition (i) fails to hold at $a = 1$ and $a = 4$. Therefore, f is discontinuous at 1 and 4.

10. $H(x) = \begin{cases} 1 + x & \text{if } x \leqslant -2 \\ 2 - x & \text{if } -2 < x \leqslant 2 \\ 2x - 1 & \text{if } 2 < x \end{cases}$

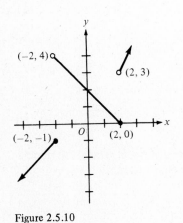

Figure 2.5.10

SOLUTION: A sketch of the graph of H is shown in Fig. 2.5.10. There are breaks in the graph at the points where $x = -2$ and $x = 2$. We consider the points separately.

Since $H(-2) = -1$, condition (i) of Definition 2.5.1 is satisfied when $a = -2$. Because

$$\lim_{x \to -2^+} H(x) = \lim_{x \to -2^+} (2 - x) = 4$$

and

$$\lim_{x \to -2^-} H(x) = \lim_{x \to -2^-} (1 + x) = -1$$

$\lim_{x \to -2} H(x)$ does not exist. Therefore condition (ii) of Definition 2.5.1 is not satisfied when $a = -2$, and thus H is discontinuous at -2.

Since $H(2) = 0$, condition (i) of Definition 2.5.1 is satisfied when $a = 2$. Because

$$\lim_{x \to 2^+} H(x) = \lim_{x \to 2^+} (2x - 1) = 3$$

and

$$\lim_{x \to 2^-} H(x) = \lim_{x \to 2^-} (2 - x) = 0$$

$\lim_{x \to 2} H(x)$ does not exist. Thus condition (ii) in Definition 2.5.1 is not satisfied when $a = 2$, and thus H is discontinuous at 2.

14. $F(x) = \begin{cases} \dfrac{x - 3}{|x - 3|} & \text{if } x \neq 3 \\ 0 & \text{if } x = 3 \end{cases}$

SOLUTION: Because $|x - 3| = x - 3$ if $x > 3$ and $|x - 3| = -(x - 3)$ if $x < 3$, then

$$F(x) = \begin{cases} \dfrac{x - 3}{x - 3} & \text{if } x > 3 \\ 0 & \text{if } x = 3 \\ \dfrac{x - 3}{-(x - 3)} & \text{if } x < 3 \end{cases}$$

or, equivalently,

$$F(x) = \begin{cases} 1 & \text{if } x > 3 \\ 0 & \text{if } x = 3 \\ -1 & \text{if } x < 3 \end{cases}$$

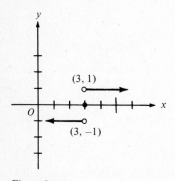

(3, 1)

(3, −1)

Figure 2.5.14

A sketch of the graph of F is shown in Fig. 2.5.14. There is a break in the graph at the point where $x = 3$. Since $F(3) = 0$, condition (i) of Definition 2.5.1 is satisfied when $a = 3$.

Because

$$\lim_{x \to 3^+} F(x) = 1 \quad \text{and} \quad \lim_{x \to 3^-} F(x) = -1$$

$\lim_{x \to 3} F(x)$ does not exist. Hence, condition (ii) of Definition 2.5.1 is not satisfied when $a = 3$. Thus, F is discontinuous at 3.

In Exercises 23-30, prove that the function is discontinuous at the number a. Then determine if the discontinuity is removable or essential. If the discontinuity is removable, define $f(a)$ so that the discontinuity is removed.

24. $f(s) = \begin{cases} \dfrac{1}{s + 5} & \text{if } s \neq -5 \\ 0 & \text{if } s = -5 \end{cases} \; ; a = -5$

$$\lim_{s \to -5^+} f(s) = \lim_{s \to -5^+} \frac{1}{s + 5} = +\infty$$

$$\lim_{s \to -5^-} f(s) = \lim_{s \to -5^-} \frac{1}{s + 5} = -\infty$$

Condition (ii) of Definition 2.5.1 is not satisfied when $a = -5$. Thus, f is discontinuous at -5. Since $\lim_{s \to -5} f(s)$ does not exist, the discontinuity is essential.

26. $f(x) = \begin{cases} \dfrac{x^2 - 4x + 3}{x - 3} & \text{if } x \neq 3 \\ 5 & \text{if } x = 3 \end{cases} \; ; a = 3$

SOLUTION: We are given that $f(3) = 5$. Furthermore,

$$\lim_{x \to 3} f(x) = \lim_{x \to 3} \frac{(x-3)(x-1)}{x-3}$$

$$= \lim_{x \to 3} (x-1)$$

$$= 2$$

Because $\lim_{x \to 3} f(x) \neq f(3)$, condition (iii) of Definition 2.5.1 is not satisfied when $a = 3$. Thus, f is discontinuous at 3. Since $\lim_{x \to 3} f(x)$ exists and is a finite number, the discontinuity is removable. We remove the discontinuity by defining $f(3) = 2$.

30. $f(x) = \dfrac{\sqrt[3]{x+1} - 1}{x}$; $a = 0$

SOLUTION: Because $f(0)$ is not defined, condition (i) of Definition 2.5.1 is not satisfied when $a = 0$. Thus, f is discontinuous at 0. To determine whether the discontinuity is removable or essential, we must determine whether or not $\lim_{x \to 0} f(x)$ exists.

To rationalize the numerator we use the factoring formula: $(a - b)(a^2 + ab + b^2) = a^3 - b^3$, with $a = \sqrt[3]{x+1}$ and $b = 1$. Thus,

$$\lim_{x \to 0} f(x) = \lim_{x \to 0} \frac{(\sqrt[3]{x+1} - 1)(\sqrt[3]{(x+1)^2} + \sqrt[3]{x+1} + 1)}{x(\sqrt[3]{(x+1)^2} + \sqrt[3]{x+1} + 1)}$$

$$= \lim_{x \to 0} \frac{(x+1) - 1}{x(\sqrt[3]{(x+1)^2} + \sqrt[3]{x+1} + 1)}$$

$$= \lim_{x \to 0} \frac{1}{\sqrt[3]{(x+1)^2} + \sqrt[3]{x+1} + 1}$$

$$= \tfrac{1}{3}$$

Because $\lim_{x \to 0} f(x)$ exists and is a finite number, the discontinuity at 0 is removable. We remove the discontinuity by defining $f(0) = \tfrac{1}{3}$.

2.6 THEOREMS ON CONTINUITY

The following theorems are used to determine the numbers at which a function is continuous.

2.6.1 Theorem   If f and g are two functions that are continuous at the number a, then

    (i) $f + g$ is continuous at a
    (ii) $f - g$ is continuous at a
    (iii) $f \circ g$ is continuous at a
    (iv) f/g is continuous at a, provided $g(a) \neq 0$.

2.6.2 Theorem   A polynomial function is continuous at every number.

2.6.3 Theorem   A rational function is continuous at every number in its domain.

2.6.5 Theorem   If $\lim_{x \to a} g(x) = b$ and if the function f is continuous at b,

$$\lim_{x \to a} (f \circ g)(x) = f(b)$$

or, equivalently,

$$\lim_{x \to a} f(g(x)) = f(\lim_{x \to a} g(x))$$

2.6.6 Theorem If the function g is continuous at a and the function f is continuous at $g(a)$, then the composite function $f \circ g$ is continuous at a.

Furthermore, Theorem 2.2.12 and Definition 2.5.1 imply the following: If $f(x) = x^{1/n}$, then

(i) f is continuous at all x if n is an odd positive integer
(ii) f is continuous at all $x > 0$ if n is an even positive integer.

In particular, $\sqrt[3]{x}$ is continuous at all x, and $\sqrt{x}$ is continuous at all $x > 0$.

Exercises 2.6

---

8. Show the application of Theorem 2.6.5 to find the limit.

$$\lim_{x \to 2} \sqrt{x^3 + 1}$$

SOLUTION: Let f and g be defined by

$$f(x) = \sqrt{x} \quad \text{and} \quad g(x) = x^3 + 1$$

The composite function $f \circ g$ is defined by $f(g(x)) = \sqrt{x^3 + 1}$. Since

$$\lim_{x \to 2} g(x) = \lim_{x \to 2} (x^3 + 1) = 9$$

and because f is continuous at 9, we may apply Theorem 2.6.5 to obtain

$$\lim_{x \to 2} \sqrt{x^3 + 1} = \lim_{x \to 2} f(g(x))$$
$$= f(\lim_{x \to 2} g(x)) \quad \text{(by Theorem 2.6.5)}$$
$$= f(9)$$
$$= \sqrt{9}$$
$$= 3$$

In Exercises 11–26, determine all values of x for which the given function is continuous. Indicate which theorems you apply.

14. $h(x) = \dfrac{x + 2}{x^3 - 7x - 6}$

SOLUTION: Because h is a rational function, by Theorem 2.6.3 h is continuous at every number in its domain, that is, at every x for which $x^3 - 7x - 6 \neq 0$. Dividing $x^3 - 7x - 6$ by the trial divisor $x + 2$, we obtain the factors

$$x^3 - 7x - 6 = (x + 1)(x + 2)(x - 3)$$

Because $x^3 - 7x - 6 = 0$ only when $x = -1$, $x = -2$, or $x = 3$, we conclude that h is continuous for all x, except -1, -2, and 3.

16. $g(x) = \sqrt{x^2 + 4}$

SOLUTION: Let $f(x) = \sqrt{x}$ and $h(x) = x^2 + 4$. Then $g(x) = f(h(x))$. Because h is a polynomial function, it follows from Theorem 2.6.2 that h is continuous at every x. Furthermore, f is continuous at every positive number. Therefore, by Theorem 2.6.6, g is continuous at every number x for which $h(x) > 0$, that is, whenever $x^2 + 4 > 0$. Since $x^2 + 4 > 0$ for all real values of x, we conclude that g is continuous at every real number.

20. $f(x) = \sqrt{\dfrac{4-x}{4+x}}$

SOLUTION: Let g and h be defined by $g(x) = \sqrt{x}$ and $h(x) = (4-x)/(4+x)$. Then $f(x) = g(h(x))$. By Theorem 2.6.3, h is continuous at all $x \neq -4$. Because g is continuous at all positive numbers, then by Theorem 2.6.6 f is continuous at all x for which $x \neq -4$ and $h(x) > 0$, that is, at all x such that

$$\frac{4-x}{4+x} > 0$$

or, equivalently,

$$\frac{x-4}{x+4} < 0 \tag{1}$$

Table 20 indicates that the factor $x - 4$ changes sign at $x = 4$ and the factor $x + 4$ changes sign at $x = -4$. Since inequality (1) is satisfied whenever the quotient of the factors is negative, Table 20 indicates that the function f is continuous for all x in $(-4, 4)$.

Table 20

	$x < -4$	$x = -4$	$-4 < x < 4$	$x = 4$	$x > 4$
$x - 4$	$-$	$-$	$-$	0	$+$
$x + 4$	$-$	0	$+$	$+$	$+$
$\dfrac{x-4}{x+4}$	$+$	does not exist	$-$	0	$+$

24. $G(x) = \left(\dfrac{x^2}{x^2 - 4} - \dfrac{1}{x}\right)^{1/3}$

SOLUTION: Let F and H be defined by

$$F(x) = x^{1/3} \quad \text{and} \quad H(x) = \frac{x^2}{x^2 - 4} - \frac{1}{x}$$

Then $G(x) = F(H(x))$. Because

$$\frac{x^2}{x^2 - 4} \quad \text{is not defined when } x = \pm 2$$

and

$$\frac{1}{x} \quad \text{is not defined when } x = 0$$

by Theorem 2.6.1 (ii) H is continuous at all x except 2, -2, and 0. Furthermore, since F is continuous at every real number, then by Theorem 2.6.6 G is continuous at all x except 2, -2, and 0.

28. Prove that if f is continuous at a and g is discontinuous at a, then $f + g$ is discontinuous at a.

SOLUTION: The proof is indirect. We assume the conclusion is false and show that this leads to a contradiction of the hypothesis. Suppose that $f + g$ is continuous at a. Then because f is continuous at a, by Theorem 2.6.1(ii) $(f + g) - f$ is continuous at a. Since

$$((f+g) - f)(x) = (f+g)(x) - f(x)$$
$$= f(x) + g(x) - f(x)$$
$$= g(x)$$

the function $(f+g) - f$ is the function g. But by hypothesis, g is discontinuous at a. The contradiction proves that our assumption is false. Hence, $f+g$ is discontinuous at a.

30. Give an example of two functions that are both discontinuous at a number a, but the sum of the two functions is continuous at a.

SOLUTION: There are many examples. If f and g are any two functions that are discontinuous at a, then $f+g$ is continuous at a if and only if

$$\lim_{x \to a^+} (f+g)(x) = \lim_{x \to a^-} (f+g)(x) = (f+g)(a)$$

or, equivalently, if and only if

$$\lim_{x \to a^+} f(x) + \lim_{x \to a^+} g(x) = \lim_{x \to a^-} f(x) + \lim_{x \to a^-} g(x) = f(a) + g(a)$$

For example, let f and g be defined as follows:

$$f(x) = \begin{cases} x+1 & \text{if } x \geq 0 \\ x-1 & \text{if } x < 0 \end{cases} \qquad g(x) = \begin{cases} x-1 & \text{if } x \geq 0 \\ x+1 & \text{if } x < 0 \end{cases}$$

Because

$$\lim_{x \to 0^+} f(x) = 1 \quad \text{and} \quad \lim_{x \to 0^-} f(x) = -1$$

$\lim_{x \to 0} f(x)$ does not exist. Hence f is discontinuous at 0.

Because

$$\lim_{x \to 0^+} g(x) = -1 \quad \text{and} \quad \lim_{x \to 0^-} g(x) = 1$$

$\lim_{x \to 0} g(x)$ does not exist. Therefore g is discontinuous at 0. However, $f+g$ is defined by

$$(f+g)(x) = f(x) + g(x) = \begin{cases} (x+1) + (x-1) & \text{if } x \geq 0 \\ (x-1) + (x+1) & \text{if } x < 0 \end{cases}$$

Therefore,

$$(f+g)(x) = 2x \quad \text{for all } x$$

and thus $f+g$ is continuous at 0.

32. If the function g is continuous at a number a and the function f is discontinuous at a, is it possible for the quotient of the two functions, f/g, to be continuous at a? Prove your answer.

SOLUTION: No. Suppose that f/g is continuous at a. Then by Definition 2.5.1

$$\lim_{x \to a} \frac{f(x)}{g(x)} = \frac{f(a)}{g(a)} \tag{1}$$

Hence, $g(a) \neq 0$. Because g is continuous at a, then

$$\lim_{x \to a} g(x) = g(a) \neq 0 \tag{2}$$

We show that (1) and (2) imply that f is continuous at a; thus the hypothesis is contradicted. Since $g(a) \neq 0$, we may multiply and divide by $g(a)$ and obtain

$$\lim_{x \to a} f(x) = \frac{\lim_{x \to a} f(x)}{g(a)} g(a)$$

$$= \frac{\lim_{x \to a} f(x)}{\lim_{x \to a} g(x)} g(a) \quad \text{[By (2)]}$$

$$= \lim_{x \to a} \frac{f(x)}{g(x)} g(a) \quad \text{[By L.T. 9 and (2)]}$$

$$= \frac{f(a)}{g(a)} g(a) \quad \text{[By (1)]}$$

$$= f(a)$$

Because $\lim_{x \to a} f(x) = f(a)$, f is continuous at a. But this is a contradiction of the hypothesis. Hence, our assumption is false, and it is not possible for f/g to be continuous at a.

Review Exercises

8. Draw a sketch of the graph and discuss the continuity of the function

$$F(x) = \frac{|x^2 - 4|}{x + 2}$$

SOLUTION: Because $x^2 - 4$ changes sign at $x = 2$ and $x = -2$, we consider three cases:

Case 1: $x \geqslant 2$
Because $x^2 - 4 \geqslant 0$, then $|x^2 - 4| = x^2 - 4$. Thus

$$\frac{|x^2 - 4|}{x + 2} = \frac{x^2 - 4}{x + 2} = \frac{(x + 2)(x - 2)}{x + 2} = x - 2$$

Case 2: $-2 < x < 2$
Because $x^2 - 4 < 0$, then $|x^2 - 4| = -(x^2 - 4)$. Thus,

$$\frac{|x^2 - 4|}{x + 2} = \frac{-(x^2 - 4)}{x + 2} = \frac{-(x + 2)(x - 2)}{x + 2} = -x + 2$$

Case 3: $x < -2$
Because $x^2 - 4 > 0$, then as in Case 1, we have

$$\frac{|x^2 - 4|}{x + 2} = x - 2$$

Note that we do not consider $x = -2$ because F is not defined when $x = -2$. Then we have shown that,

$$F(x) = \begin{cases} x - 2 & \text{if } x < -2 \\ -x + 2 & \text{if } -2 < x < 2 \\ x - 2 & \text{if } 2 \leqslant x \end{cases}$$

A sketch of the graph of F is shown in Fig. 2.8R. F is continuous at all $x \neq -2$.

14. Establish the limit by using Definition 2.1.1; that is, for any $\epsilon > 0$ find a $\delta > 0$ such that $|f(x) - L| < \epsilon$ whenever $0 < |x - a| < \delta$.

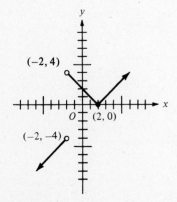

$(-2, 4)$

$(2, 0)$

$(-2, -4)$

Figure 2.8R

$$\lim_{x \to -2} x^3 = -8$$

SOLUTION: For any $\epsilon > 0$ we must find a $\delta > 0$ such that

$$|x^3 + 8| < \epsilon \quad \text{whenever} \quad 0 < |x + 2| < \delta$$

or, equivalently, such that

$$|x + 2| \, |x^2 - 2x + 4| < \epsilon \quad \text{whenever} \quad 0 < |x + 2| < \delta \qquad (1)$$

We find an upper bound for $|x^2 - 2x + 4|$. Suppose that $\delta \leqslant 1$. Then whenever $0 < |x + 2| < \delta$, we have

$$|x + 2| < 1$$
$$-1 < x + 2 \quad < 1$$
$$-3 < x \qquad < -1$$
$$1 < -x \qquad < 3 \qquad (2)$$
$$1 < x^2 \qquad < 9 \qquad (3)$$

And from (2) we obtain

$$2 < -2x \qquad < 6$$
$$6 < -2x + 4 < 10 \qquad (4)$$

Thus, by adding corresponding members of (3) and (4), we have

$$7 < x^2 - 2x + 4 \quad < 19$$
$$7 < |x^2 - 2x + 4| < 19 \qquad (5)$$

Because $|x + 2| < \delta$, from (5) we obtain

$$|x + 2| \cdot |x^2 - 2x + 4| < 19\delta \qquad (6)$$

And from (6) and (1) we have

$$19\delta \leqslant \epsilon$$

$$\delta \leqslant \frac{\epsilon}{19}$$

The proof follows.

PROOF: For any $\epsilon > 0$ choose $\delta = \min(1, \frac{1}{19}\epsilon)$. Then $\delta > 0$, $\delta \leqslant 1$, and $\delta \leqslant \frac{1}{19}\epsilon$. Whenever $0 < |x + 2| < \delta$, it follows that

$$|x + 2| < \frac{\epsilon}{19} \qquad (7)$$

Moreover, since $|x + 2| < 1$, inequality (5) follows. By (5) and (7) we obtain

$$|x + 2| \cdot |x^2 - 2x + 4| < \frac{\epsilon}{19} \, 19$$

or, equivalently,

$$|x^3 + 8| < \epsilon$$

Therefore,

$$\lim_{x \to -2} x^3 = -8$$

16. If $f(x) = (|x| - x)/x$, evaluate:

    (a) $\displaystyle\lim_{x \to 0^-} f(x)$ (b) $\displaystyle\lim_{x \to 0^+} f(x)$ (c) $\displaystyle\lim_{x \to 0} f(x)$

SOLUTION:

(a) Because $x \to 0^-$, then $x < 0$ and $|x| = -x$. Thus,

$$\lim_{x \to 0^-} f(x) = \lim_{x \to 0^-} \frac{-x - x}{x} = \lim_{x \to 0^-} (-2) = -2$$

(b) Because $x \to 0^+$, then $x > 0$ and $|x| = x$. Thus,

$$\lim_{x \to 0^+} f(x) = \lim_{x \to 0^+} \frac{x - x}{x} = \lim_{x \to 0^+} (0) = 0$$

(c) Because $\lim_{x \to 0^-} f(x) \neq \lim_{x \to 0^+} f(x)$, then $\lim_{x \to 0} f(x)$ does not exist.

In Exercises 17-24, evaluate the limit if it exists.

20. $\displaystyle \lim_{y \to 5^-} \frac{\sqrt{25 - y^2}}{y - 5}$

SOLUTION: Limit Theorem 9 cannot be used because the limit of the denominator is 0. Limit Theorem 12 cannot be used because the limit of both the numerator and denominator is 0. However, since $y \to 5^-$, then $y < 5$ and $5 - y > 0$. Thus, $y - 5 = -(5 - y) = -\sqrt{(5 - y)^2}$. Therefore,

$$\lim_{y \to 5^-} \frac{\sqrt{25 - y^2}}{y - 5} = \lim_{y \to 5^-} \frac{\sqrt{(5 - y)(5 + y)}}{-\sqrt{(5 - y)^2}}$$

$$= \lim_{y \to 5^-} \frac{\sqrt{5 - y} \cdot \sqrt{5 + y}}{-\sqrt{5 - y} \cdot \sqrt{5 - y}}$$

$$= \lim_{y \to 5^-} \frac{\sqrt{5 + y}}{-\sqrt{5 - y}}$$

Because
$$\lim_{y \to 5^-} \sqrt{5 + y} = \sqrt{10}$$

and
$$\lim_{y \to 5^-} -\sqrt{5 - y} = 0$$

where $-\sqrt{5 - y}$ is approaching 0 through negative values, it follows from Limit Theorem 12(ii) (2.4.4) that

$$\lim_{y \to 5^-} \frac{\sqrt{25 - y^2}}{y - 5} = -\infty$$

22. $\displaystyle \lim_{x \to 8} \frac{\sqrt{7 + \sqrt[3]{x}} - 3}{x - 8}$

SOLUTION: The limit of both the numerator and denominator is 0. We rationalize the numerator

$$\lim_{x \to 8} \frac{\sqrt{7 + \sqrt[3]{x}} - 3}{x - 8} = \lim_{x \to 8} \frac{(\sqrt{7 + \sqrt[3]{x}} - 3)(\sqrt{7 + \sqrt[3]{x}} + 3)}{(x - 8)(\sqrt{7 + \sqrt[3]{x}} + 3)}$$

$$= \lim_{x \to 8} \frac{\sqrt[3]{x} - 2}{(x - 8)(\sqrt{7 + \sqrt[3]{x}} + 3)}$$

$$= \lim_{x \to 8} \frac{(\sqrt[3]{x} - 2)(\sqrt[3]{x^2} + 2\sqrt[3]{x} + 4)}{(x - 8)(\sqrt{7 + \sqrt[3]{x}} + 3)(\sqrt[3]{x^2} + 2\sqrt[3]{x} + 4)}$$

$$= \lim_{x \to 8} \frac{x - 8}{(x - 8)(\sqrt{7 + \sqrt[3]{x}} + 3)(\sqrt[3]{x^2} + 2\sqrt[3]{x} + 4)}$$

$$= \lim_{x \to 8} \frac{1}{(\sqrt{7 + \sqrt[3]{x}} + 3)(\sqrt[3]{x^2} + 2\sqrt[3]{x} + 4)}$$

$$= \frac{1}{(\sqrt{7 + \sqrt[3]{8}} + 3)(\sqrt[3]{8^2} + 2\sqrt[3]{8} + 4)}$$

$$= \tfrac{1}{72}$$

24. $\displaystyle \lim_{x \to 1^+} \frac{[\![x^2]\!] - [\![x]\!]^2}{x^2 - 1}$

SOLUTION: Because the limit of both the numerator and denominator is 0, we cannot find the limit of the fraction by using the limit theorems. However, if $1 < x < \sqrt{2}$, then $1 < x^2 < 2$; so $[\![x^2]\!] = 1$, $[\![x]\!]^2 = 1$, and $x^2 - 1 > 0$. Therefore,

$$\frac{[\![x^2]\!] - [\![x]\!]^2}{x^2 - 1} = 0 \quad \text{if} \quad 1 < x < \sqrt{2} \tag{1}$$

By the definition of a right-hand limit and (1) it follows that

$$\lim_{x \to 1^+} \frac{[\![x^2]\!] - [\![x]\!]^2}{x^2 - 1} = 0$$

28. Give an example of a function f that is discontinuous at 1 for which

(a) $\displaystyle\lim_{x \to 1} f(x)$ exists but $f(1)$ does not exist

(b) $f(1)$ exists but $\displaystyle\lim_{x \to 1} f(x)$ does not exist

(c) $\displaystyle\lim_{x \to 1} f(x)$ and $f(1)$ both exist but are not equal.

SOLUTION: There are many examples.

(a) Let $f(x) = (x^2 - 1)/(x - 1)$. Then $f(1)$ does not exist, but

$$\lim_{x \to 1} f(x) = \lim_{x \to 1} \frac{(x + 1)(x - 1)}{x - 1} = \lim_{x \to 1} (x + 1) = 2$$

(b) Let $f(x) = \begin{cases} x & \text{if } x < 1 \\ 0 & \text{if } x \geqslant 1 \end{cases}$

Then $f(1) = 0$, but since

$$\lim_{x \to 1^+} f(x) = 0 \quad \text{and} \quad \lim_{x \to 1^-} f(x) = 1$$

then $\displaystyle\lim_{x \to 1} f(x)$ does not exist.

(c) Let

$$f(x) = \begin{cases} \dfrac{x^2 - 1}{x - 1} & \text{if } x \neq 1 \\[2mm] 0 & \text{if } x = 1 \end{cases}$$

In part (a) we showed that $\displaystyle\lim_{x \to 1} f(x) = 2$. Because $f(1) = 0$, then $\displaystyle\lim_{x \to 1} f(x) \neq f(1)$.

32. If the domain of f is the set of all real numbers and f is continuous at 0, prove that if $f(a + b) = f(a) \cdot f(b)$ for all a and b, then f is continuous at every number.

SOLUTION: Because f is continuous at 0, then

$$\lim_{x \to 0} f(x) = f(0) \tag{1}$$

Moreover, by hypothesis

$$f(a + b) = f(a) \cdot f(b) \tag{2}$$

To show that f is continuous at every number, we show that for every real number a,

$$\lim_{x \to a} f(x) = f(a) \tag{3}$$

Let a be any real number and let $x = z + a$. Then $x \to a$ if and only if $z \to 0$. Thus (3) is equivalent to

$$\lim_{z \to 0} f(z + a) = f(a) \tag{4}$$

Now,

$$
\begin{aligned}
\lim_{z \to 0} f(z + a) &= \lim_{z \to 0} \left[f(z) \cdot f(a) \right] && \text{[By (2)]} \\
&= \lim_{z \to 0} f(z) \cdot \lim_{z \to 0} f(a) && \text{(By L.T.6)} \\
&= f(0) \cdot f(a) && \text{[By (1) and L.T. 2]} \\
&= f(0 + a) && \text{[By (2)]} \\
&= f(a)
\end{aligned}
$$

We have shown that (4) is true, and therefore that f is continuous at every real number a.

3

The derivative

[handwritten notes:]

$$f(t) = \lim_{\Delta x \to 0} \frac{f(t + \Delta x) - f(x_1)}{\Delta t}$$

$$m(t_1) = \lim_{\Delta x \to 0} \frac{f(t + \Delta x) - f(x_1)}{\Delta x}$$

$$v(t_1) = \lim_{t \to t_1} \frac{f(t) - f(t_1)}{t - t_1}$$

$$v(t_1) = \lim_{\Delta t \to 0} \frac{f(t_1 + \Delta t) - f(t_1)}{t}$$

$$\Rightarrow f(t) = \lim_{\Delta x \to 0} \frac{f(x + \Delta x - f(t))}{\Delta x}$$

3.1 THE TANGENT LINE

3.1.1 Definition

If the function f is continuous at x_1, then the *tangent line* to the graph of f at the point $P(x_1, f(x_1))$ is

(i) the line through P having slope $m(x_1)$, given by

$$m(x_1) = \lim_{\Delta x \to 0} \frac{f(x_1 + \Delta x) - f(x_1)}{\Delta x}$$

if this limit exists;

(ii) the line $x = x_1$ if $\displaystyle\lim_{\Delta x \to 0} \frac{f(x_1 + \Delta x) - f(x_1)}{\Delta x} = +\infty$ or $-\infty$

The limit indicated in Definition 3.1.1 will always fail to exist if f is discontinuous at x_1. Moreover, for certain functions that are continuous at x_1

$$\lim_{\Delta x \to 0^+} \frac{f(x_1 + \Delta x) - f(x_1)}{\Delta x} \neq \lim_{\Delta x \to 0^-} \frac{f(x_1 + \Delta x) - f(x_1)}{\Delta x}$$

In either of these cases there is no tangent line to the graph at x_1.

We note that $(x_1, f(x_1))$ must be a point that lies on the graph of f. If we are interested in finding the tangent line at only one point and if the coordinates of the point of tangency are known, it is easier to evaluate the limit if we make the indicated replacement for x_1 before finding the limit. This is illustrated in Exercises 10, 16, and 18. However, if we want to find the slope of the tangent line to the curve at more than one point, as in Exercise 6, we find the limit and then make the indicated replacements for x_1.

3.1.2 Definition   The *normal line* to a curve at a given point is the line perpendicular to the tangent line at that point.

Exercises 3.1

6.   Find the slope of the tangent line to the graph at the point (x_1, y_1). Make a table of values of $x, y,$ and m at the various points on the graph, and include in the table all points where the graph has a horizontal tangent. Draw a sketch of the graph.

$$y = x^3 - x^2 - x + 10$$

SOLUTION: We use Definition 3.1.1 with

$$f(x) = x^3 - x^2 - x + 10$$

$$m(x_1) = \lim_{\Delta x \to 0} \frac{f(x_1 + \Delta x) - f(x_1)}{\Delta x}$$

$$= \lim_{\Delta x \to 0} \frac{[(x_1 + \Delta x)^3 - (x_1 + \Delta x)^2 - (x_1 + \Delta x) + 10] - [x_1^3 - x_1^2 - x_1 + 10]}{\Delta x}$$

$$= \lim_{\Delta x \to 0} \frac{x_1^3 + 3x_1^2\Delta x + 3x_1(\Delta x)^2 + (\Delta x)^3 - x_1^2 - 2x_1\Delta x - (\Delta x)^2 - x_1 - \Delta x + 10 - x_1^3 + x_1^2 + x_1 - 10}{\Delta x}$$

$$= \lim_{\Delta x \to 0} \frac{3x_1^2\Delta x + 3x_1(\Delta x)^2 + (\Delta x)^3 - 2x_1\Delta x - (\Delta x)^2 - \Delta x}{\Delta x}$$

$$= \lim_{\Delta x \to 0} \frac{\Delta x(3x_1^2 + 3x_1\Delta x + (\Delta x)^2 - 2x_1 - \Delta x - 1)}{\Delta x}$$

$$= \lim_{\Delta x \to 0} (3x_1^2 + 3x_1\Delta x + (\Delta x)^2 - 2x_1 - \Delta x - 1)$$

$$= 3x_1^2 - 2x_1 - 1 \tag{1}$$

To find the points where the graph has a horizontal tangent, we solve the equation $m(x_1) = 0$. Thus, by (1) we have

$$3x_1^2 - 2x_1 - 1 = 0$$
$$(3x_1 + 1)(x_1 - 1) = 0$$

Therefore, the graph has a horizontal tangent at the points where $x = -\frac{1}{3}$ and $x = 1$. Table 6 gives values of $x, y,$ and m. In the table we choose x first, calculate y from the given equation, and calculate m from Eq. (1). Fig. 3.1.6 shows a sketch of the graph.

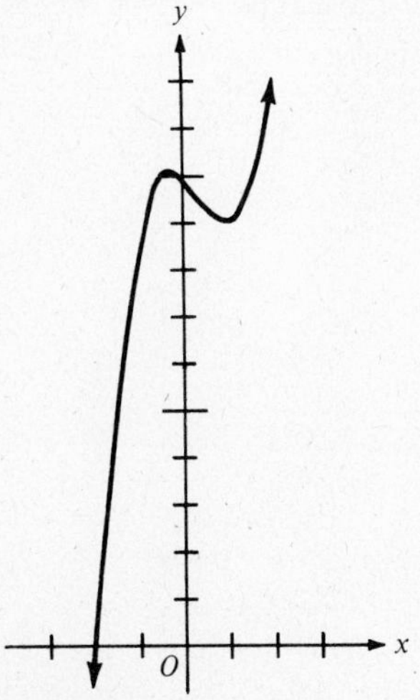

Table 6

x	y	m
2	12	7
1	9	0
0	10	−1
$-\frac{1}{3}$	$10\frac{5}{27}$	0
−1	9	4
−2	0	15

Figure 3.1.6

In Exercises 9–18, find an equation of the tangent line and an equation of the normal line to the given curve at the indicated point. Draw a sketch of the curve together with the resulting tangent line and normal line.

10. $y = x^2 + 2x + 1 ; (1, 4)$

SOLUTION: To find the slope of the tangent line to the given curve at the point $(1, 4)$, we use Definition 3.1.1 with $f(x) = x^2 + 2x + 1$ and $x_1 = 1$. Thus,

$$m(x_1) = \lim_{\Delta x \to 0} \frac{f(x_1 + \Delta x) - f(x_1)}{\Delta x}$$

$$m(1) = \lim_{\Delta x \to 0} \frac{f(1 + \Delta x) - f(1)}{\Delta x}$$

$$= \lim_{\Delta x \to 0} \frac{[(1 + \Delta x)^2 + 2(1 + \Delta x) + 1] - 4}{\Delta x}$$

$$= \lim_{\Delta x \to 0} \frac{1 + 2\Delta x + (\Delta x)^2 + 2 + 2\Delta x + 1 - 4}{\Delta x}$$

$$= \lim_{\Delta x \to 0} \frac{4\Delta x + (\Delta x)^2}{\Delta x}$$

$$= \lim_{\Delta x \to 0} (4 + \Delta x)$$
$$= 4$$

We use the point-slope form with $m = 4$ and $(x_1, y_1) = (1, 4)$ to find an equation of the tangent line to the curve at $(1, 4)$.

$$y - 4 = 4(x - 1)$$
$$4x - y = 0$$

Because the normal line is perpendicular to the tangent line, the slope of the normal line is $-\frac{1}{4}$ (the negative reciprocal of the slope of the tangent line). Thus, the point-slope form of the equation of the normal line is

$$y - 4 = -\tfrac{1}{4}(x - 1)$$
$$x + 4y - 17 = 0$$

A sketch of the curve together with the tangent and normal lines is shown in Fig. 3.1.10.

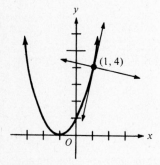

Figure 3.1.10

16. $y = -\dfrac{8}{\sqrt{x}} ; (4, -4)$

SOLUTION: To find the slope of the tangent line to the curve at $(4, -4)$ we use Definition 3.1.1 with $f(x) = -8/\sqrt{x}$ and $x_1 = 4$.

$$m(x_1) = \lim_{\Delta x \to 0} \frac{f(x_1 + \Delta x) - f(x_1)}{\Delta x}$$

$$m(4) = \lim_{\Delta x \to 0} \frac{f(4 + \Delta x) - f(4)}{\Delta x}$$

$$= \lim_{\Delta x \to 0} \frac{\dfrac{-8}{\sqrt{4 + \Delta x}} - (-4)}{\Delta x}$$

$$= \lim_{\Delta x \to 0} \frac{\left(\dfrac{-8}{\sqrt{4 + \Delta x}} + 4\right)\sqrt{4 + \Delta x}}{\Delta x \sqrt{4 + \Delta x}}$$

$$= \lim_{\Delta x \to 0} \frac{-8 + 4\sqrt{4 + \Delta x}}{\Delta x \sqrt{4 + \Delta x}}$$

$$= \lim_{\Delta x \to 0} \frac{-4(2 - \sqrt{4 + \Delta x})}{\Delta x \sqrt{4 + \Delta x}} \cdot \frac{2 + \sqrt{4 + \Delta x}}{2 + \sqrt{4 + \Delta x}}$$

$$= \lim_{\Delta x \to 0} \frac{-4[4 - (4 + \Delta x)]}{\Delta x \sqrt{4 + \Delta x} \,(2 + \sqrt{4 + \Delta x})}$$

$$= \lim_{\Delta x \to 0} \frac{4\Delta x}{\Delta x \sqrt{4 + \Delta x} \,(2 + \sqrt{4 + \Delta x})}$$

$$= \lim_{\Delta x \to 0} \frac{4}{\sqrt{4 + \Delta x} \,(2 + \sqrt{4 + \Delta x})}$$

$$= \frac{4}{\sqrt{4} \,(2 + \sqrt{4})} = \frac{1}{2}$$

We use the point-slope form with $m = \frac{1}{2}$ to find an equation of the tangent line to the curve at the point $(4, -4)$.

$$y - (-4) = \tfrac{1}{2}(x - 4)$$
$$x - 2y - 12 = 0$$

Because the normal line is perpendicular to the tangent line, the slope of the normal line is -2 (the negative reciprocal of the slope of the tangent line). The point-slope form of the equation of the normal line to the curve at the point $(4, -4)$ is

$$y - (-4) = -2(x - 4)$$
$$2x + y - 4 = 0$$

A sketch of the graph of the curve and the tangent and normal lines at the point $(4, -4)$ is shown in Fig. 3.1.16.

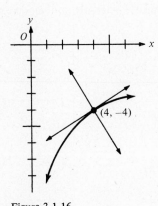

Figure 3.1.16

18. $y = \sqrt[3]{5 - x}$; $(-3, 2)$

SOLUTION: To find the slope of the tangent line to the curve at $(-3, 2)$, we use Definition 3.1.1 with $f(x) = \sqrt[3]{5 - x}$ and $x_1 = -3$.

$$m(x_1) = \lim_{\Delta x \to 0} \frac{f(x_1 + \Delta x) - f(x_1)}{\Delta x}$$

$$m(-3) = \lim_{\Delta x \to 0} \frac{f(-3 + \Delta x) - f(-3)}{\Delta x}$$

$$= \lim_{\Delta x \to 0} \frac{\sqrt[3]{8 - \Delta x} - 2}{\Delta x}$$

$$= \lim_{\Delta x \to 0} \frac{(\sqrt[3]{8 - \Delta x} - 2)(\sqrt[3]{(8 - \Delta x)^2} + 2\sqrt[3]{8 - \Delta x} + 4)}{\Delta x (\sqrt[3]{(8 - \Delta x)^2} + 2\sqrt[3]{8 - \Delta x} + 4)}$$

$$= \lim_{\Delta x \to 0} \frac{(8 - \Delta x) - 8}{\Delta x (\sqrt[3]{(8 - \Delta x)^2} + 2\sqrt[3]{8 - \Delta x} + 4)}$$

$$= \lim_{\Delta x \to 0} \frac{-1}{\sqrt[3]{(8 - \Delta x)^2} + 2\sqrt[3]{8 - \Delta x} + 4}$$

$$= \frac{-1}{\sqrt[3]{8^2} + 2\sqrt[3]{8} + 4} = -\frac{1}{12}$$

Therefore, the slope of the tangent line to the curve at $(-3, 2)$ is $-\frac{1}{12}$ and an equation of the line is

$$y - 2 = -\tfrac{1}{12}(x + 3)$$
$$x + 12y - 21 = 0$$

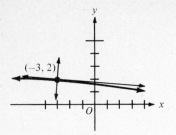

Figure 3.1.18

Because the normal line is perpendicular to the tangent line, the slope of the normal line is 12. Hence, an equation of the normal line is

$$y - 2 = 12(x + 3)$$
$$12x - y + 38 = 0$$

A sketch of the graph of the curve and the tangent and normal lines at the point $(-3, 2)$ is shown in Fig. 3.1.18. Because the curve is almost straight at the point $(-3, 2)$, it is difficult to distinguish between the curve and the tangent line.

22. Find an equation of each line through the point $(3, -2)$ that is tangent to the curve $y = x^2 - 7$.

SOLUTION: Let L_1 be one of the required lines and let $P_1(x_1, y_1)$ be the point at which L_1 is tangent to the curve, as shown in Fig. 3.1.22. We must find the slope of L_1. Because the point $(3, -2)$ does not lie on the curve, we cannot use Definition 3.1.1 with $x_1 = 3$ to find the slope of L_1. However, since P_1 does lie on the curve, we may use the definition with $f(x) = x^2 - 7$.

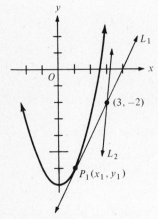

Figure 3.1.22

$$
\begin{aligned}
m(x_1) &= \lim_{\Delta x \to 0} \frac{f(x_1 + \Delta x) - f(x_1)}{\Delta x} \\
&= \lim_{\Delta x \to 0} \frac{[(x_1 + \Delta x)^2 - 7] - [x_1^2 - 7]}{\Delta x} \\
&= \lim_{\Delta x \to 0} \frac{x_1^2 + 2x_1\Delta x + (\Delta x)^2 - 7 - x_1^2 + 7}{\Delta x} \\
&= \lim_{\Delta x \to 0} \frac{2x_1\Delta x + (\Delta x)^2}{\Delta x} \\
&= \lim_{\Delta x \to 0} (2x_1 + \Delta x) \\
&= 2x_1 \tag{1}
\end{aligned}
$$

Therefore, $2x_1$ is the slope of L_1. Using Definition 1.5.1 for the slope of a line, the two points (x_1, y_1), and $(3, -2)$, which lie on L_1, we have

$$2x_1 = \frac{y_1 + 2}{x_1 - 3} \tag{2}$$

Because P_1 lies on the curve, (x_1, y_1) must satisfy the equation of the curve, and thus

$$y_1 = x_1^2 - 7 \tag{3}$$

Solving Eqs. (2) and (3) simultaneously, we obtain the pairs $(1, -6)$ and $(5, 18)$. Therefore, $P_1 = (1, -6)$, and by substituting $x_1 = 1$ in Eq. (1) we find that the slope of L_1 is 2. Thus, an equation of L_1 is

$$y + 6 = 2(x - 1)$$
$$2x - y - 8 = 0$$

Because we found two solutions to the system of equations (2) and (3), we know there is a second line, L_2, that passes through $(3, -2)$ and is tangent to the curve at the point $P_2 = (5, 18)$. By substituting $x_1 = 5$ in Eq. (1), we find the slope of L_2 to be 10. An equation of line L_2 is

$$y - 18 = 10(x - 5)$$
$$10x - y - 32 = 0$$

3.2 INSTANTANEOUS VELOCITY IN RECTILINEAR MOTION

3.2.1 Definition If f is a function given by the equation $s = f(t)$ and a particle is moving along a straight line so that s is the number of units in the directed distance of the particle from a fixed point on the line at t units of time, then the *instantaneous velocity* of the particle at t_1 units of time is $v(t_1)$ units of velocity where

$$v(t_1) = \lim_{t \to t_1} \frac{f(t) - f(t_1)}{t - t_1}$$

if this limit exists.

Note that the limit in Definition 3.2.1 may be expressed as

$$v(t_1) = \lim_{\Delta t \to 0} \frac{f(t_1 + \Delta t) - f(t_1)}{t}$$

Therefore, the formula for instantaneous velocity is of the same form as the formula for the slope of the tangent line to a curve given in Definition 3.1.1.

Exercises 3.2

In Exercises 3-8, a particle is moving along a horizontal line according to the given equation of motion, where s ft is the directed distance of the particle from a point 0 at t sec. Find the instantaneous velocity, $v(t_1)$ ft/sec, at t_1 sec, and then find $v(t_1)$ for the particular value of t_1 given.

4. $s = 8 - t^2$; $t_1 = 5$

SOLUTION: We use Definition 3.2.1 with $f(t) = 8 - t^2$.

$$v(t_1) = \lim_{t \to t_1} \frac{f(t) - f(t_1)}{t - t_1}$$

$$= \lim_{t \to t_1} \frac{(8 - t^2) - (8 - t_1^2)}{t - t_1}$$

$$= \lim_{t \to t_1} \frac{-(t + t_1)(t - t_1)}{t - t_1}$$

$$= \lim_{t \to t_1} -(t + t_1)$$

$$= -2t_1$$

Thus, $v(5) = -10$.

8. $s = \sqrt[3]{t + 2}$; $t_1 = 6$

SOLUTION: We use Definition 3.2.1 with $f(t) = \sqrt[3]{t + 2}$.

$$v(t_1) = \lim_{t \to t_1} \frac{f(t) - f(t_1)}{t - t_1}$$

$$= \lim_{t \to t_1} \frac{\sqrt[3]{t + 2} - \sqrt[3]{t_1 + 2}}{t - t_1}$$

$$= \lim_{t \to t_1} \frac{(\sqrt[3]{t + 2} - \sqrt[3]{t_1 + 2})(\sqrt[3]{(t + 2)^2} + \sqrt[3]{t + 2}\sqrt[3]{t_1 + 2} + \sqrt[3]{(t_1 + 2)^2})}{(t - t_1)(\sqrt[3]{(t + 2)^2} + \sqrt[3]{t + 2}\sqrt[3]{t_1 + 2} + \sqrt[3]{(t_1 + 2)^2})}$$

$$= \lim_{t \to t_1} \frac{(t + 2) - (t_1 + 2)}{(t - t_1)(\sqrt[3]{(t + 2)^2} + \sqrt[3]{t + 2}\sqrt[3]{t_1 + 2} + \sqrt[3]{(t_1 + 2)^2})}$$

$$= \lim_{t \to t_1} \frac{1}{\sqrt[3]{(t + 2)^2} + \sqrt[3]{t + 2}\sqrt[3]{t_1 + 2} + \sqrt[3]{(t_1 + 2)^2}}$$

$$= \frac{1}{3\sqrt[3]{(t_1 + 2)^2}}$$

Thus,

$$v(6) = \frac{1}{3\sqrt[3]{(6+2)^2}} = \frac{1}{12}$$

12. The motion of a particle is along a horizontal line, according to the given equation of motion, where s ft is the directed distance of the particle from a point 0 at t sec. The positive direction is to the right. Determine the intervals of time when the particle is moving to the right and when it is moving to the left. Also determine when the particle reverses direction. Show the behavior of the motion with a figure similar to Fig. 3.2.2 (in the text), choosing values of t at random but including the values of t when the particle reverses its direction.

$$s = \frac{t}{1+t^2}$$

SOLUTION: We find the instantaneous velocity by using Definition 3.2.1 with $f(t) = t/(1+t^2)$.

$$v(t_1) = \lim_{t \to t_1} \frac{f(t) - f(t_1)}{t - t_1}$$

$$= \lim_{t \to t_1} \frac{\dfrac{t}{1+t^2} - \dfrac{t_1}{1+t_1^2}}{t - t_1}$$

$$= \lim_{t \to t_1} \frac{\left(\dfrac{t}{1+t^2} - \dfrac{t_1}{1+t_1^2}\right)(1+t^2)(1+t_1^2)}{(t-t_1)(1+t^2)(1+t_1^2)}$$

$$= \lim_{t \to t_1} \frac{t(1+t_1^2) - t_1(1+t^2)}{(t-t_1)(1+t^2)(1+t_1^2)}$$

$$= \lim_{t \to t_1} \frac{t + tt_1^2 - t_1 - t_1t^2}{(t-t_1)(1+t^2)(1+t_1^2)}$$

$$= \lim_{t \to t_1} \frac{(t-t_1) - tt_1(t-t_1)}{(t-t_1)(1+t^2)(1+t_1^2)}$$

$$= \lim_{t \to t_1} \frac{(t-t_1)(1-tt_1)}{(t-t_1)(1+t^2)(1+t_1^2)}$$

$$= \lim_{t \to t_1} \frac{1-tt_1}{(1+t^2)(1+t_1^2)}$$

$$= \frac{1-t_1^2}{(1+t_1^2)^2}$$

$$= \frac{(1-t_1)(1+t_1)}{(1+t_1^2)^2} \tag{1}$$

Because $v(t_1) = 0$ when $t_1 = 1$ and $t_1 = -1$, the particle reverses direction at each of these times. Table 12a indicates the sign of each factor of $v(t_1)$ when $t_1 < -1$, $-1 < t_1 < 1$, and $t_1 > 1$. The signs for $v(t_1)$ are found by using the rules for multiplying and dividing with positive and negative numbers. Because the particle is moving to the right when $v(t_1) > 0$ and is moving to the left when $v(t_1) < 0$, Table 12a also indicates that the motion is to the right when $-1 < t_1 < 1$ and is to the left when $t_1 > 1$ and when $t_1 < -1$. The behavior of the motion is illustrated in Fig. 3.2.12. Table 12b gives values of s and v for specific replacements of t, where we use the original equation of motion to calculate s and Eq. (1) to calculate v.

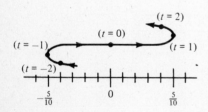

Figure 3.2.12

Table 12a

	$1 - t_1$	$1 + t_1$	$(1 + t_1{}^2)^2$	$v(t_1)$
$t_1 < -1$	+	−	+	−
$t_1 = -1$	+	0	+	0
$-1 < t_1 < 1$	+	+	+	+
$t_1 = 1$	0	+	+	0
$1 < t_1$	−	+	+	−

Table 12b

t	s	v
-2	$-\frac{2}{5}$	$-\frac{3}{25}$
-1	$-\frac{1}{2}$	0
0	0	1
1	$\frac{1}{2}$	0
2	$\frac{2}{5}$	$-\frac{3}{25}$

16. A rocket is fired vertically upward, and it is s ft above the ground t sec after being fired, where $s = 560t - 16t^2$ and the positive direction is upward. Find: (a) the velocity of the rocket 2 sec after being fired, and (b) how long it takes for the rocket to reach its maximum height.

SOLUTION: We find the instantaneous velocity by using Definition 3.2.1 with $f(t) = 560t - 16t^2$.

$$v(t_1) = \lim_{t \to t_1} \frac{f(t) - f(t_1)}{t - t_1}$$

$$= \lim_{t \to t_1} \frac{(560t - 16t^2) - (560t_1 - 16t_1{}^2)}{t - t_1}$$

$$= \lim_{t \to t_1} \frac{560(t - t_1) - 16(t^2 - t_1{}^2)}{t - t_1}$$

$$= \lim_{t \to t_1} \frac{(t - t_1)[560 - 16(t + t_1)]}{t - t_1}$$

$$= \lim_{t \to t_1} [560 - 16(t + t_1)]$$

$$= 560 - 32t_1 \tag{1}$$

(a) We use Eq. (1) with $t_1 = 2$.

$$v(2) = 560 - 32 \cdot 2 = 496$$

The velocity of the rocket 2 sec after being fired is 496 ft/sec.

(b) Because the instantaneous velocity is zero at the moment when the rocket reaches its maximum height, we use Eq. (1) with $v(t_1) = 0$.

$$0 = 560 - 32t_1$$
$$t_1 = 17.5$$

Thus, it takes 17.5 sec for the rocket to reach its maximum height.

3.3 THE DERIVATIVE OF A FUNCTION

The formula for the slope of the tangent line to the graph of f, which was given by Definition 3.1.1, and the formula for the instantaneous velocity of a particle in rectilinear motion, which was given by Definition 3.2.1, are both special cases of the formula for the derivative of a function. Following is one of the most important definitions in the book.

3.3.1 Definition

The *derivative* of the function f is that function, denoted by f', such that its value at any number x in the domain of f is given by

$$f'(x) = \lim_{\Delta x \to 0} \frac{f(x + \Delta x) - f(x)}{\Delta x} \tag{3}$$

if this limit exists. (f' is read "f prime," and $f'(x)$ is read "f prime of x.")

We also use the symbol $D_x f(x)$ to represent $f'(x)$. If $y = f(x)$, then the symbols $D_x y$, y', and dy/dx are sometimes used to represent $f'(x)$. If Δy is defined by

$$\Delta y = f(x + \Delta x) - f(x)$$

then

$$D_x y = \lim_{\Delta x \to 0} \frac{\Delta y}{\Delta x}$$

We note that Formula (3) yields a function of x whenever the limit exists. To find the value of this function at some particular number x_1 in its domain [that is, to find $f'(x_1)$], we may use either of the following formulas, which are equivalent.

$$f'(x_1) = \lim_{\Delta x \to 0} \frac{f(x_1 + \Delta x) - f(x_1)}{\Delta x} \tag{4}$$

$$f'(x_1) = \lim_{x \to x_1} \frac{f(x) - f(x_1)}{x - x_1} \tag{5}$$

The domain of f always includes every element in the domain of f'. However, sometimes the domains are not the same; that is, the domain of f' may be a proper subset of the domain of f. Hence, the following two definitions are needed.

3.3.2 Definition

The function f is said to be *differentiable at* x_1 if $f'(x_1)$ exists.

3.3.3 Definition

A function is said to be *differentiable* if it is differentiable at every number in its domain.

Exercises 3.3

In Exercises 1-10, find $f'(x)$ for the given function by applying Formula (3) of this section.

4. $f(x) = \sqrt{3x + 5}$

SOLUTION:

$$f'(x) = \lim_{\Delta x \to 0} \frac{f(x + \Delta x) - f(x)}{\Delta x}$$

$$= \lim_{\Delta x \to 0} \frac{\sqrt{3(x + \Delta x) + 5} - \sqrt{3x + 5}}{\Delta x}$$

$$= \lim_{\Delta x \to 0} \frac{(\sqrt{3(x + \Delta x) + 5} - \sqrt{3x + 5})(\sqrt{3(x + \Delta x) + 5} + \sqrt{3x + 5})}{\Delta x(\sqrt{3(x + \Delta x) + 5} + \sqrt{3x + 5})}$$

$$= \lim_{\Delta x \to 0} \frac{[3(x + \Delta x) + 5] - (3x + 5)}{\Delta x(\sqrt{3(x + \Delta x) + 5} + \sqrt{3x + 5})}$$

$$= \lim_{\Delta x \to 0} \frac{3\Delta x}{\Delta x(\sqrt{3(x + \Delta x) + 5} + \sqrt{3x + 5})}$$

$$= \lim_{\Delta x \to 0} \frac{3}{\sqrt{3(x + \Delta x) + 5} + \sqrt{3x + 5}}$$

$$= \frac{3}{\sqrt{3x + 5} + \sqrt{3x + 5}}$$

$$= \frac{3}{2\sqrt{3x + 5}}$$

8. $f(x) = \dfrac{3}{1 + x^2}$

SOLUTION:

$$f'(x) = \lim_{\Delta x \to 0} \frac{f(x + \Delta x) - f(x)}{\Delta x}$$

$$= \lim_{\Delta x \to 0} \frac{\dfrac{3}{1 + (x + \Delta x)^2} - \dfrac{3}{1 + x^2}}{\Delta x}$$

$$= \lim_{\Delta x \to 0} \frac{3(1 + x^2) - 3[1 + (x + \Delta x)^2]}{\Delta x[1 + (x + \Delta x)^2](1 + x^2)}$$

$$= \lim_{\Delta x \to 0} \frac{3[1 + x^2 - 1 - x^2 - 2x\Delta x - (\Delta x)^2]}{\Delta x[1 + (x + \Delta x)^2](1 + x^2)}$$

$$= \lim_{\Delta x \to 0} \frac{3\Delta x(-2x - \Delta x)}{\Delta x[1 + (x + \Delta x)^2](1 + x^2)}$$

$$= \lim_{\Delta x \to 0} \frac{3(-2x - \Delta x)}{[1 + (x + \Delta x)^2](1 + x^2)}$$

$$= \frac{-6x}{(1 + x^2)^2}$$

16. Find $f'(a)$ for the given value of a by applying Formula (4) of this section.

$$f(x) = \frac{1}{x} + x + x^2; \quad a = -3$$

SOLUTION:

$$f'(x_1) = \lim_{\Delta x \to 0} \frac{f(x_1 + \Delta x) - f(x_1)}{\Delta x}$$

$$f'(-3) = \lim_{\Delta x \to 0} \frac{f(-3 + \Delta x) - f(-3)}{\Delta x}$$

$$= \lim_{\Delta x \to 0} \frac{\left[\dfrac{1}{-3 + \Delta x} + (-3 + \Delta x) + (-3 + \Delta x)^2\right] - \left[-\dfrac{1}{3} - 3 + 9\right]}{\Delta x}$$

$$= \lim_{\Delta x \to 0} \frac{\dfrac{1}{-3 + \Delta x} + \dfrac{1}{3}}{\Delta x} + \lim_{\Delta x \to 0} \frac{(-3 + \Delta x) + (-3 + \Delta x)^2 - 6}{\Delta x}$$

$$= \lim_{\Delta x \to 0} \frac{3 - 3 + \Delta x}{3\Delta x(-3 + \Delta x)} + \lim_{\Delta x \to 0} \frac{-3 + \Delta x + 9 - 6\Delta x + (\Delta x)^2 - 6}{\Delta x}$$

$$= \lim_{\Delta x \to 0} \frac{\Delta x}{3\Delta x(-3 + \Delta x)} + \lim_{\Delta x \to 0} \frac{\Delta x(-5 + \Delta x)}{\Delta x}$$

$$= \lim_{\Delta x \to 0} \frac{1}{3(-3 + \Delta x)} + \lim_{\Delta x \to 0} (-5 + \Delta x)$$

$$= -\tfrac{1}{9} + (-5)$$

$$= -\tfrac{46}{9}$$

22. Find $f'(a)$ for the given value of a by applying Formula (5) of this section.

$$f(x) = \frac{1}{\sqrt[3]{x}} - x; \quad a = -8$$

SOLUTION:

$$f'(x_1) = \lim_{x \to x_1} \frac{f(x) - f(x_1)}{x - x_1}$$

$$f'(-8) = \lim_{x \to -8} \frac{f(x) - f(-8)}{x - (-8)}$$

$$= \lim_{x \to -8} \frac{\left(\dfrac{1}{\sqrt[3]{x}} - x\right) - \left(-\dfrac{1}{2} + 8\right)}{x + 8}$$

$$= \lim_{x \to -8} \frac{\dfrac{1}{\sqrt[3]{x}} + \dfrac{1}{2}}{x + 8} + \lim_{x \to -8} \frac{-x - 8}{x + 8}$$

$$= \lim_{x \to -8} \frac{2 + \sqrt[3]{x}}{2\sqrt[3]{x}\,(x + 8)} + \lim_{x \to -8} (-1)$$

$$= \lim_{x \to -8} \frac{(2 + \sqrt[3]{x})(4 - 2\sqrt[3]{x} + \sqrt[3]{x^2})}{2\sqrt[3]{x}\,(x + 8)(4 - 2\sqrt[3]{x} + \sqrt[3]{x^2})} + (-1)$$

$$= \lim_{x \to -8} \frac{8 + x}{2\sqrt[3]{x}\,(x + 8)(4 - 2\sqrt[3]{x} + \sqrt[3]{x^2})} - 1$$

$$= \lim_{x \to -8} \frac{1}{2\sqrt[3]{x}\,(4 - 2\sqrt[3]{x} + \sqrt[3]{x^2})} - 1$$

$$= \tfrac{1}{-48} - 1 = -\tfrac{49}{48}$$

24. Given $f(x) = \sqrt[3]{(4x - 3)^2}$, find $f'(x)$. Is f differentiable at $\tfrac{3}{4}$? Draw a sketch of the graph of f.

SOLUTION:

$$f(x) = (4x - 3)^{2/3}$$

$$f'(x) = \lim_{\Delta x \to 0} \frac{f(x + \Delta x) - f(x)}{\Delta x}$$

$$= \lim_{\Delta x \to 0} \frac{(4x + 4\Delta x - 3)^{2/3} - (4x - 3)^{2/3}}{\Delta x}$$

$$= \lim_{\Delta x \to 0} \frac{[(4x + 4\Delta x - 3)^{2/3} - (4x - 3)^{2/3}]\,[(4x + 4\Delta x - 3)^{4/3} + (4x + 4\Delta x - 3)^{2/3}(4x - 3)^{2/3} + (4x - 3)^{4/3}]}{\Delta x[(4x + 4\Delta x - 3)^{4/3} + (4x + 4\Delta x - 3)^{2/3}(4x - 3)^{2/3} + (4x - 3)^{4/3}]}$$

$$= \lim_{\Delta x \to 0} \frac{(4x + 4\Delta x - 3)^2 - (4x - 3)^2}{\Delta x[(4x + 4\Delta x - 3)^{4/3} + (4x + 4\Delta x - 3)^{2/3}(4x - 3)^{2/3} + (4x - 3)^{4/3}]}$$

$$= \lim_{\Delta x \to 0} \frac{16x^2 + 16(\Delta x)^2 + 9 + 32x\Delta x - 24x - 24\Delta x - 16x^2 + 24x - 9}{\Delta x[(4x + 4\Delta x - 3)^{4/3} + (4x + 4\Delta x - 3)^{2/3}(4x - 3)^{2/3} + (4x - 3)^{4/3}]}$$

$$= \lim_{\Delta x \to 0} \frac{16\Delta x + 32x - 24}{(4x + 4\Delta x - 3)^{4/3} + (4x + 4\Delta x - 3)^{2/3}(4x - 3)^{2/3} + (4x - 3)^{4/3}}$$

$$= \frac{32x - 24}{(4x - 3)^{4/3} + (4x - 3)^{2/3}(4x - 3)^{2/3} + (4x - 3)^{4/3}}$$

$$= \frac{32x - 24}{3(4x - 3)^{4/3}}$$

$$= \frac{8(4x - 3)}{3(4x - 3)^{4/3}}$$

$$f'(x) = \frac{8}{3(4x - 3)^{1/3}} = \frac{8}{3\sqrt[3]{4x - 3}}$$

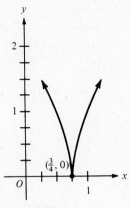

Figure 3.3.24

Because the denominator of $f'(x)$ is zero when $x = \frac{3}{4}$, then $f'(\frac{3}{4})$ is not defined; thus, f is not differentiable at $\frac{3}{4}$. A sketch of the graph of f is shown in Fig. 3.3.24. Note that the tangent line to the curve at the point where $x = \frac{3}{4}$ is parallel to the y-axis.

26. Let f be a function whose domain is the set of all real numbers and $f(a + b) = f(a) \cdot f(b)$ for all a and b. Furthermore, suppose that $f(0) = 1$ and $f'(0)$ exists. Prove that $f'(x)$ exists for all x and that $f'(x) = f'(0) \cdot f(x)$.

SOLUTION: Because $f'(0)$ exists, then $f'(x_1)$ exists when $x_1 = 0$. Thus,

$$f'(x_1) = \lim_{\Delta x \to 0} \frac{f(x_1 + \Delta x) - f(x_1)}{\Delta x}$$

$$f'(0) = \lim_{\Delta x \to 0} \frac{f(\Delta x) - f(0)}{\Delta x}$$

Since $f(0) = 1$, we have

$$f'(0) = \lim_{\Delta x \to 0} \frac{f(\Delta x) - 1}{\Delta x} \qquad (1)$$

Because $f(a + b) = f(a) \cdot f(b)$, then $f(x + \Delta x) = f(x) \cdot f(\Delta x)$. Hence,

$$f'(x) = \lim_{\Delta x \to 0} \frac{f(x + \Delta x) - f(x)}{\Delta x}$$

$$= \lim_{\Delta x \to 0} \frac{f(x) \cdot f(\Delta x) - f(x)}{\Delta x}$$

$$= \lim_{\Delta x \to 0} \frac{[f(\Delta x) - 1]}{\Delta x} \cdot f(x) \qquad (2)$$

We use Limit Theorem 6 (the limit of a product). Thus, (2) becomes

$$f'(x) = \lim_{\Delta x \to 0} \frac{f(\Delta x) - 1}{\Delta x} \cdot \lim_{\Delta x \to 0} f(x) \qquad (3)$$

By Limit Theorem 2 (the limit of a constant) and by substituting from Eq. (1) into (3), we obtain

$$f'(x) = f'(0) \cdot f(x)$$

Because $f(x)$ exists for all x and $f'(0)$ exists, it follows that $f'(x)$ exists for all x.

3.4 DIFFERENTIABILITY AND CONTINUITY

3.4.1 Theorem   If a function f is differentiable at x_1, then f is continuous at x_1.

The theorem implies that if $f'(x_1)$ exists, then there must be no break in the graph of f at the point where $x = x_1$. The converse of Theorem 3.4.1 is not true. That is, a function that is continuous at x_1 may not be differentiable at x_1. For example, the function whose graph is shown in Fig. 3.4.4 is continuous at -2 but not differentiable at -2. Note that the graph has a "corner" at $(-2, 1)$. If a function is differentiable at a point, then the graph of the function must be "smooth" at that point. Furthermore, if a function is differentiable at a point, then the tangent line to the graph of the function at that point must be a line that is not vertical. That is, if the tangent line to the graph of f at x_1 is a vertical line, then f is not differentiable at x_1.

One-sided derivatives are defined as follows.

3.4.2 Definition   If the function f is defined at x_1, then the *derivative from the right* of f at x_1, denoted by $f'_+(x_1)$, is defined by

$$f'_+(x_1) = \lim_{\Delta x \to 0^+} \frac{f(x_1 + \Delta x) - f(x_1)}{\Delta x}$$

or, equivalently,

$$f'_+(x_1) = \lim_{x \to x_1^+} \frac{f(x) - f(x_1)}{x - x_1}$$

if the limit exists.

3.4.3 Definition   If the function f is defined at x_1, then the *derivative from the left* of f at x_1, denoted by $f'_-(x_1)$, is defined by

$$f'_-(x_1) = \lim_{\Delta x \to 0^-} \frac{f(x_1 + \Delta x) - f(x_1)}{\Delta x}$$

or, equivalently,

$$f'_-(x_1) = \lim_{x \to x_1^-} \frac{f(x) - f(x_1)}{x - x_1}$$

if the limit exists.

Exercises 3.4

In Exercises 1-14, do each of the following:

 (a) Draw a sketch of the graph of the function.
 (b) Determine if f is continuous at x_1.
 (c) Find $f'_-(x_1)$ and $f'_+(x_1)$ if they exist.
 (d) Determine if f is differentiable at x_1.

4. $f(x) = 1 + |x + 2|$; $x_1 = -2$

SOLUTION: Because $|x + 2| = x + 2$ if $x \geqslant -2$ and $|x + 2| = -(x + 2)$ if $x < -2$,

$$f(x) = \begin{cases} -x - 1 & \text{if} \quad x < -2 \\ x + 3 & \text{if} \quad x \geqslant -2 \end{cases}$$

 (a) A sketch of the graph of f is shown in Fig. 3.4.4.
 (b) Because $f(-2) = 1$, and

$$\lim_{x \to -2^-} f(x) = \lim_{x \to -2^-} (-x - 1) = 1$$

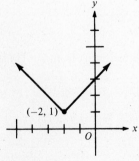

$(-2, 1)$

Figure 3.4.4

$$\lim_{x \to -2^+} f(x) = \lim_{x \to -2^+} (x + 3) = 1$$

then f is continuous at -2.

(c) By Definition 3.4.3

$$f'_-(-2) = \lim_{x \to -2^-} \frac{f(x) - f(-2)}{x - (-2)}$$

$$= \lim_{x \to -2^-} \frac{(-x - 1) - 1}{x + 2}$$

$$= \lim_{x \to -2^-} \frac{-(x + 2)}{x + 2}$$

$$= -1$$

And by Definition 3.4.2

$$f'_+(-2) = \lim_{x \to -2^+} \frac{f(x) - f(-2)}{x - (-2)}$$

$$= \lim_{x \to -2^+} \frac{(x + 3) - 1}{x + 2}$$

$$= \lim_{x \to -2^+} \frac{x + 2}{x + 2}$$

$$= 1$$

(d) Because $f'_-(-2) \neq f'_+(-2)$, then $f'(-2)$ does not exist. Thus, f is not differentiable at -2.

8. $f(x) = \begin{cases} x^2 - 4 & \text{if} \quad x < 2 \\ \sqrt{x - 2} & \text{if} \quad x \geq 2 \end{cases}$; $x_1 = 2$

SOLUTION:

(a) A sketch of the graph is shown in Fig. 3.4.8.

(b) Because $f(2) = 0$, and

$$\lim_{x \to 2^-} f(x) = \lim_{x \to 2^-} (x^2 - 4) = 0$$

$$\lim_{x \to 2^+} f(x) = \lim_{x \to 2^+} \sqrt{x - 2} = 0$$

then f is continuous at 2.

(c) By Definition 3.4.3

$$f'_-(2) = \lim_{x \to 2^-} \frac{f(x) - f(2)}{x - 2}$$

$$= \lim_{x \to 2^-} \frac{(x^2 - 4) - 0}{x - 2}$$

$$= \lim_{x \to 2^-} \frac{(x + 2)(x - 2)}{x - 2}$$

$$= \lim_{x \to 2^-} (x + 2)$$

$$= 4$$

By Definition 3.4.2

$$f'_+(2) = \lim_{x \to 2^+} \frac{f(x) - f(2)}{x - 2}$$

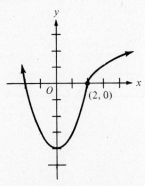

Figure 3.4.8

$$= \lim_{x \to 2^+} \frac{\sqrt{x-2} - 0}{x-2}$$

$$= \lim_{x \to 2^+} \frac{\sqrt{x-2}}{\sqrt{x-2}\sqrt{x-2}}$$

$$= \lim_{x \to 2^+} \frac{1}{\sqrt{x-2}}$$

$$= +\infty$$

(d) Because $f'_+(2) = +\infty$, $f'(2)$ does not exist, and f is not differentiable at 2.

12. $f(x) = (x-2)^{-2}$; $x_1 = 2$

SOLUTION:

   (a) A sketch of the graph is shown in Fig. 3.4.12.

   (b) Because $f(2)$ is not defined, f is discontinuous at 2.

   (c) Because $f(2)$ is not defined, then neither $f'_-(2)$ nor $f'_+(2)$ is defined.

   (d) Because $f(2)$ is not defined, then $f'(2)$ is not defined, and f is not differentiable at 2.

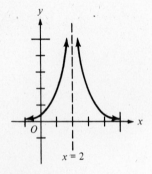

Figure 3.4.12

20. Given $f(x) = [\![x]\!]$, find $f'(x_1)$ if x_1 is not an integer. Prove by applying Theorem 3.4.1 that $f'(x_1)$ does not exist if x_1 is an integer. If x_1 is an integer, what can you say about $f'_-(x_1)$ and $f'_+(x_1)$?

SOLUTION: For a sketch of the graph of f see Fig. 1.7.11 in the text. If x_1 is not an integer, then there is an integer N such that $N < x_1 < N + 1$. Hence, by definition

$$f(x_1) = N$$

Furthermore, for all x such that $N < x < N + 1$,

$$f(x) = N$$

Thus, whenever $N < x < N + 1$,

$$f(x) - f(x_1) = 0$$

Therefore, by Definition 3.3.1

$$f'(x_1) = \lim_{x \to x_1} \frac{f(x) - f(x_1)}{x - x_1}$$

$$= \lim_{x \to x_1} \frac{0}{x - x_1}$$

$$= 0$$

Thus, if x_1 is not an integer, $f'(x_1) = 0$. Furthermore, if x_1 is an integer, then

$$\lim_{x \to x_1^+} f(x) = x_1 \quad \text{and} \quad \lim_{x \to x_1^-} f(x) = x_1 - 1$$

Hence, $\lim_{x \to x_1} f(x)$ does not exist, and thus, f is discontinuous at x_1. By Theorem 3.4.1 this proves that $f'(x_1)$ does not exist if x_1 is an integer.

   However, if x_1 is an integer, then by Definition 3.4.3

$$f'_-(x_1) = \lim_{x \to x_1^-} \frac{f(x) - f(x_1)}{x - x_1}$$

$$= \lim_{x \to x_1^-} \frac{(x_1 - 1) - x_1}{x - x_1}$$

$$= \lim_{x \to x_1^-} \frac{-1}{x - x_1}$$

$$= +\infty$$

And if x_1 is an integer, by Definition 3.4.2 we have

$$f'_+(x_1) = \lim_{x \to x_1^+} \frac{f(x) - f(x_1)}{x - x_1}$$

$$= \lim_{x \to x_1^+} \frac{x_1 - x_1}{x - x_1}$$

$$= 0$$

24. Let the function f be defined by

$$f(x) = \begin{cases} \dfrac{g(x) - g(a)}{x - a} & \text{if} \quad x \neq a \\ g'(a) & \text{if} \quad x = a \end{cases}$$

Prove that if $g'(a)$ exists, f is continuous at a.

SOLUTION: Because $g'(a)$ exists, then

$$g'(a) = \lim_{x \to a} \frac{g(x) - g(a)}{x - a} \tag{1}$$

Now from the definition of f we have

$$\lim_{x \to a} f(x) = \lim_{x \to a} \frac{g(x) - g(a)}{x - a} \tag{2}$$

And by substitution from Eq. (1) in (2) we obtain

$$\lim_{x \to a} f(x) = g'(a) \tag{3}$$

Because $f(a)$ is defined to be $g'(a)$, Eq. (3) yields

$$\lim_{x \to a} f(x) = f(a)$$

Therefore, by Definition 2.5.1, f is continuous at a.

3.5 SOME THEOREMS ON DIFFERENTIATION OF ALGEBRAIC FUNCTIONS

The differentiation formulas that are proved in this section are among the most useful in the book and should be memorized.

3.5.1 Theorem (Formula for the derivative of a constant)

If c is any constant,

$$D_x(c) = 0$$

The derivative of a constant is zero.

3.5.3 Theorem (Formula for the derivative of a constant times a function)

If c is any constant and if $D_x f(x)$ exists,

$$D_x[c \cdot f(x)] = c \cdot D_x f(x)$$

The derivative of a constant times a function is the constant times the derivative of the function, if this derivative exists.

3.5.4 Theorem (Formula for the derivative of the sum of two functions)

If $D_x f(x)$ and $D_x g(x)$ exist,

$$D_x[f(x) + g(x)] = D_x f(x) + D_x g(x)$$

The derivative of the sum of two functions is the sum of their derivatives, if these derivatives exist.

3.5.6 Theorem (Formula for the derivative of a product) If $D_x f(x)$ and $D_x g(x)$ exist,

$$D_x[f(x) \cdot g(x)] = f(x) \cdot D_x g(x) + g(x) \cdot D_x f(x)$$

The derivative of the product of two functions is the first function times the derivative of the second function plus the second function times the derivative of the first function, if these derivatives exist.

3.5.7 Theorem (Formula for the derivative of a quotient) If $g(x) \neq 0$ and $D_x f(x)$ and $D_x g(x)$ exist, then

$$D_x\left[\frac{f(x)}{g(x)}\right] = \frac{g(x) \cdot D_x f(x) - f(x) \cdot D_x g(x)}{[g(x)]^2}$$

The derivative of the quotient of two functions is the denominator times the derivative of the numerator minus the numerator times the derivative of the denominator, all divided by the square of the denominator, if the derivatives exist.

Theorem (Formula for the derivative of the power function for an integer exponent) If r is any positive or negative integer, then

$$D_x(x^r) = rx^{r-1}$$

We have proved the formula for the derivative of the power function only for the case in which the exponent r is an integer. In Section 3.7 we show that the formula holds when r is any rational number, and in Section 9.5 we show that the formula is valid when r is any real number.

When it is convenient to do so, we replace a given product by an equivalent sum before differentiating, because the formula for the derivative of a sum is easier to apply than the formula for the derivative of a product. This is illustrated in Exercise 14. Also, if possible we replace a given quotient by an equivalent sum before differentiating, as illustrated in Exercise 22.

Exercises 3.5

In Exercises 1-26, differentiate the given function by applying the theorems of this section.

4. $g(x) = x^7 - 2x^5 + 5x^3 - 7x$

SOLUTION:

$$\begin{aligned} g'(x) &= 7x^{7-1} - 2(5x^{5-1}) + 5(3x^{3-1}) - 7(x^{1-1}) \\ &= 7x^6 - 10x^4 + 15x^2 - 7 \end{aligned}$$

10. $f(x) = x^4 - 5 + x^{-2} + 4x^{-4}$

SOLUTION:

$$\begin{aligned} f'(x) &= 4x^{4-1} - 0 + (-2)x^{-2-1} + 4 \cdot (-4)x^{-4-1} \\ &= 4x^3 - 2x^{-3} - 16x^{-5} \end{aligned}$$

12. $H(x) = \dfrac{5}{6x^5}$

SOLUTION: We first write $H(x)$ in the form cx^r and then use the formula for the derivative of a constant times a function.

$$H(x) = \tfrac{5}{6}x^{-5}$$
$$H'(x) = \tfrac{5}{6}(-5)x^{-5-1}$$
$$= -\tfrac{25}{6}x^{-6}$$
$$= -\frac{25}{6x^6}$$

Note that we could also use the formula for the derivative of a quotient to find $H'(x)$.

14. $g(x) = (2x^2 + 5)(4x - 1)$

SOLUTION: We express $g(x)$ as a sum and use the formula for the derivative of a sum.

$$g(x) = 8x^3 - 2x^2 + 20x - 5$$
$$g'(x) = 24x^2 - 4x + 20$$

ALTERNATE SOLUTION: We use the formula for the derivative of a product.

$$g'(x) = (2x^2 + 5) \cdot (D_x(4x - 1) + (4x - 1) \cdot D_x(2x^2 + 5)$$
$$= (2x^2 + 5)(4) + (4x - 1)(4x)$$
$$= 8x^2 + 20 + 16x^2 - 4x$$
$$= 24x^2 - 4x + 20$$

18. $F(y) = \dfrac{2y + 1}{3y + 4}$

SOLUTION: We use the formula for the derivative of a quotient.

$$F'(y) = \frac{(3y + 4) \cdot D_y(2y + 1) - (2y + 1) \cdot D_y(3y + 4)}{(3y + 4)^2}$$

$$= \frac{(3y + 4)(2) - (2y + 1)(3)}{(3y + 4)^2}$$

$$= \frac{5}{(3y + 4)^2}$$

22. $g(x) = \dfrac{x^4 - 2x^2 + 5x + 1}{x^4}$

SOLUTION: Because the denominator of $g(x)$ is a monomial, we may divide the numerator by the denominator and thus express $g(x)$ as a sum. Then we differentiate by the formula for the derivative of a sum.

$$g(x) = 1 - 2x^{-2} + 5x^{-3} + x^{-4}$$
$$g'(x) = 4x^{-3} - 15x^{-4} - 4x^{-5}$$
$$= x^{-5}(4x^2 - 15x^1 - 4)$$
$$= \frac{4x^2 - 15x - 4}{x^5}$$

26. $g(x) = \dfrac{x^3 + 1}{x^2 + 3}(x^2 - 2x^{-1} + 1)$

SOLUTION: First, we simplify $g(x)$

$$g(x) = \frac{(x^3 + 1)(x^2 - 2x^{-1} + 1)}{x^2 + 3}$$

$$= \frac{x^5 + x^3 - x^2 - 2x^{-1} + 1}{x^2 + 3} \cdot \frac{x}{x}$$

$$= \frac{x^6 + x^4 - x^3 + x - 2}{x^3 + 3x}$$

Now we use the formula for the derivative of a quotient. Note that we cannot proceed as in Exercise 22 because the denominator of $g(x)$ is not a monomial.

$$g'(x) = \frac{(x^3 + 3x)(6x^5 + 4x^3 - 3x^2 + 1) - (x^6 + x^4 - x^3 + x - 2)(3x^2 + 3)}{(x^3 + 3x)^2}$$

And if we simplify $g'(x)$, the result is

$$g'(x) = \frac{3x^8 + 16x^6 + 9x^4 - 8x^3 + 6x^2 + 6}{x^2(x^2 + 3)^2}$$

32. Find an equation of the tangent line to the curve $y = 8/(x^2 + 4)$ at the point $(2, 1)$.

SOLUTION: Because the formula for the slope of the tangent line to the graph of f at the point where $x = x_1$ is a special case of $f'(x_1)$, that is, because $m(x_1) = f'(x_1)$, we differentiate f, where

$$f(x) = \frac{8}{x^2 + 4}$$

to find a formula for the slope. We use the formula for the derivative of a quotient.

$$f'(x) = \frac{(x^2 + 4) \cdot 0 - 8 \cdot (2x)}{(x^2 + 4)^2}$$

$$= \frac{-16x}{(x^2 + 4)^2}$$

Then $m(2) = f'(2)$, so

$$m(2) = -\frac{16 \cdot 2}{(2^2 + 4)^2} = -\frac{1}{2}$$

We use the point-slope form of an equation of a line with $m = -\frac{1}{2}$ and $(x_1, y_1) = (2, 1)$. Thus, we have

$$y - 1 = -\tfrac{1}{2}(x - 2)$$
$$x + 2y - 4 = 0$$

34. Find an equation of each of the tangent lines to the curve $3y = x^3 - 3x^2 + 6x + 4$, which is parallel to the line $2x - y + 3 = 0$.

SOLUTION: Let L_1 be one of the required lines and let (x_1, y_1) be the point at which L_1 is tangent to the curve. As in Exercise 32, the slope of L_1 is given by $m(x_1) = f'(x_1)$, where $y = f(x)$ is the equation of the curve. Dividing both members of the equation of the curve by 3, we obtain

$$f(x) = \tfrac{1}{3} x^3 - x^2 + 2x + \tfrac{4}{3} \qquad (1)$$

Differentiating, we have

$$f'(x) = x^2 - 2x + 2$$

Thus,

$$m(x_1) = x_1{}^2 - 2x_1 + 2 \qquad (2)$$

Because the slope-intercept form of the equation of the given line is $y = 2x + 3$, the slope of this line is 2. Because L_1 is parallel to the given line, the slope of L_1 is also 2. Therefore, $m(x_1) = 2$, and by Eq. (2) we have

$$2 = x_1{}^2 - 2x_1 + 2$$
$$x_1 = 0 \quad \text{or} \quad x_1 = 2$$

If $x_1 = 0$, then by Eq. (1) $f(x_1) = \frac{4}{3}$. Thus, L_1 contains $(0, \frac{4}{3})$ and has slope 2. An equation of L_1 is

$$y = 2x + \tfrac{4}{3}$$
$$6x - 3y + 4 = 0$$

If $x_1 = 2$, then $f(x_1) = 4$. Therefore, the point-slope form of the equation of the second of the required lines is

$$y - 4 = 2(x - 2)$$
$$2x - y = 0$$

36. An object is moving along a straight line according to the equation of motion $s = 3t/(t^2 + 9)$, with $t \geq 0$, where s ft is the directed distance of the object from the starting point at t sec.

(a) What is the instantaneous velocity of the object at t_1 sec?
(b) What is the instantaneous velocity at 1 sec?
(c) At what time is the instantaneous velocity zero?

SOLUTION:
(a) The value for $v(t_1)$ (where $v(t_1)$ ft/sec is the instantaneous velocity at t_1 sec) is $f'(t_1)$ where

$$f(t) = \frac{3t}{t^2 + 9}$$

$$f'(t) = \frac{(t^2 + 9)(3) - (3t)(2t)}{(t^2 + 9)^2}$$

$$= \frac{-3t^2 + 27}{(t^2 + 9)^2}$$

Then $v(t_1) = f'(t_1)$; thus,

$$v(t_1) = \frac{-3t_1{}^2 + 27}{(t_1{}^2 + 9)^2} \tag{1}$$

(b) We let $t_1 = 1$ in Eq. (1).

$$v(1) = \frac{-3 + 27}{10^2} = \frac{6}{25}$$

The instantaneous velocity is $\frac{6}{25}$ ft/sec at 1 sec.

(c) We let $v(t_1) = 0$ in Eq. (1).

$$0 = \frac{-3t_1{}^2 + 27}{(t_1{}^2 + 9)^2}$$

$$0 = -3t_1{}^2 + 27$$

$$t_1{}^2 = 9$$

$$t_1 = 3 \quad \text{because} \quad t_1 \geq 0$$

The instantaneous velocity is zero at 3 sec.

3.6 THE DERIVATIVE OF A COMPOSITE FUNCTION

3.6.1 Theorem (Chain Rule) If y is a function of u, defined by $y = f(u)$, and $D_u y$ exists, and if u is a function of x, defined by $u = g(x)$, and $D_x u$ exists, then y is a function of x and $D_x y$ exists and is given by

$$D_x y = D_u y \cdot D_x u$$
$$= f'(u) \cdot g'(x)$$

The following differentiation formula is a special case of the chain rule.

Theorem (Formula for the derivative of a composite power function) If $f'(x)$ exists and n is any integer,

$$D_x [f(x)]^n = n[f(x)]^{n-1} f'(x)$$

As the exercises illustrate, we usually want to write the derivative in completely factored form. Sometimes it is easier to differentiate if we first replace a given quotient by an equivalent product and then use the chain rule. This technique is illustrated in Exercise 20.

Exercises 3.6

In Exercises 1-20, find the derivative of the given function.

4. $g(r) = (2r^4 + 8r^2 + 1)^5$

SOLUTION: We use the special case of the chain rule called the "formula for the derivative of a composite power function."

$$g'(r) = 5(2r^4 + 8r^2 + 1)^4 \cdot (8r^3 + 16r)$$
$$= 40r(2r^4 + 8r^2 + 1)^4 (r^2 + 2)$$

8. $f(x) = (4x^2 + 7)^2 (2x^3 + 1)^4$

SOLUTION: Because the last operation in the expression is multiplication, this expression is a product. We use the formula for the derivative of a product.

$$f'(x) = (4x^2 + 7)^2 \cdot D_x (2x^3 + 1)^4 + (2x^3 + 1)^4 \cdot D_x (4x^2 + 7)^2 \tag{1}$$

By the formula for the derivative of a composite power function, we have

$$D_x (2x^3 + 1)^4 = 4(2x^3 + 1)^3 (6x^2) \tag{2}$$

and

$$D_x (4x^2 + 7)^2 = 2(4x^2 + 7)^1 (8x) \tag{3}$$

By substituting from (2) and (3) into (1), we obtain

$$f'(x) = (4x^2 + 7)^2 \cdot 24x^2 (2x^3 + 1)^3 + (2x^3 + 1)^4 \cdot 16x(4x^2 + 7)$$

We factor $f'(x)$ by removing the repeated factors.

$$f'(x) = (4x^2 + 7) \cdot 8x \cdot (2x^3 + 1)^3 \cdot [(4x^2 + 7) \cdot 3x + (2x^3 + 1) \cdot 2]$$
$$= 8x(4x^2 + 7)(2x^3 + 1)^3 (16x^3 + 21x + 2)$$

12. $g(t) = \left(\dfrac{2t^2 + 1}{3t^3 + 1} \right)^2$

SOLUTION: We use the formula for the derivative of a composite power function.

$$g'(t) = 2 \left(\frac{2t^2 + 1}{3t^3 + 1} \right)^1 \cdot D_t \left(\frac{2t^2 + 1}{3t^3 + 1} \right) \tag{1}$$

We now use the formula for the derivative of a quotient.

$$D_t \left(\frac{2t^2 + 1}{3t^3 + 1} \right) = \frac{(3t^3 + 1)(4t) - (2t^2 + 1)(9t^2)}{(3t^3 + 1)^2}$$

$$= \frac{-6t^4 - 9t^2 + 4t}{(3t^3 + 1)^2} \tag{2}$$

Substituting from (2) into Eq. (1), we obtain

$$g'(t) = \frac{2(2t^2 + 1)}{3t^3 + 1} \cdot \frac{-t(6t^3 + 9t - 4)}{(3t^3 + 1)^2}$$

$$= -\frac{2t(2t^2 + 1)(6t^3 + 9t - 4)}{(3t^3 + 1)^3}$$

16. $f(y) = (y + 3)^3 (5y + 1)^2 (3y^2 - 4)$

SOLUTION: We use the formula of Exercise 27 in Exercises 3.5. If
$\phi(x) = f(x) \cdot g(x) \cdot h(x)$, then

$$\phi'(x) = f(x) \cdot g(x) \cdot h'(x) + f(x) \cdot g'(x) \cdot h(x) + f'(x) \cdot g(x) \cdot h(x)$$

provided $f'(x), g'(x),$ and $h'(x)$ exist. Thus, by this formula and the chain rule,

$$\begin{aligned}
f'(y) &= (y + 3)^3 (5y + 1)^2 (6y) + (y + 3)^3 [2(5y + 1)^1 (5)] (3y^2 - 4) \\
&\quad + [3(y + 3)^2 (1)] (5y + 1)^2 (3y^2 - 4) \\
&= (y + 3)^2 (5y + 1)[(y + 3)(5y + 1)(6y) + (y + 3)(10)(3y^2 - 4) \\
&\quad\quad + 3(5y + 1)(3y^2 - 4)] \\
&= (y + 3)^2 (5y + 1)[(5y^2 + 16y + 3)(6y) + 10(3y^3 + 9y^2 - 4y - 12) \\
&\quad\quad + 3(15y^3 + 3y^2 - 20y - 4)] \\
&= (y + 3)^2 (5y + 1)(105y^3 + 195y^2 - 82y - 132)
\end{aligned}$$

20. $F(x) = \dfrac{(5x - 8)^{-2}}{(x^2 + 3)^{-3}}$

SOLUTION: Before differentiating, we replace the given quotient by an equivalent product.

$$F(x) = (x^2 + 3)^3 (5x - 8)^{-2}$$

As in Exercise 8, we use the formula for the derivative of a product and the chain rule. We obtain

$$F'(x) = (x^2 + 3)^3 [(-2)(5x - 8)^{-3}(5)] + (5x - 8)^{-2} [3(x^2 + 3)^2 (2x)]$$

We factor $F'(x)$ by removing the repeated factors. Because the exponent -3 is less than the exponent -2, we choose to remove $(5x - 8)^{-3}$ rather than $(5x - 8)^{-2}$.

$$\begin{aligned}
F'(x) &= (x^2 + 3)^2 (2)(5x - 8)^{-3} [(x^2 + 3)(-5) + (5x - 8)(3x)] \\
&= 2(x^2 + 3)^2 (5x - 8)^{-3} (10x^2 - 24x - 15) \\
&= \frac{2(x^2 + 3)^2 (10x^2 - 24x - 15)}{(5x - 8)^3}
\end{aligned}$$

26. Given $f(u) = u^2 + 5u + 5$ and $g(x) = (x + 1)/(x - 1)$, find the derivative of $f \circ g$ in two ways:

    (a) by first finding $(f \circ g)(x)$ and then finding $(f \circ g)'(x)$
    (b) by using the chain rule.

SOLUTION:
    (a) Because $(f \circ g)(x) = f(g(x)) = (g(x))^2 + 5(g(x)) + 5$, then

$$(f \circ g)(x) = \left(\frac{x+1}{x-1}\right)^2 + 5\left(\frac{x+1}{x-1}\right) + 5$$

$$= \frac{(x+1)^2}{(x-1)^2} + \frac{5(x+1)(x-1)}{(x-1)(x-1)} + \frac{5(x-1)^2}{(x-1)^2}$$

$$= \frac{x^2 + 2x + 1 + 5x^2 - 5 + 5x^2 - 10x + 5}{(x-1)^2}$$

$$= \frac{11x^2 - 8x + 1}{x^2 - 2x + 1}$$

We use the formula for the derivative of a quotient to find $(f \circ g)'(x)$.

$$(f \circ g)'(x) = \frac{(x^2 - 2x + 1)(22x - 8) - (11x^2 - 8x + 1)(2x - 2)}{(x^2 - 2x + 1)^2}$$

$$= \frac{2(x-1)[(x-1)(11x-4) - (11x^2 - 8x + 1)]}{(x-1)^4}$$

$$= \frac{2(-7x + 3)}{(x-1)^3}$$

$$= \frac{-2(7x - 3)}{(x-1)^3}$$

(b) Let $y = f(u)$ and $u = g(x)$. That is, let

$$y = u^2 + 5u + 5$$

and

$$u = \frac{x+1}{x-1} \tag{1}$$

Then

$$D_u y = 2u + 5$$

and

$$D_x u = \frac{(x-1)(1) - (x+1)(1)}{(x-1)^2}$$

$$= \frac{-2}{(x-1)^2}$$

By the chain rule, $D_x y = D_u y \cdot D_x u$. Thus,

$$D_x y = (2u + 5) \cdot \frac{-2}{(x-1)^2} \tag{2}$$

Because $D_x y = (f \circ g)'(x)$, by substituting from (1) into Eq. (2) we obtain

$$(f \circ g)'(x) = \left[2\left(\frac{x+1}{x-1}\right) + 5\right] \cdot \frac{-2}{(x-1)^2}$$

$$= \frac{7x - 3}{x - 1} \cdot \frac{-2}{(x-1)^2}$$

$$= \frac{-2(7x - 3)}{(x-1)^3}$$

28. Use the chain rule to prove that (a) the derivative of an even function is an odd function, and (b) the derivative of an odd function is an even function, provided these derivatives exist.

SOLUTION:

(a) Let f be an even function. Then

$$f(-x) = f(x)$$

We must show that f' is an odd function; that is, we must show

$$f'(-x) = -f'(x)$$

Let $y = f(u)$ and $u = -x$. Because we are given that f' exists, then by the chain rule,

$$\begin{aligned} D_x y &= D_u y \cdot D_x u \\ &= f'(u) \cdot (-1) \\ &= -f'(-x) \end{aligned}$$ (1)

And because $y = f(u) = f(-x)$ and we are given that $f(-x) = f(x)$, then $y = f(x)$ and $D_x y = f'(x)$. Thus, by Eq. (1) we obtain

$$f'(x) = -f'(-x)$$

or, equivalently,

$$f'(-x) = -f'(x)$$

Therefore, f' is an odd function.

(b) If f is an odd function, then $f(-x) = -f(x)$. Let $y = f(u)$ and $u = -x$. Then $y = f(-x) = -f(x)$ and $D_x y = -f'(x)$. And by substituting in Eq. (1), we obtain

$$-f'(x) = -f'(-x)$$

or, equivalently,

$$f'(-x) = f'(x)$$

Therefore, f' is an even function.

30. Suppose that f and g are functions such that $f'(x) = 1/x$ and $f(g(x)) = x$. Prove that if $g'(x)$ exists, then $g'(x) = g(x)$.

SOLUTION: Let $y = f(u)$ and $u = g(x)$. Then $y = f(g(x))$ and because we are given that $f(g(x)) = x$, then $D_x y = 1$. Moreover, since f' and g' exist, we have from the chain rule

$$\begin{aligned} D_x y &= D_u y \cdot D_x u \\ &= f'(u) \cdot g'(x) \end{aligned}$$

Since we are given that $f'(x) = 1/x$, then $f'(u) = 1/u$, and hence

$$D_x y = \frac{1}{u} \cdot g'(x)$$ (1)

And because $D_x y = 1$ and $u = g(x)$, Eq. (1) is equivalent to

$$1 = \frac{g'(x)}{g(x)}$$

$$g'(x) = g(x)$$

3.7 THE DERIVATIVE OF THE POWER FUNCTION FOR RATIONAL EXPONENTS

The two theorems proved in this section extend the formulas for the derivative of a power function and for the derivative of a composite power function to include those powers with rational number exponents.

3.7.1 Theorem

If f is the power function where r is any rational number (i.e., $f(x) = x^r$), then

$$f'(x) = rx^{r-1}$$

3.7.2 Theorem If f and g are functions such that $f(x) = [g(x)]^r$, where r is any rational number, and if $g'(x)$ exists, then

$$f'(x) = r[g(x)]^{r-1}g'(x)$$

Exercises 3.7

In Exercises 1-18, find the derivative of the given function.

2. $f(s) = \sqrt{2 - 3s^2}$

SOLUTION: We replace the radical sign by a fractional exponent and use the formula for the derivative of a power function.

$$f(s) = (2 - 3s^2)^{1/2}$$
$$f'(x) = \tfrac{1}{2}(2 - 3s^2)^{-1/2} \cdot (-6s)$$
$$= \frac{-3s}{\sqrt{2 - 3s^2}}$$

6. $g(y) = (y^2 + 3)^{1/3}(y^3 - 1)^{1/2}$

SOLUTION: By the formula for the derivative of a product, we have

$$g'(y) = (y^2 + 3)^{1/3}D_y(y^3 - 1)^{1/2} + (y^3 - 1)^{1/2}D_y(y^2 + 3)^{1/3} \tag{1}$$

We now apply the formula for the derivative of a composite power function and obtain

$$D_y(y^3 - 1)^{1/2} = \tfrac{1}{2}(y^3 - 1)^{-1/2}(3y^2) \tag{2}$$
$$D_y(y^2 + 3)^{1/3} = \tfrac{1}{3}(y^2 + 3)^{-2/3}(2y) \tag{3}$$

By substituting (2) and (3) in (1), we obtain

$$g'(y) = (y^2 + 3)^{1/3}(\tfrac{3}{2}y^2)(y^3 - 1)^{-1/2} + (y^3 - 1)^{1/2}(\tfrac{2}{3}y)(y^2 + 3)^{-2/3}$$

We factor $g'(y)$ by removing the repeated factors. Because $-\tfrac{2}{3}$, the exponent for the factor $(y^2 + 3)^{-2/3}$, is less than $\tfrac{1}{3}$, the exponent for the factor $(y^2 + 3)^{1/3}$, we remove the factor $(y^2 + 3)^{-2/3}$.

Similarly, because $-\tfrac{1}{2}$, the exponent for the factor $(y^3 - 1)^{-1/2}$, is less than $\tfrac{1}{2}$, the exponent for the factor $(y^3 - 1)^{1/2}$, we remove the factor $(y^3 - 1)^{-1/2}$. And because 6 is the least common denominator for the fractional coefficients, $\tfrac{3}{2}$ and $\tfrac{2}{3}$, we remove $\tfrac{1}{6}$. Thus

$$g'(y) = (\tfrac{1}{6}y)(y^2 + 3)^{-2/3}(y^3 - 1)^{-1/2}[(y^2 + 3)(9y) + (y^3 - 1)4]$$
$$= \frac{y(13y^3 + 27y - 4)}{6(y^2 + 3)^{2/3}(y^3 - 1)^{1/2}}$$

12. $G(x) = \dfrac{4x + 6}{\sqrt{x^2 + 3x + 4}}$

SOLUTION: We replace the given quotient by an equivalent product.

$$G(x) = (4x + 6)(x^2 + 3x + 4)^{-1/2}$$

As in Exercise 6, we use the formula for the derivative of a product and the formula for the derivative of a composite power function to obtain

$$G'(x) = (4x + 6)(-\tfrac{1}{2})(x^2 + 3x + 4)^{-3/2}(2x + 3) + (x^2 + 3x + 4)^{-1/2}(4)$$

$$= (x^2 + 3x + 4)^{-3/2} [2(2x + 3)(-\tfrac{1}{2})(2x + 3) + (x^2 + 3x + 4)(4)]$$

$$= (x^2 + 3x + 4)^{-3/2} [(-4x^2 - 12x - 9) + (4x^2 + 12x + 16)]$$

$$= \frac{7}{(x^2 + 3x + 4)^{3/2}}$$

16. $G(t) = \sqrt{\dfrac{5t + 6}{5t - 4}}$

SOLUTION:

$$G(t) = \left(\frac{5t + 6}{5t - 4}\right)^{1/2}$$

We use the formula for the derivative of a composite power function and the formula for the derivative of a quotient.

$$G'(t) = \frac{1}{2}\left(\frac{5t + 6}{5t - 4}\right)^{-1/2} \cdot \frac{(5t - 4)(5) - (5t + 6)(5)}{(5t - 4)^2}$$

$$= \frac{\tfrac{1}{2}(5t + 6)^{-1/2}(-50)}{(5t - 4)^{-1/2}(5t - 4)^2}$$

$$= \frac{-25}{(5t + 6)^{1/2}(5t - 4)^{3/2}}$$

22. Find an equation of the tangent line to the curve $y = 1/\sqrt[3]{7x - 6}$ which is perpendicular to the line $12x - 7y + 2 = 0$.

SOLUTION: Let ℓ_1 be the required line, and let (x_1, y_1) be the point at which ℓ_1 is tangent to the curve. Because the slope of ℓ_1 is given by $m(x_1) = f'(x_1)$, where f is the function whose graph is the given curve, we find the derivative of f with

$$f(x) = \frac{1}{\sqrt[3]{7x - 6}} = (7x - 6)^{-1/3} \tag{1}$$

$$f'(x) = -\tfrac{1}{3}(7x - 6)^{-4/3}(7)$$

$$= \frac{-7}{3(7x - 6)^{4/3}}$$

Therefore,

$$m(x_1) = \frac{-7}{3(7x_1 - 6)^{4/3}} \tag{2}$$

An equation of the given line ℓ in the slope-intercept form is

$$y = \frac{12}{7}x + \frac{2}{7}$$

Thus, the slope of ℓ is $\frac{12}{7}$. Because ℓ_1 is perpendicular to ℓ , the slope of ℓ_1 is $-\frac{7}{12}$. We substitute $m(x_1) = -\frac{7}{12}$ in Eq. (2) and solve for x_1 .

$$-\frac{7}{12} = \frac{-7}{3(7x_1 - 6)^{4/3}}$$

$$(7x_1 - 6)^{4/3} = 4$$

$$7x_1 - 6 = 4^{3/4}$$

$$7x_1 - 6 = 2^{3/2} \tag{3}$$

$$x_1 = \frac{6 + 2\sqrt{2}}{7}$$

By Eq. (1) we have

$$y_1 = f(x_1) = (7x_1 - 6)^{-1/3} \qquad \text{(4)}$$

By substituting from Eq. (3) into (4) we have

$$y_1 = (2^{3/2})^{-1/3} = \frac{\sqrt{2}}{2}$$

Therefore, the point-slope form of the equation of ℓ_1 is

$$y - \frac{\sqrt{2}}{2} = -\frac{7}{12}\left(x - \frac{6 + 2\sqrt{2}}{7}\right)$$

$$12y - 6\sqrt{2} = -7x + 6 + 2\sqrt{2}$$

$$7x + 12y - 8\sqrt{2} - 6 = 0$$

30. Suppose that $g(x) = \sqrt{9 - x^2}$ and $h(x) = f(g(x))$ where f is differentiable at 3. Prove that $h'(0) = 0$.

SOLUTION: Let $y = f(u)$ and $u = g(x)$. Because $h(x) = f(g(x))$, then $h(x) = y$, and the chain rule,

$$D_x y = D_u y \cdot D_x u$$

is equivalent to

$$h'(x) = f'(u) \cdot g'(x) \qquad \text{(1)}$$

provided the derivatives exist. Because $u = \sqrt{9 - x^2}$, then if $x = 0$, $u = 3$. By substituting $x = 0$ and $u = 3$ in Eq. (1), we have

$$h'(0) = f'(3) \cdot g'(0) \qquad \text{(2)}$$

provided $f'(3)$ and $g'(0)$ exist. We are given that f is differentiable at 3, that is, that $f'(3)$ exists. Furthermore,

$$g(x) = (9 - x^2)^{1/2}$$

Thus,

$$g'(x) = \tfrac{1}{2}(9 - x^2)^{-1/2}(-2x)$$
$$g'(0) = 0$$

By substituting in Eq. (2) we have

$$h'(0) = f'(3) \cdot 0$$
$$h'(0) = 0$$

3.8 IMPLICIT DIFFERENTIATION

With implicit differentiation we must be careful to distinguish the independent variable from the dependent variable. If x is the independent variable, then we regard y as a function of x. Thus,

$$D_x x = 1$$

but $D_x y$ is not known. Similarly,

$$D_x(x^r) = rx^{r-1}$$

but to find $D_x(y^r)$ we must use the chain rule, since y^r is a composite function of x. Thus,

$$D_x(y^r) = ry^{r-1}D_x y$$

If we regard y as the independent variable and differentiate implicitly with respect to y, then the roles of x and y are interchanged. Thus,

$$D_y(x^r) = rx^{r-1}D_y x$$
$$D_y(y^r) = ry^{r-1}$$

Exercises 3.8

In Exercises 1-16, find $D_x y$ by implicit differentiation.

2. $2x^3y + 3xy^3 = 5$

SOLUTION: We differentiate with respect to x, and have

$$D_x(2x^3y) + D_x(3xy^3) = 0 \tag{1}$$

Because $2x^3y$ is the product of two functions of x, namely $2x^3$ and the function represented by y, to find $D_x(2x^3y)$ we use the formula for the derivative of a product.

$$D_x(2x^3y) = 2x^3 D_x y + y \cdot 6x^2 \tag{2}$$

Because $3xy^3$ is the product of two functions of x and because y^3 is a composite function of x, to find $D_x(3xy^3)$ we use the formula for the derivative of a product and the chain rule.

$$D_x(3xy^3) = 3x \cdot 3y^2 \cdot D_x y + y^3 \cdot 3 \tag{3}$$

By substituting in Eq. (1) from Eq. (2) and Eq. (3), we obtain

$$2x^3 D_x y + 6x^2 y + 9xy^2 D_x y + 3y^3 = 0$$
$$(2x^3 + 9xy^2) D_x y = -6x^2y - 3y^3$$
$$D_x y = \frac{-6x^2y - 3y^3}{2x^3 + 9xy^2}$$

6. $\dfrac{x}{y} - 4y = x$

SOLUTION: Differentiating with respect to x, we have

$$\frac{y \cdot 1 - xD_x y}{y^2} - 4D_x y = 1$$
$$y - x D_x y - 4y^2 D_x y = y^2$$
$$(-x - 4y^2)D_x y = y^2 - y$$
$$D_x y = \frac{y^2 - y}{-x - 4y^2} = \frac{-y^2 + y}{x + 4y^2} \tag{1}$$

ALTERNATE SOLUTION: We eliminate the fraction and then differentiate implicitly with respect to x.

$$x - 4y^2 = xy$$
$$1 - 8y D_x y = x D_x y + y \cdot 1$$
$$(-8y - x) D_x y = y - 1$$
$$D_x y = \frac{y - 1}{-x - 8y}$$
$$D_x y = \frac{-y + 1}{x + 8y} \tag{2}$$

We show that Eq. (1) is equivalent to Eq. (2). Dividing the numerator and denominator of the fraction in Eq. (1) by y, we have

$$D_x y = \frac{-y + 1}{\dfrac{x}{y} + 4y} \tag{3}$$

From the given equation we have

$$\frac{x}{y} = x + 4y \tag{4}$$

Substituting from Eq. (4) to Eq. (3), we obtain

$$D_x y = \frac{-y + 1}{(x + 4y) + 4y} = \frac{-y + 1}{x + 8y}$$

which agrees with Eq. (2).

10. $\quad y\sqrt{2 + 3x} + x\sqrt{1 + y} = x$

SOLUTION: Replacing the radicals by fractional exponents, we have

$$y(2 + 3x)^{1/2} + x(1 + y)^{1/2} = x$$

Differentiating with respect to x, we obtain

$$y \cdot \tfrac{1}{2}(2 + 3x)^{-1/2} \cdot 3 + (2 + 3x)^{1/2} D_x y + x \cdot \tfrac{1}{2}(1 + y)^{-1/2} D_x y + (1 + y)^{1/2} \cdot 1 = 1$$

Separating the terms that contain $D_x y$ from those that do not, we get

$$[(2 + 3x)^{1/2} + \tfrac{1}{2}x(1 + y)^{-1/2}]D_x y = 1 - \tfrac{3}{2}y(2 + 3x)^{-1/2} - (1 + y)^{1/2}$$

$$D_x y = \frac{1 - \tfrac{3}{2}y(2 + 3x)^{-1/2} - (1 + y)^{1/2}}{(2 + 3x)^{1/2} + \tfrac{1}{2}x(1 + y)^{-1/2}}$$

Multiplying the numerator and denominator of the fraction by $2(2 + 3x)^{1/2}(1 + y)^{1/2}$, we get

$$D_x y = \frac{2(2 + 3x)^{1/2}(1 + y)^{1/2} - 3y(1 + y)^{1/2} - 2(1 + y)(2 + 3x)^{1/2}}{2(2 + 3x)(1 + y)^{1/2} + x(2 + 3x)^{1/2}}$$

18. Consider y as the independent variable and find $D_y x$.

$$y = 2x^3 - 5x$$

SOLUTION: Because y is the independent variable, we regard x as a function of y and differentiate with respect to y. Thus,

$$D_y y = 6x^2 D_y x - 5 D_y x$$
$$1 = (6x^2 - 5) D_y x$$

$$D_y x = \frac{1}{6x^2 - 5}$$

22. There are two lines through the point $(-1, 3)$ that are tangent to the curve

$$x^2 + 4y^2 - 4x - 8y + 3 = 0 \tag{1}$$

Find an equation of each of these lines.

SOLUTION: Let L_1 be one of the required lines and let (x_1, y_1) be the point at which L_1 is tangent to the curve. To find m_1, the slope of line L_1, we use Eq. (1) and differentiate implicitly with respect to x. Thus,

$$2x + 8y\,D_x y - 4 - 8\,D_x y = 0$$

$$D_x y = \frac{-2x + 4}{8y - 8}$$

$$= \frac{-x + 2}{4y - 4}$$

Hence,

$$m_1 = \frac{-x_1 + 2}{4y_1 - 4} \tag{2}$$

Because L_1 contains $(-1, 3)$ and (x_1, y_1), by definition of slope we have

$$m_1 = \frac{y_1 - 3}{x_1 + 1} \tag{3}$$

Using Eqs. (2) and (3) and eliminating m_1, we obtain

$$x_1{}^2 + 4y_1{}^2 - x_1 - 16y_1 + 10 = 0 \tag{4}$$

Because the curve contains (x_1, y_1), this pair must satisfy Eq. (1). That is,

$$x_1{}^2 + 4y_1{}^2 - 4x_1 - 8y_1 + 3 = 0 \tag{5}$$

By subtracting terms of Eq. (5) from corresponding terms of Eq. (4), we get

$$3x_1 - 8y_1 + 7 = 0$$

$$x_1 = \frac{8y_1 - 7}{3} \tag{6}$$

Substituting from Eq. (6) into Eq. (4), we have

$$\frac{64y_1{}^2 - 112y_1 + 49}{9} + 4y_1{}^2 - \frac{8y_1 - 7}{3} - 16y_1 + 10 = 0$$

The solutions for this equation are

$$y_1 = 2 \quad \text{and} \quad y_1 = \frac{4}{5}$$

From Eq. (6) we find that the solutions for the system of Eqs. (4) and (5) are $(3, 2)$ and $(-\frac{1}{5}, \frac{4}{5})$. If $(x_1, y_1) = (3, 2)$, by Eq. (2) we have $m_1 = -\frac{1}{4}$. Thus, an equation for line L_1 is

$$y - 2 = -\frac{1}{4}(x - 3)$$

$$x + 4y - 11 = 0$$

To find an equation of the second line we use $(x_1, y_1) = (-\frac{1}{5}, \frac{4}{5})$, and from Eq. (2) we have $m_1 = -\frac{11}{4}$. Hence, an equation for the second line is

$$y - \frac{4}{5} = -\frac{11}{4}\left(x + \frac{1}{5}\right)$$

$$11x + 4y - 1 = 0$$

24. If $x^n y^m = (x + y)^{n+m}$, prove that $x \cdot D_x y = y$.

SOLUTION: Differentiating implicitly with respect to x, we have

$$x^n \cdot m y^{m-1} \cdot D_x y + y^m \cdot n x^{n-1} = (n + m)(x + y)^{n+m-1}(1 + D_x y)$$

$$x^n m y^{m-1} \cdot D_x y + y^m \cdot n x^{n-1} = (n + m)(x + y)^{n+m-1}$$

$$+ (n + m)(x + y)^{n+m-1} \cdot D_x y$$

$$[m x^n y^{m-1} - (n + m)(x + y)^{n+m-1}]\,D_x y = (n + m)(x + y)^{n+m-1} - n y^m x^{n-1}$$

$$D_x y = \frac{(n+m)(x+y)^{n+m-1} - nx^{n-1}y^m}{mx^n y^{m-1} - (n+m)(x+y)^{n+m-1}}$$

Multiplying the numerator and denominator of the fraction by $x + y$, we obtain

$$D_x y = \frac{(n+m)(x+y)^{n+m} - nx^{n-1}y^m(x+y)}{mx^n y^{m-1}(x+y) - (n+m)(x+y)^{n+m}}$$

If we replace $(x+y)^{n+m}$ by $x^n y^m$ (since these expressions are equal by hypothesis), we have

$$D_x y = \frac{(n+m)x^n y^m - nx^{n-1}y^m(x+y)}{mx^n y^{m-1}(x+y) - (n+m)x^n y^m}$$

Next, we factor both the numerator and denominator.

$$D_x y = \frac{x^{n-1}y^m[(n+m)x - n(x+y)]}{x^n y^{m-1}[m(x+y) - (n+m)y]}$$

$$= \frac{x^{n-1}y^m(mx - ny)}{x^n y^{m-1}(mx - ny)}$$

$$= \frac{y}{x}$$

Therefore,

$$x \cdot D_x y = y$$

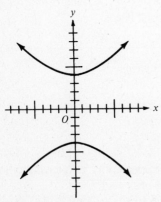

Figure 3.8.28(a)

28. Given the equation $y^2 - x^2 = 16$,

(a) find two functions defined by the equation and state their domains
(b) draw a sketch of the graph of each of the functions obtained in part (a)
(c) draw a sketch of the graph of the equation
(d) find the derivative of each of the functions obtained in part (a) and state the domains of the derivatives
(e) find $D_x y$ by implicit differentiation from the given equation, and verify that the result so obtained agrees with the results in part (d)
(f) find an equation of each tangent line at $x_1 = -3$.

Figure 3.8.28(b)

SOLUTION:
(a) Because $y = \pm\sqrt{x^2 + 16}$, the functions are f_1 and f_2 with $f_1(x) = \sqrt{x^2 + 16}$ and $f_2(x) = -\sqrt{x^2 + 16}$. The domain of each function is $(-\infty, +\infty)$.
(b) A sketch of the graph of f_1 is shown in Fig. 3.8.28(a), and a sketch of the graph of f_2 is shown in Fig. 3.8.28(b).
(c) A sketch of the graph of the given equation is shown in Fig. 3.8.28(c).
(d) $f_1(x) = (x^2 + 16)^{1/2}$

$$f_1'(x) = \frac{1}{2}(x^2 + 16)^{-1/2}(2x)$$

$$f_1'(x) = \frac{x}{\sqrt{x^2 + 16}}$$

Similarly,

$$f_2'(x) = \frac{-x}{\sqrt{x^2 + 16}}$$

The domain of each derivative is $(-\infty, +\infty)$.
(e) Differentiating the given equation implicitly with respect to x, we have

$$2y\, D_x y - 2x = 0$$

Figure 3.8.28(c)

$$D_x y = \frac{x}{y}$$

For $y = f_1(x) = \sqrt{x^2 + 16}$, we found in part **(d)** that

$$D_x y = f_1'(x) = \frac{x}{\sqrt{x^2 + 16}} = \frac{x}{y}$$

which agrees with the result found by implicit differentiation. For $y = f_2(x) = -\sqrt{x^2 + 16}$, we found in part **(d)** that

$$D_x y = f_2'(x) = \frac{-x}{\sqrt{x^2 + 16}} = \frac{-x}{-y} = \frac{x}{y}$$

which also agrees with the result found by implicit differentiation.

(f) For $y = f_1(x)$, if $x_1 = -3$, we have

$$y_1 = f_1(x_1) = \sqrt{(-3)^2 + 16} = 5$$

$$m(x_1) = f_1'(x_1) = \frac{-3}{\sqrt{(-3)^2 + 16}} = \frac{-3}{5}$$

Thus, an equation of the tangent line to the graph of f_1 at $(-3, 5)$ is

$$y - 5 = -\frac{3}{5}(x + 3)$$

$$3x + 5y - 16 = 0$$

For $y = f_2(x)$, if $x_1 = -3$, we have

$$y_1 = f_2(x_1) = -\sqrt{(-3)^2 + 16} = -5$$

$$m(x_1) = f_2'(x_1) = \frac{-(-3)}{\sqrt{(-3)^2 + 16}} = \frac{3}{5}$$

An equation of the tangent line to the graph of f_2 at $(-3, -5)$ is

$$y + 5 = \frac{3}{5}(x + 3)$$

$$3x - 5y - 16 = 0$$

3.9 THE DERIVATIVE AS A RATE OF CHANGE

3.9.1 Definition If $y = f(x)$, *the instantaneous rate of change of* y *per unit change in* x at x_1 is $f'(x_1)$ or, equivalently, the derivative of y with respect to x at x_1, if it exists there.

3.9.2 Definition If $y = f(x)$, the *relative rate of change of* y *per unit change in* x at x_1 is given by $f'(x_1)/f(x_1)$ or, equivalently, $D_x y/y$ evaluated at $x = x_1$.

We use the symbol

$$D_x y \Big]_{x = x_1}$$

to represent the value of $D_x y$ when $x = x_1$.

When the equation that defines the functional relationship between the variables is not given, we must first find this equation. Sometimes the equation is a formula from geometry, as in Exercise 8. Sometimes even the variables are not explicitly identified, as in Exercise 12. By a careful reading of the given information and the question asked, we must first choose appropriate variables and then find an equation that expresses the functional relationship between these variables.

4. Let s be the principal square root of a number x. Find the instantaneous rate of change of s with respect to x and the relative rate of change of s per unit change in x when x is
(a) 9   (b) 4

SOLUTION:

$$s = \sqrt{x} = x^{1/2}$$

$$D_x s = \frac{1}{2} x^{-1/2} = \frac{1}{2\sqrt{x}}$$

(a) $D_x s \Big]_{x=9} = \frac{1}{2\sqrt{9}} = \frac{1}{6}$

$\dfrac{D_x s}{s} \Big]_{x=9} = \dfrac{1/6}{\sqrt{9}} = \dfrac{1}{18}$

Therefore, when $x = 9$ the instantaneous rate of change of s per unit change in x is $\frac{1}{6}$, and the relative rate of change of s per unit change in x is $\frac{1}{18}$.

(b) $D_x s \Big]_{x=4} = \frac{1}{2\sqrt{4}} = \frac{1}{4}$

$\dfrac{D_x s}{s} \Big]_{x=4} = \dfrac{1/4}{\sqrt{4}} = \dfrac{1}{8}$

Therefore, when $x = 4$ the instantaneous rate of change of s and the relative rate of change of s per unit change in x are $\frac{1}{4}$ and $\frac{1}{8}$, respectively.

6. The supply equation for a certain kind of pencil is $x = 3p^2 + 2p$ where p cents is the price per pencil when $1000x$ pencils are supplied.

(a) Find the average rate of change of the supply per 1 cent change in the price when the price is increased from 10 cents to 11 cents.
(b) Find the instantaneous (or marginal) rate of change of the supply per 1 cent change in the price when the price is 10 cents.

SOLUTION: Because $1000x = 1000(3p^2 + 2p)$, let

$$f(p) = 1000(3p^2 + 2p)$$

(a) We find $\Delta f / \Delta p$ when $p = 10$ and $p + \Delta p = 11$. Hence $\Delta p = 1$ and

$$\frac{\Delta f}{\Delta p} = \frac{f(p + \Delta p) - f(p)}{\Delta p}$$

$$= \frac{f(11) - f(10)}{1}$$

$$= 1000(3 \cdot 11^2 + 2 \cdot 11) - 1000(3 \cdot 10^2 + 2 \cdot 10)$$
$$= 65{,}000$$

The average rate of change of the supply when the price is increased from 10 cents to 11 cents is 65,000 pencils per 1 cent change in the price.
(b) We find $f'(10)$.

$$f'(p) = 1000(6p + 2)$$
$$f'(10) = 1000(6 \cdot 10 + 2) = 62{,}000$$

Therefore, when the price is 10 cents the instantaneous rate of change of the supply is 62,000 pencils per 1 cent change in the price.

8. A balloon maintains the shape of a sphere as it is being inflated. Find the rate of change of the surface area with respect to the radius at the instant when the radius is 2 in.

SOLUTION: Let S sq in be the surface area of a sphere whose radius is r in. Then $S = 4\pi r^2$. We find $D_r S$ when $r = 2$.

$$D_r S = 8\pi r$$

$$D_r S \Big]_{r=2} = 16\pi$$

Therefore, when the radius of the balloon is 2 in, the rate of change of the surface area is 16π sq in per inch change in the radius.

12. At 8 A.M. a ship sailing due north at 24 knots (nautical miles per hour) is at a point P. At 10 A.M. a second ship sailing due east at 32 knots is at P. At what rate is the distance between the two ships changing at

(a) 9 A.M.   (b) 11 A.M.?

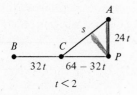

Figure 3.9.12(a)

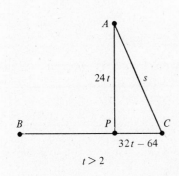

Figure 3.9.12(b)

SOLUTION: Let s nautical miles be the distance between the ships at the moment when it is t hours after 8 A.M. We must find an equation that defines s as a function of t. Refer to Fig. 3.9.12(a) and Fig. 3.9.12(b). The ship that is sailing north at 24 knots is at point P at 8 A.M. and at point A when it is t hours after 8 A.M. Thus, $24t$ is the number of nautical miles in the distance $|\overline{PA}|$. The ship that is sailing east at 32 knots is at point B at 8 A.M. and at point C when it is t hours after 8 A.M. Thus $|\overline{BC}| = 32t$. Because this ship is at point P at 10 A.M., then $|\overline{PB}| = 64$. If $t < 2$, point C is between B and P, and thus $|\overline{CP}| = 64 - 32t$, as illustrated in Fig. 3.9.12(a). And if $t > 2$, point P is between B and C [see Fig. 3.9.12(b)], and $|\overline{CP}| = 32t - 64$. Because $s = |\overline{AC}|$, and triangle APC is a right triangle, by the Pythagorean theorem we have for all $t \geqslant 0$

$$s^2 = (24t)^2 + (32t - 64)^2 \tag{1}$$

Differentiating implicitly with respect to t, we get

$$2s\, D_t s = 2(24t) \cdot 24 + 2(32t - 64) \cdot 32$$

$$D_t s = \frac{(24)^2 t + (32)^2 (t - 2)}{s} \tag{2}$$

(a) We find $D_t s$ when $t = 1$. From Eq. (1) we obtain $s = 40$ when $t = 1$. Thus by Eq. (2), we get

$$D_t s \Big]_{t=1} = \frac{(24)^2 + (32)^2 (-1)}{40} = -\frac{56}{5}$$

Therefore, at 9 A.M. the rate of change of the distance between the ships is $-\frac{56}{5}$ knots; that is, the distance is decreasing at the rate of $\frac{56}{5}$ knots, or 11.2 knots.

(b) We find $D_t s$ when $t = 3$. By Eq. (1) we get $s = 8\sqrt{97}$ when $t = 3$. Thus Eq. (2) yields

$$D_t s \Big]_{t=3} = \frac{(24)^2(3) + (32)^2}{8\sqrt{97}} = \frac{344}{\sqrt{97}}$$

Therefore, at 11 A.M. the distance between the ships is increasing at the rate of $344/\sqrt{97}$ knots, or approximately 34.9 knots.

3.10 RELATED RATES If x is a variable that is a function of time, which is represented by t, then the rate of change of x with respect to time is given by $D_t x$. The following steps should be followed to solve problems involving the rate of change with respect to time for two or more related variables.

1. Select a letter to represent each variable.
2. Identify the constant rates of change that are given.
3. Identify the rate of change that must be found.
4. Find an equation that expresses the functional relationship between the variables.
5. Differentiate with respect to t on both sides of the equation.
6. Replace the given rates of change with their constant values.
7. Replace the variables by their values at the particular time of interest.
8. Solve the resulting equation for the unknown rate of change.

Often it is helpful to draw a figure in order to find the equation in step 4. Be careful to distinguish the *variables,* which represent length, area, volume, etc., from the *rates of change* of these variables. Although a variable that represents length may appear as a dimension in the figure, the rate of change of this variable does not appear in the figure. To find the equation you may use any formulas from geometry for length, area, and volume. The Pythagorean theorem may be used whenever there is a right triangle in the figure. If a and b are the measures of the legs of the right triangle and c is the measure of the hypoteneuse, then

$$a^2 + b^2 = c^2$$

Also, the fact that corresponding sides of similar triangles are proportional may be used. Sometimes, it may be helpful to introduce an additional variable in order to find the equation, as is illustrated in Exercise 16.

Following are some of the formulas from geometry.

1. Circumference of a circle: $C = 2\pi r$
2. Area formulas for plane figures:
   a. Rectangle: $A = lw$
   b. Triangle: $A = \frac{1}{2}bh$
   c. Parallelogram: $A = bh$
   d. Trapezoid: $A = \frac{1}{2}(b + B)h$
   e. Circle: $A = \pi r^2$
3. Surface area formulas for solids:
   a. Right circular cylinder:
      (i) Lateral area: $A = 2\pi rh$
      (ii) Total area: $A = 2\pi rh + 2\pi r^2$
   b. Right circular cone:
      (i) Lateral area: $A = \pi r\sqrt{r^2 + h^2}$
      (ii) Total area: $A = \pi r\sqrt{r^2 + h^2} + \pi r^2$
   c. Sphere: $A = 4\pi r^2$
4. Volume formulas for solids:
   a. Right circular cylinder: $V = \pi r^2 h$
   b. Right circular cone: $V = \frac{1}{3}\pi r^2 h$
   c. Sphere: $V = \frac{4}{3}\pi r^3$
   d. Prism: Volume equals the area of the base times the distance between bases.
   e. Pyramid: Volume equals one-third the area of the base times the altitude.

Exercises 3.10

---

2. A spherical balloon is being inflated so that its volume is increasing at the rate of 5 ft³/min. At what rate is the diameter increasing when the diameter is 12 ft?

SOLUTION: First we identify the variables.

t = the number of minutes in the time that has elapsed since the balloon began to be inflated

x = the number of feet in the diameter of the balloon at t min

V = the number of cubic feet in the volume of the balloon at t min

Because the volume is increasing at the rate of 5 ft³/min, we are given that $D_t V = 5$. Because $D_t x$ represents the rate of change of the diameter with respect to time, we want to find $D_t x$ when $x = 12$. Since the balloon is a sphere, we use the formula for the volume of a sphere, $V = \frac{4}{3} \pi r^3$, where r is the number of feet in the radius. Because $r = x/2$, we have

$$V = \frac{4}{3} \pi \left(\frac{x}{2}\right)^3$$

$$V = \frac{\pi x^3}{6}$$

Differentiating with respect to t, we get

$$D_t V = \frac{1}{2} \pi x^2 D_t x$$

Substituting 5 for $D_t V$ and solving for $D_t x$, we get

$$D_t x = \frac{10}{\pi x^2}$$

$$D_t x \Big]_{x=12} = \frac{10}{\pi 12^2} = \frac{5}{72\pi}$$

Therefore the diameter is increasing at the rate of $5/(72\pi)$ ft/min, or approximately 0.022 ft/min, when the diameter is 12 ft.

8. A man 6 ft tall is walking toward a building at the rate of 5 ft/sec. If there is a light on the ground 50 ft from the building, how fast is the man's shadow on the building growing shorter when he is 30 ft from the building?

SOLUTION: We identify the variables.

t = the number of seconds in the time that has elapsed since the man began to walk toward the building

x = the number of feet in the distance between the man and the light at t sec

y = the number of feet in the distance between the man and the building at t sec

z = the number of feet in the length of the man's shadow on the building at t sec

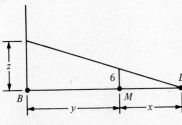

Figure 3.10.8

Fig. 3.10.8 illustrates the case when the man is at point M, between point L (the light), and point B (the building). Because the man is walking at the rate of 5 ft/sec, we are given that $D_t x = 5$. Because $D_t z$ is the rate of change of the length of the shadow, we want to find $D_t z$ when $y = 30$. Because we are given that $x + y = 50$, by similar triangles we have

$$\frac{z}{50} = \frac{6}{x}$$

$$z = 300x^{-1}$$

Differentiating on both sides with respect to t, we obtain

$$D_t z = -300x^{-2} D_t x$$

Replacing $D_t x$ by 5, we have

$$D_t z = \frac{-1500}{x^2}$$

When $y = 30$, $x = 20$, and

$$D_t z \Big]_{x=20} = \frac{-1500}{20^2} = -\frac{15}{4}$$

We conclude that the shadow is growing shorter at the rate of $\frac{15}{4}$ ft/sec when the man is 30 ft from the building.

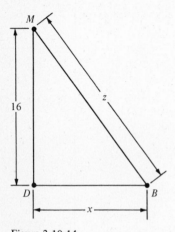

Figure 3.10.14

14. A man on a dock is pulling in a boat at the rate of 50 ft/min by means of a rope attached to the boat at water level. If the man's hands are 16 ft above the water level, how fast is the boat approaching the dock when the amount of rope out is 20 ft?

SOLUTION: We identify the variables.

t = the number of minutes in the time that has elapsed since the man began to pull in the boat
x = the number of feet in the distance between the boat and the dock at t min
z = the number of feet in the amount of rope out at t min

In Fig. 3.10.14 the boat is at point B, the dock enters the water at point D, and the man's hands are at point M. Because the man is pulling in the rope at the rate of 50 ft/min, then z is decreasing, and $D_t z = -50$. Since $D_t x$ is the rate of change of the distance between the boat and the dock, we want to find $D_t x$ when $z = 20$. By the Pythagorean theorem we have

$$z^2 = x^2 + 16^2 \tag{1}$$

Differentiating on both sides of Eq. (1) with respect to t, we obtain

$$2z \, D_t z = 2x \, D_t x$$

Substituting -50 for $D_t z$ and solving for $D_t x$, gives

$$D_t x = \frac{-50z}{x}$$

When $z = 20$, it follows from Eq. (1) that $x = 12$. Thus, from Eq. (2) we have

$$D_t x \Big]_{\substack{x=12 \\ z=20}} = \frac{-50 \cdot 20}{12} = \frac{-250}{3}$$

Therefore, x is decreasing, and thus the boat is approaching the dock at the rate of $\frac{250}{3}$ ft/min when 20 ft of rope are out.

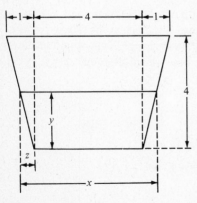

Figure 3.10.16

16. A horizontal trough is 16 ft long, and its ends are isosceles trapezoids with an altitude of 4 ft, a lower base of 4 ft, and an upper base of 6 ft. Water is being poured into the trough at the rate of 10 ft³/min. How fast is the water level rising when the water is 2 ft deep?

SOLUTION: Refer to Fig. 3.10.16, which illustrates one end of the trough.

t = the number of minutes that have elapsed since water began pouring into the trough
y = the number of feet in the depth of the water at t min
x = the number of feet in the width of the surface of the water at t min
V = the number of cubic feet in the volume of the water at t min

Because water is being poured into the trough at the rate of 10 ft³/min, we are given that $D_t V = 10$. Since $D_t y$ is the rate at which the depth of the water is changing, we want to find $D_t y$ when $y = 2$. The water that is in the trough is in the shape of a prism with altitude 16 ft and base a trapezoid. Because the volume of a prism is the area of its base times its altitude and the area of a trapezoid is given by the formula $A = \frac{1}{2}(b + B) \cdot h$, we have

$$V = \frac{1}{2}(4 + x) \cdot y \cdot 16$$

$$V = 8y(x + 4) \tag{1}$$

We find an equation that expresses x as a function of y. Let z be the number of feet in the distance indicated in Fig. 3.10.16. Because the trapezoid is isosceles,

$$x = 2z + 4 \tag{2}$$

By similar triangles

$$\frac{z}{y} = \frac{1}{4} \tag{3}$$

Substituting from Eq. (3) into Eq. (2) gives

$$x = \frac{y}{2} + 4 \tag{4}$$

Substituting the value for x in Eq. (4) into Eq. (1), we obtain

$$V = 8y\left(\frac{y}{2} + 8\right)$$

$$V = 4y^2 + 64y$$
$$D_t V = 8y\, D_t y + 64\, D_t y$$

Substituting $D_t V = 10$ and $y = 2$, and we have

$$10 = 16\, D_t y + 64\, D_t y$$

$$D_t y = \frac{1}{8}$$

We conclude that the depth of the water is increasing at the rate of $\frac{1}{8}$ ft/min at the moment when the water is 2 ft deep.

3.11 DERIVATIVES OF HIGHER ORDER

The first derivative of the function f is represented by f'; the second derivative of f is the first derivative of f' and is represented by f''; the third derivative of f is the first derivative of f'' and is represented by f'''; and so forth, provided these derivatives exist. If n is a positive integer greater than 1, the nth derivative of f is represented by $f^{(n)}$, and $f^{(n)}$ is the first derivative of $f^{(n-1)}$. Furthermore if $y = f(x)$, then $D_x y = f'(x)$, $D_x{}^2 y = f''(x)$, $D_x{}^3 y = f'''(x)$, and so on. In general, $D_x{}^n y = f^{(n)}(x)$.

If $s = f(t)$ is the equation of motion for a particle in rectilinear motion, then the instantaneous velocity at t sec is given by the formula $v = D_t s$, and the instantaneous acceleration at t seconds is given by $a = D_t v = D_t{}^2 s$. That is, the first derivative of f gives the velocity, and the second derivative of f gives the acceleration of the particle.

Exercises 3.11

In Exercises 1-10, find the first and second derivative of the function defined by the given equation.

6. $h(y) = \sqrt[3]{2y^3 + 5}$

SOLUTION:

$$h(y) = (2y^3 + 5)^{1/3}$$

$$h'(y) = \frac{1}{3}(2y^3 + 5)^{-2/3}(6y^2)$$

$$= 2y^2(2y^3 + 5)^{-2/3} \qquad \text{(1)}$$

$$= \frac{2y^2}{(2y^3 + 5)^{2/3}}$$

To find $h''(y)$ we differentiate $h'(y)$ by using the derivative of a product formula. Thus, from (1) we obtain

$$h''(y) = 2y^2\left(-\frac{2}{3}\right)(2y^3 + 5)^{-5/3}(6y^2) + (2y^3 + 5)^{-2/3}(4y)$$

$$= \frac{4}{3}y(2y^3 + 5)^{-5/3}[-y(6y^2) + 3(2y^3 + 5)]$$

$$= 20y(2y^3 + 5)^{-5/3}$$

$$= \frac{20y}{(2y^3 + 5)^{5/3}}$$

10. $f(x) = \dfrac{2 - \sqrt{x}}{2 + \sqrt{x}}$

SOLUTION:

$$f(x) = \frac{2 - x^{1/2}}{2 + x^{1/2}}$$

$$f'(x) = \frac{(2 + x^{1/2})(-\frac{1}{2}x^{-1/2}) - (2 - x^{1/2})(\frac{1}{2}x^{-1/2})}{(2 + x^{1/2})^2}$$

$$= \frac{\frac{1}{2}x^{-1/2}(-2 - x^{1/2} - 2 + x^{1/2})}{(2 + x^{1/2})^2}$$

$$= \frac{-2}{x^{1/2}(2 + x^{1/2})^2}$$

To find $f''(x)$, we express $f'(x)$ as a product and differentiate

$$f'(x) = -2x^{-1/2}(2 + x^{1/2})^{-2}$$

$$f''(x) = -2x^{-1/2}(-2)(2 + x^{1/2})^{-3}\left(\frac{1}{2}x^{-1/2}\right) + (2 + x^{1/2})^{-2}(x^{-3/2})$$

$$= x^{-3/2}(2 + x^{1/2})^{-3}[2x^{1/2} + (2 + x^{1/2})]$$

$$= \frac{2 + 3x^{1/2}}{x^{3/2}(2 + x^{1/2})^3}$$

14. Find $f^{(4)}(x)$ if $f(x) = \dfrac{2}{x - 1}$

SOLUTION:

$$f(x) = 2(x - 1)^{-1}$$
$$f'(x) = -2(x - 1)^{-2}$$
$$f''(x) = 4(x - 1)^{-3}$$
$$f'''(x) = -12(x - 1)^{-4}$$

$$f^{(4)}(x) = 48(x - 1)^{-5} = \frac{48}{(x - 1)^5}$$

18.   Given $x^{1/2} + y^{1/2} = 2$, show that $D_x{}^2y = 1/x^{3/2}$.

SOLUTION: Differentiating implicitly with respect to x on both sides of the given equation, we have

$$\frac{1}{2}x^{-1/2} + \frac{1}{2}y^{-1/2}D_xy = 0$$

$$D_xy = -x^{-1/2}y^{1/2} \qquad (1)$$

Differentiating implicitly with respect to x on both sides of Eq. (1), we obtain

$$D_x{}^2y = -x^{-1/2}\left(\frac{1}{2}\right)y^{-1/2}D_xy + y^{1/2}\left(\frac{1}{2}\right)x^{-3/2} \qquad (2)$$

Substituting the value for D_xy from Eq. (1) into (2), we get

$$D_x{}^2y = -x^{-1/2}\left(\frac{1}{2}\right)y^{-1/2}(-x^{-1/2}y^{1/2}) + \frac{1}{2}x^{-3/2}y^{1/2}$$

$$= \frac{1}{2}x^{-3/2}(x^{1/2} + y^{1/2}) \qquad (3)$$

Because the given equation is $x^{1/2} + y^{1/2} = 2$, Eq. (3) is equivalent to

$$D_x{}^2y = \frac{1}{2}x^{-3/2}(2) = \frac{1}{x^{3/2}}$$

20.   Given $b^2x^2 - a^2y^2 = a^2b^2$ (a and b are constants), find $D_x{}^2y$ in simplest form.

SOLUTION: We differentiate implicitly with respect to x.

$$b^2(2x) - a^2(2y)D_xy = 0$$

$$D_xy = \frac{b^2}{a^2} \cdot \frac{x}{y} \qquad (1)$$

Differentiating implicitly with respect to x on both sides of Eq. (1), we have

$$D_x{}^2y = \frac{b^2}{a^2}\left[\frac{y(1) - x D_xy}{y^2}\right] \qquad (2)$$

Substituting the value of D_xy from Eq. (1) into Eq. (2), we get

$$D_x{}^2y = \frac{b^2}{a^2}\left[\frac{y - x\left(\dfrac{b^2x}{a^2y}\right)}{y^2}\right]$$

$$= \frac{b^2}{a^2}\left[\frac{a^2y^2 - b^2x^2}{a^2y^3}\right]$$

$$= \frac{-b^2(b^2x^2 - a^2y^2)}{a^4y^3} \qquad (3)$$

Because we are given that $b^2x^2 - a^2y^2 = a^2b^2$, Eq. (3) can be simplified.

$$D_x{}^2y = \frac{-b^2(a^2b^2)}{a^4y^3} = \frac{-b^4}{a^2y^3}$$

24.   A particle is moving along a straight line according to the given equation of motion, where s ft is the directed distance of the particle from the origin at t sec. Find the time when the instantaneous acceleration is zero, and then find the directed distance of the particle from the origin and the instantaneous velocity at this instant.

$$s = 2t^3 - 6t^2 + 3t - 4, \ t \geqslant 0$$

SOLUTION:

$$v = D_t s = 6t^2 - 12t + 3 \tag{1}$$

Because the acceleration is the rate of change of the velocity, we find $D_t v$.

$$a = D_t v = 12t - 12 \tag{2}$$

We replace a by 0 in Eq. (2)

$$0 = 12t - 12$$
$$t = 1$$

The instantaneous acceleration is zero at 1 sec.

We replace t by 1 in the given equation of motion, and also replace t by 1 in Eq. (1).

$$s = 2(1)^3 - 6(1)^2 + 3(1) - 4 = -5$$
$$v = 6(1)^2 - 12(1) + 3 \qquad = -3$$

At 1 sec the directed distance of the particle from the origin is -5 ft, and the instantaneous velocity is -3 ft/sec.

30. Find the formula for $f'(x)$ and $f''(x)$ and state the domain of f' and f''.

$$f(x) = |x|^3$$

SOLUTION: Because $|x| = \sqrt{x^2}$,

$$f(x) = (\sqrt{x^2})^3 = (x^2)^{3/2}$$

$$f'(x) = \frac{3}{2}(x^2)^{1/2}(2x)$$

$$= 3x(x^2)^{1/2}$$
$$= 3x\,|x| \tag{1}$$

The domain of f' is $(-\infty, +\infty)$. Note that we do not replace $(x^2)^{3/2}$ by x^3, because x^3 has a negative value if $x < 0$, whereas $(x^2)^{3/2}$ is nonnegative for all real x. To find f'' we differentiate on both sides of Eq. (1).

$$f''(x) = 3x\left[\frac{1}{2}(x^2)^{-1/2}(2x)\right] + (x^2)^{1/2}(3)$$

$$= \frac{3x^2}{\sqrt{x^2}} + 3\sqrt{x^2}$$

$$= \frac{3x^2}{|x|} + 3|x|$$

$$= 3|x| + 3|x|$$
$$= 6|x| \quad \text{if} \ \ |x| \neq 0$$

The domain of f'' is $\{x \,|\, x \neq 0\}$.

34. Show that if $xy = 1$, then $D_x^2 y \cdot D_y^2 x = 4$.

SOLUTION: Because $xy = 1$, then

$$y = x^{-1}$$
$$D_x y = -x^{-2}$$
$$D_x^2 y = 2x^{-3} \tag{1}$$

Because $xy = 1$, then

$$x = y^{-1}$$
$$D_y x = -y^{-2}$$
$$D_y{}^2 x = 2y^{-3} \tag{2}$$

By (1) and (2) we have

$$D_x{}^2 y \cdot D_y{}^2 x = (2x^{-3})(2y^{-3})$$

$$= \frac{4}{(xy)^3} \tag{3}$$

Because $xy = 1$, Eq. (3) is equivalent to

$$D_x{}^2 y \cdot D_y{}^2 x = 4$$

38. If $y = 1/(1 - 2x)$, prove by mathematical induction that

$$D_x{}^n y = \frac{2^n n!}{(1 - 2x)^{n+1}} \tag{1}$$

SOLUTION: We must show (a) Eq. (1) is true for $n = 1$, and (b) whenever Eq. (1) is true for $n = k$, then Eq. (1) is also true for $n = k + 1$.

  (a) We are given that

$$y = (1 - 2x)^{-1}$$

Then

$$D_x y = -(1 - 2x)^{-2}(-2) = \frac{2}{(1 - 2x)^2} \tag{2}$$

Because Eq. (1) becomes Eq. (2) if $n = 1$, we conclude that Eq. (1) holds when $n = 1$.

  (b) Suppose that Eq. (1) holds when $n = k$. That is, suppose that

$$D_x{}^k y = \frac{2^k k!}{(1 - 2x)^{k+1}} = 2^k k!(1 - 2x)^{-k-1}$$

Differentiating with respect to x, we obtain

$$D_x{}^{k+1} y = 2^k k!(-k - 1)(1 - 2x)^{(-k-1)-1} \cdot (-2)$$
$$= 2^{k+1} k!(k + 1)(1 - 2x)^{-k-2} \tag{3}$$

Because $(k + 1) \cdot k! = (k + 1)!$, Eq. (3) is equivalent to

$$D_x{}^{k+1} y = \frac{2^{k+1}(k + 1)!}{(1 - 2x)^{k+2}} \tag{4}$$

If n is replaced by $k + 1$ in Eq. (1), the result is Eq. (4). Therefore, we have shown that whenever Eq. (1) holds for $n = k$, it also holds for $n = k + 1$.

Review Exercises

In Exercises 1-14, find $D_x y$.

6. $y = \dfrac{x^2}{(x + 2)^2(4x - 5)}$

SOLUTION:

$$D_x y = \frac{(x+2)^2(4x-5)(2x) - x^2[(x+2)^2(4) + (4x-5) \cdot 2(x+2)]}{[(x+2)^2(4x-5)]^2}$$

$$= \frac{2x(x+2)[(x+2)(4x-5) - 2x(x+2) - x(4x-5)]}{(x+2)^4(4x-5)^2}$$

$$= \frac{2x(-2x^2 + 4x - 10)}{(x+2)^3(4x-5)^2}$$

$$= \frac{-4x(x^2 - 2x + 5)}{(x+2)^3(4x-5)^2}$$

8. $xy^2 + 2y^3 = x - 2y$

SOLUTION: We differentiate implicitly with respect to x on both sides of the given equation

$$x(2y\, D_x y) + y^2(1) + 6y^2\, D_x y = 1 - 2\, D_x y$$
$$(2xy + 6y^2 + 2)\, D_x y = 1 - y^2$$

$$D_x y = \frac{1 - y^2}{2xy + 6y^2 + 2}$$

14. $y = \dfrac{x\sqrt{3 + 2x}}{4x - 1}$

SOLUTION:

$$y = \frac{x(2x+3)^{1/2}}{4x-1}$$

$$D_x y = \frac{(4x-1)[x \cdot \frac{1}{2}(2x+3)^{-1/2}(2) + (2x+3)^{1/2}(1)] - x(2x+3)^{1/2}(4)}{(4x-1)^2}$$

Multiplying the numerator and denominator by $(2x+3)^{1/2}$, we get

$$D_x y = \frac{(4x-1)(x) + (4x-1)(2x+3) - 4x(2x+3)}{(4x-1)^2(2x+3)^{1/2}}$$

$$= \frac{4x^2 - 3x - 3}{(4x-1)^2(2x+3)^{1/2}}$$

20. Using only the definition of a derivative, find $f'(5)$ if $f(x) = \sqrt[3]{3x+1}$.

SOLUTION: We use the definition

$$f'(x_1) = \lim_{x \to x_1} \frac{f(x) - f(x_1)}{x - x_1}$$

with $f(x) = (3x+1)^{1/3}$ and $x_1 = 5$. Therefore,

$$f'(5) = \lim_{x \to 5} \frac{(3x+1)^{1/3} - 16^{1/3}}{x - 5}$$

$$= \lim_{x \to 5} \frac{[(3x+1)^{1/3} - 16^{1/3}][(3x+1)^{2/3} + (3x+1)^{1/3} \cdot 16^{1/3} + 16^{2/3}]}{(x-5)[(3x+1)^{2/3} + (3x+1)^{1/3} \cdot 16^{1/3} + 16^{2/3}]}$$

$$= \lim_{x \to 5} \frac{(3x+1) - 16}{(x-5)[(3x+1)^{2/3} + (3x+1)^{1/3} \cdot 16^{1/3} + 16^{2/3}]}$$

$$= \lim_{x \to 5} \frac{3}{(3x+1)^{2/3} + (3x+1)^{1/3} \cdot 16^{1/3} + 16^{2/3}}$$

$$= \frac{3}{16^{2/3} + 16^{1/3} \cdot 16^{1/3} + 16^{2/3}}$$

$$= \frac{3}{3 \cdot 16^{2/3}} = \frac{1}{2^{8/3}} = \frac{\sqrt[3]{2}}{8}$$

26. Find $f'(-3)$ if $f(x) = (|x| - x)\sqrt[3]{9x}$

SOLUTION: Because $|x| = \sqrt{x^2}$, then

$$f(x) = [(x^2)^{1/2} - x](9x)^{1/3}$$

Thus,

$$f'(x) = [(x^2)^{1/2} - x]\frac{1}{3}(9x)^{-2/3}(9) + (9x)^{1/3}\left[\frac{1}{2}(x^2)^{-1/2}(2x) - 1\right]$$

$$= \frac{3(|x| - x)}{(9x)^{2/3}} + (9x)^{1/3}\left(\frac{x}{|x|} - 1\right)$$

Therefore

$$f'(-3) = \frac{3(|-3| + 3)}{(-27)^{2/3}} + (-27)^{1/3}\left(\frac{-3}{|-3|} - 1\right)$$

$$= \frac{18}{9} + (-3)(-2) = 8$$

28. Find an equation of the normal line to the curve $x - y = \sqrt{x + y}$ at the point $(3, 1)$.

SOLUTION: Differentiating implicitly with respect to x on both sides of the given equation, we get

$$1 - D_x y = \frac{1}{2}(x + y)^{-1/2}(1 + D_x y)$$

Replacing x by 3 and y by 1, we have

$$1 - D_x y = \frac{1}{2}(4)^{-1/2}(1 + D_x y)$$

Solving for $D_x y$, we obtain

$$D_x y = \frac{3}{5}$$

Hence, the slope of the tangent line to the curve at the point $(3, 1)$ is $\frac{3}{5}$. We conclude that the slope of the normal line to the curve at $(3, 1)$ is $-\frac{5}{3}$. Thus, an equation for the normal line is

$$y - 1 = -\frac{5}{3}(x - 3)$$
$$5x + 3y - 18 = 0$$

32. Suppose

$$f(x) = \begin{cases} x^3 & \text{if } x < 1 \\ ax^2 + bx + c & \text{if } x \geqslant 1 \end{cases} \qquad (1)$$

Find the values of a, b, and c so that $f''(1)$ exists.

SOLUTION: If $x \neq 1$, we may differentiate f as follows.

$$f'(x) = \begin{cases} 3x^2 & \text{if } x < 1 \\ 2ax + b & \text{if } x > 1 \end{cases} \tag{2}$$

$$f''(x) = \begin{cases} 6x & \text{if } x < 1 \\ 2a & \text{if } x > 1 \end{cases} \tag{3}$$

If $f''(1)$ exists, then $f''_+(1) = f''_-(1)$, or equivalently,

$$\lim_{x \to 1^+} f''(x) = \lim_{x \to 1^-} f''(x) \tag{4}$$

From (3) we have

$$\lim_{x \to 1^+} f''(x) = \lim_{x \to 1^+} 2a = 2a \tag{5}$$

and

$$\lim_{x \to 1^-} f''(x) = \lim_{x \to 1^-} 6x = 6 \tag{6}$$

Substituting from (5) and (6) into (4), we have

$$2a = 6$$
$$a = 3$$

Because $f''(1)$ exists, then $f'(1)$ exists. Thus $f'_+(1) = f'_-(1)$ or, equivalently,

$$\lim_{x \to 1^+} f'(x) = \lim_{x \to 1^-} f'(x) \tag{7}$$

From (2), we have

$$\lim_{x \to 1^+} f'(x) = \lim_{x \to 1^+} (2ax + b) = 2a + b \tag{8}$$

and

$$\lim_{x \to 1^-} f'(x) = \lim_{x \to 1^-} 3x^2 = 3 \tag{9}$$

Substituting from (8) and (9) into (7), we have

$$2a + b = 3$$

Thus, because $a = 3$, we obtain $b = -3$.

Furthermore, because $f'(1)$ exists, then f is continuous at 1. Then

$$\lim_{x \to 1^+} f(x) = \lim_{x \to 1^-} f(x) \tag{10}$$

From (1), we have

$$\lim_{x \to 1^+} f(x) = \lim_{x \to 1^+} (ax^2 + bx + c) = a + b + c \tag{11}$$

and

$$\lim_{x \to 1^-} f(x) = \lim_{x \to 1^-} x^3 = 1 \tag{12}$$

Substituting from (11) and (12) into (10) we get

$$a + b + c = 1$$

Because $a = 3$ and $b = -3$, then $c = 1$.

34. A particle is moving in a straight line according to the equation of motion $s = \sqrt{a + bt^2}$, where a and b are positive constants. Prove that the measure of the acceleration of the particle is inversely proportional to s^3 for any t.

SOLUTION: Because the acceleration is given by $a = D_t{}^2 s$, we must show that

$$D_t{}^2 s = \frac{c}{s^3} \tag{1}$$

for some constant c. We are given that

$$s = (a + bt^2)^{1/2} \tag{2}$$

Hence,

$$D_t s = \frac{1}{2}(a + bt^2)^{-1/2}(2bt)$$

$$= bt(a + bt^2)^{-1/2}$$

Differentiating $D_t s$ with respect to t, we have

$$D_t{}^2 s = bt\left(-\frac{1}{2}\right)(a + bt^2)^{-3/2}(2bt) + (a + bt^2)^{-1/2}(b)$$

$$= b(a + bt^2)^{-3/2}[-bt^2 + (a + bt^2)]$$

$$= \frac{ab}{(a + bt^2)^{3/2}} \tag{3}$$

By cubing both sides of Eq. (2), we have

$$s^3 = (a + bt^2)^{3/2} \tag{4}$$

And by substituting from Eq. (4) into Eq. (3), we obtain

$$D_t{}^2 s = \frac{ab}{s^3}$$

which becomes Eq. (1) if we let $c = ab$.

36. As the last car of a train passes under a bridge, an automobile crosses the bridge on a roadway perpendicular to the track and 30 ft above it. The train is traveling at the rate of 80 ft/sec and the automobile is traveling at the rate of 40 ft/sec. How fast are the train and the automobile separating after 2 sec?

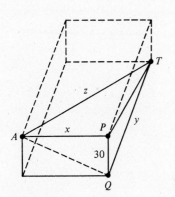

Figure 3.36R

SOLUTION: Refer to Fig. 3.36R. The roadway is in the line that contains points P and A, and the railway track is in the line that contains Q and T. We are given that line PA is perpendicular to line QT. The last car of the train is at point Q when the automobile is at point P. After t seconds the last car of the train is at point T and the automobile is at point A. Let the variables be defined as follows.

t = the number of seconds that have elapsed since the automobile was at point P and the last car of the train was at Q

x = the number of feet in the distance of the automobile from point P at t sec

y = the number of feet in the distance of the last car of the train from point Q at t sec

z = the number of feet in the distance between the automobile and the last car of the train at t sec

Because the train is traveling at 80 ft/sec, we are given that $D_t y = 80$. Because the automobile is traveling at 40 ft/sec, we are given that $D_t x = 40$. Since $D_t z$ represents the rate at which the automobile and the train are separating, we want to find $D_t z$ at the moment when $t = 2$. Hence, we must find an equation with variables x, y, and z. Because triangle AQT is a right triangle, we have

$$z^2 = |\overline{AQ}|^2 + y^2 \tag{1}$$

Because triangle APQ is a right triangle, we have

$$|\overline{AQ}|^2 = x^2 + (30)^2 \tag{2}$$

Substituting from Eq. (2) into Eq. (1), we obtain

$$z^2 = x^2 + (30)^2 + y^2 \tag{3}$$

Differentiating with respect to t on both sides of Eq. (3), we get

$$2z\, D_t z = 2x\, D_t x + 2y\, D_t y$$

$$D_t z = \frac{x\, D_t x + y\, D_t y}{z} \tag{4}$$

Substituting 40 for $D_t x$ and 80 for $D_t y$ in Eq. (4), we have

$$D_t z = \frac{40x + 80y}{z} \tag{5}$$

When $t = 2$, we have $x = 80$ because the automobile is traveling at 40 ft/sec. Similarly, when $t = 2$, we have $y = 160$ because the train is traveling at 80 ft/sec. From Eq. (3) we get $z = 10\sqrt{329}$ when $x = 80$ and $y = 160$. Thus, from Eq. (5) we obtain

$$D_t z\Big]_{t=2} = \frac{40 \cdot 80 + 80 \cdot 160}{10\sqrt{329}} = \frac{1600}{\sqrt{329}}$$

We conclude that after 2 sec the train and the automobile are separating at the rate of $1600/\sqrt{329}$ ft/sec, or approximately 88.2 ft/sec.

40. If the two functions f and g are differentiable at the number x_1, is the composite function $f \circ g$ necessarily differentiable at x_1? If your answer is yes, prove it. If your answer is no, give a counterexample.

SOLUTION: No, f must be differentiable at $g(x_1)$ rather than x_1. For a counterexample, let f and g be defined as follows

$$f(x) = \frac{1}{x} \quad \text{and} \quad g(x) = x - 1$$

and consider $x_1 = 1$. We note that

$$f'(x) = -\frac{1}{x^2} \quad \text{and} \quad g'(x) = 1$$

$$f'(1) = -1 \quad \text{and} \quad g'(1) = 1$$

Thus, both f and g are differentiable at 1. But the composite function $f \circ g$ is defined by

$$(f \circ g)(x) = f(g(x)) = \frac{1}{x - 1}$$

Hence,

$$(f \circ g)'(x) = \frac{-1}{(x - 1)^2}$$

Because $(f \circ g)'(1)$ is not defined, then $f \circ g$ is not differentiable at 1.

42. Give an example of two functions f and g for which f is differentiable at $g(0)$, g is not differentiable at 0, and $f \circ g$ is differentiable at 0.

SOLUTION: Let f and g be defined as follows:

$$f(x) = x^2 \quad \text{and} \quad g(x) = |x|$$

Because $g(x) = \sqrt{x^2}$, then

$$g'(x) = \frac{1}{2}(x^2)^{-1/2} \cdot 2x$$

$$= \frac{x}{|x|}$$

Hence, $g'(0)$ does not exist, and thus g is not differentiable at 0. Moreover, because $f'(x) = 2x$, then f is differentiable for all x, and in particular f is differentiable at $g(0) = 0$. However, $f \circ g$ is defined by

$$(f \circ g)(x) = f(g(x)) = |x|^2 = x^2$$

Thus

$$(f \circ g)'(x) = 2x$$

Therefore, $f \circ g$ is differentiable at all x, and in particular $f \circ g$ is differentiable at 0.

46. Let f and g be two functions whose domains are the set of all real numbers. Furthermore, suppose that (i) $g(x) = xf(x) + 1$; (ii) $g(a + b) = g(a) \cdot g(b)$ for all a and b; and (iii) $\lim\limits_{x \to 0} f(x) = 1$. Prove that $g'(x) = g(x)$.

SOLUTION:

$$g'(x) = \lim_{\Delta x \to 0} \frac{g(x + \Delta x) - g(x)}{\Delta x}$$

$$= \lim_{\Delta x \to 0} \frac{g(x) \cdot g(\Delta x) - g(x)}{\Delta x} \quad \text{(Hypothesis ii)}$$

$$= \lim_{\Delta x \to 0} g(x) \cdot \lim_{\Delta x \to 0} \frac{g(\Delta x) - 1}{\Delta x}$$

$$= g(x) \cdot \lim_{\Delta x \to 0} \frac{[\Delta x \cdot f(\Delta x) + 1] - 1}{\Delta x} \quad \text{(Hypothesis i)}$$

$$= g(x) \cdot \lim_{\Delta x \to 0} f(\Delta x)$$

$$= g(x) \cdot 1 \quad \text{(Hypothesis iii)}$$

Therefore, $g'(x) = g(x)$.

4

Topics on limits, continuity, and the derivative

4.1 LIMITS AT INFINITY

4.1.1 Definition Let f be a function which is defined at every number in some interval $(a, +\infty)$. The *limit of* f(x), *as* x *increases without bound*, is L, written

$$\lim_{x \to +\infty} f(x) = L$$

if for any $\epsilon > 0$, however small, there exists a number $N > 0$ such that $|f(x) - L| < \epsilon$ whenever $x > N$.

4.1.2 Definition Let f be a function which is defined at every number in some interval $(-\infty, a)$. The *limit of* f(x), *as* x *decreases without bound*, is L, written

$$\lim_{x \to -\infty} f(x) = L$$

if for any $\epsilon > 0$, however small, there exists a number $N < 0$ such that $|f(x) - L| < \epsilon$ whenever $x < N$.

4.1.3 Limit Theorem 13 If r is any positive integer, then

$$\text{(i)} \ \lim_{x \to +\infty} \frac{1}{x^r} = 0$$

$$\text{(ii)} \ \lim_{x \to -\infty} \frac{1}{x^r} = 0$$

Limit Theorems 2, 4, 5, 6, 7, 8, 9, and 10, given in Sec. 2.2, and Limit Theorems 11 and 12, given in Sec. 2.4, remain unchanged when "$x \to a$" is replaced by "$x \to +\infty$" or "$x \to -\infty$."

Suppose that f is a function defined by

$$f(x) = \frac{g(x)}{h(x)}$$

where $g(x)$ and $h(x)$ are polynomials. To find the limit of f as $x \to +\infty$ or as $x \to -\infty$, first divide both $g(x)$ and $h(x)$ by the highest power of x that appears in either the numerator or the denominator. Then you may use Limit Theorem 13 together with the other limit theorems to find the limit of f.

It can be shown that if f and g are both polynomial functions, then

(i) If the degree of $f(x)$ is less than the degree of $g(x)$, then

$$\lim_{x \to +\infty} \frac{f(x)}{g(x)} = \lim_{x \to -\infty} \frac{f(x)}{g(x)} = 0$$

(ii) If the degree of $f(x)$ and the degree of $g(x)$ are both equal to n, a is the coefficient of x^n in $f(x)$, and b is the coefficient of x^n in $g(x)$, then

$$\lim_{x \to +\infty} \frac{f(x)}{g(x)} = \lim_{x \to -\infty} \frac{f(x)}{g(x)} = \frac{a}{b}$$

(iii) If the degree of $f(x)$ is greater than the degree of $g(x)$, then the limit of $f(x)/g(x)$ as x approaches either $+\infty$ or $-\infty$ does not exist. That is,

$$\lim_{x \to +\infty} \frac{f(x)}{g(x)} = \pm\infty \quad \text{and} \quad \lim_{x \to -\infty} \frac{f(x)}{g(x)} = \pm\infty$$

Exercise 4 illustrates part (i); Exercise 2 illustrates part (ii); and Exercise 14 illustrates part (iii) of the above discussion.

Exercises 4.1

In Exercises 1-14, find the limits and when applicable, indicate the limit theorems being used.

2. $\displaystyle\lim_{s \to +\infty} \frac{4s^2 + 3}{2s^2 - 1}$

SOLUTION: We divide the numerator and denominator by s^2. Thus,

$$\lim_{s \to +\infty} \frac{4s^2 + 3}{2s^2 - 1} = \lim_{s \to +\infty} \frac{4 + \dfrac{3}{s^2}}{2 - \dfrac{1}{s^2}}$$

$$= \frac{\displaystyle\lim_{s \to +\infty} \left(4 + \dfrac{3}{s^2}\right)}{\displaystyle\lim_{s \to +\infty} \left(2 - \dfrac{1}{s^2}\right)} \quad \text{(L.T. 9)}$$

$$= \frac{\displaystyle\lim_{s \to +\infty} 4 + \lim_{s \to +\infty} 3 \cdot \lim_{s \to +\infty} \dfrac{1}{s^2}}{\displaystyle\lim_{s \to +\infty} 2 - \lim_{s \to +\infty} \dfrac{1}{s^2}} \quad \text{(L.T. 4 and L.T. 6)}$$

$$= \frac{4 + 3 \cdot 0}{2 - 0} \quad \text{(L.T. 2 and L.T. 13)}$$

$$= 2$$

4. $\displaystyle\lim_{x \to +\infty} \frac{x^2 - 2x + 5}{7x^3 + x + 1}$

SOLUTION: Dividing the numerator and denominator by x^3, the highest power of x which appears, we have

$$\lim_{x \to +\infty} \frac{x^2 - 2x + 5}{7x^3 + x + 1} = \lim_{x \to +\infty} \frac{\dfrac{1}{x} - \dfrac{2}{x^2} + \dfrac{5}{x^3}}{7 + \dfrac{1}{x^2} + \dfrac{1}{x^3}}$$

$$= \frac{\displaystyle\lim_{x \to +\infty} \left(\frac{1}{x} - \frac{2}{x^2} + \frac{5}{x^3} \right)}{\displaystyle\lim_{x \to +\infty} \left(7 + \frac{1}{x^2} + \frac{1}{x^3} \right)} \quad \text{(L.T. 9)}$$

$$= \frac{\displaystyle\lim_{x \to +\infty} \frac{1}{x} - 2 \lim_{x \to +\infty} \frac{1}{x^2} + 5 \lim_{x \to +\infty} \frac{1}{x^3}}{7 + \displaystyle\lim_{x \to +\infty} \frac{1}{x^2} + \lim_{x \to +\infty} \frac{1}{x^3}} \quad \begin{array}{l}\text{(L.T. 5, 6,}\\ \text{and 2)}\end{array}$$

$$= \frac{0 - 2 \cdot 0 + 5 \cdot 0}{7 + 0 + 0} \quad \text{(L. T. 13)}$$

$$= 0$$

6. $\displaystyle\lim_{x \to -\infty} \frac{\sqrt{x^2 + 4}}{x + 4}$

SOLUTION: We divide the numerator and denominator by x. Because $x \to -\infty$, then $x < 0$ and $\sqrt{x^2} = -x$ or, equivalently, $x = -\sqrt{x^2}$. Thus,

$$\frac{\sqrt{x^2 + 4}}{x} = \frac{\sqrt{x^2 + 4}}{-\sqrt{x^2}} = -\sqrt{\frac{x^2 + 4}{x^2}} = -\sqrt{1 + \frac{4}{x^2}} \qquad \text{(1)}$$

By Eq. (1) we have

$$\lim_{x \to -\infty} \frac{\sqrt{x^2 + 4}}{x + 4} = \lim_{x \to -\infty} \frac{\dfrac{\sqrt{x^2 + 4}}{x}}{\dfrac{x + 4}{x}}$$

$$= \lim_{x \to -\infty} \frac{-\sqrt{1 + \dfrac{4}{x^2}}}{1 + \dfrac{4}{x}}$$

$$= \frac{-\sqrt{\displaystyle\lim_{x \to -\infty} \left(1 + \frac{4}{x^2} \right)}}{\displaystyle\lim_{x \to -\infty} \left(1 + \frac{4}{x} \right)} \quad \text{(L.T. 9 and L.T. 10)}$$

$$= \frac{-\sqrt{1 + 4 \displaystyle\lim_{x \to -\infty} \frac{1}{x^2}}}{1 + 4 \displaystyle\lim_{x \to -\infty} \frac{1}{x}} \quad \text{(L.T. 4, L.T. 2, and L.T. 6)}$$

$$= \frac{-\sqrt{1 + 4 \cdot 0}}{1 + 4 \cdot 0} \quad \text{(L.T. 13)}$$

$$= -1$$

10. $\displaystyle \lim_{x \to +\infty} (\sqrt{x^2 + x} - x)$

SOLUTION: Because this expression is not a fraction, we cannot use the technique illustrated in Exercise 6. However, we may replace the given expression by a fraction with denominator 1 and then rationalize the numerator. Thus

$$\lim_{x \to +\infty} (\sqrt{x^2 + x} - x) = \lim_{x \to +\infty} \frac{\sqrt{x^2 + x} - x}{1} \cdot \frac{\sqrt{x^2 + x} + x}{\sqrt{x^2 + x} + x}$$

$$= \lim_{x \to +\infty} \frac{x}{\sqrt{x^2 + x} + x} \qquad (1)$$

Now we may divide the numerator and denominator by x. Because $x \to +\infty$, then $x > 0$ and $x = \sqrt{x^2}$. Thus,

$$\frac{\sqrt{x^2 + x} + x}{x} = \frac{\sqrt{x^2 + x}}{x} + \frac{x}{x}$$

$$= \frac{\sqrt{x^2 + x}}{\sqrt{x^2}} + 1$$

$$= \sqrt{\frac{x^2 + x}{x^2}} + 1$$

$$= \sqrt{1 + \frac{1}{x}} + 1 \qquad (2)$$

By Eq. (1) and Eq. (2), we have

$$\lim_{x \to +\infty} (\sqrt{x^2 + x} - x) = \lim_{x \to +\infty} \frac{x}{\sqrt{x^2 + x} + x} \cdot \frac{\dfrac{1}{x}}{\dfrac{1}{x}}$$

$$= \lim_{x \to +\infty} \frac{1}{\sqrt{1 + \dfrac{1}{x}} + 1}$$

$$= \frac{1}{\sqrt{1 + \displaystyle\lim_{x \to +\infty} \dfrac{1}{x}} + 1} \qquad \text{(L.T. 9, 2, 4, and 10)}$$

$$= \frac{1}{\sqrt{1 + 0} + 1} \qquad \text{(L.T. 13)}$$

$$= \frac{1}{2}$$

14. $\displaystyle \lim_{x \to -\infty} \frac{5x^3 - 12x + 7}{4x^2 - 1}$

SOLUTION: We divide the numerator and denominator by x^3, the highest power of x which appears. Thus,

$$\lim_{x \to -\infty} \frac{5x^3 - 12x + 7}{4x^2 - 1} = \lim_{x \to -\infty} \frac{5 - \dfrac{12}{x^2} + \dfrac{7}{x^3}}{\dfrac{4}{x} - \dfrac{1}{x^3}} \qquad (1)$$

We consider the limit of the numerator and the limit of the denominator separately

$$\lim_{x \to -\infty} 5 - \frac{12}{x^2} + \frac{7}{x^3} = 5 - 12 \lim_{x \to -\infty} \frac{1}{x^2} + 7 \lim_{x \to -\infty} \frac{1}{x^3} \quad \text{(L.T. 5, 6, and 2)}$$

$$= 5 - 12 \cdot 0 + 7 \cdot 0 \quad \text{(L.T. 13)}$$
$$= 5 \qquad\qquad (2)$$

$$\lim_{x \to -\infty} \frac{4}{x} - \frac{1}{x^3} = 4 \lim_{x \to -\infty} \frac{1}{x} - \lim_{x \to -\infty} \frac{1}{x^3} \quad \text{(L.T. 4, 6, and 2)}$$

$$= 4 \cdot 0 - 0 = 0 \quad \text{(L.T. 13)} \qquad\qquad (3)$$

By Eqs. (1), (2), (3), and L.T. 12, we conclude that the limit of the given function is either $+\infty$ or $-\infty$. Furthermore, by factoring the denominator we have

$$\frac{4}{x} - \frac{1}{x^3} = \frac{1}{x}\left(4 - \frac{1}{x^2}\right)$$

Because $x \to -\infty$, then $x < -1$. Thus

$$\frac{1}{x} < 0 \quad \text{and} \quad 4 - \frac{1}{x^2} > 0$$

and hence

$$\frac{1}{x}\left(4 - \frac{1}{x^2}\right) < 0 \quad \text{if } x < -1$$

Therefore the denominator approaches zero through negative values. By L.T. 12(ii) we conclude that

$$\lim_{x \to -\infty} \frac{5x^3 - 12x + 7}{4x^2 - 1} = -\infty$$

16. Prove that $\lim_{x \to +\infty} f(x) = 1$ by applying Definition 4.1.1; that is, for any $\epsilon > 0$, show that there exists a number $N > 0$ such that $|f(x) - 1| < \epsilon$ whenever $x > N$.

$$f(x) = \frac{x^2 + 2x}{x^2 - 1}$$

SOLUTION: For any $\epsilon > 0$ we must find an $N > 0$ such that

$$\left| \frac{x^2 + 2x}{x^2 - 1} - 1 \right| < \epsilon \quad \text{whenever } x > N \qquad\qquad (1)$$

Because

$$\left| \frac{x^2 + 2x}{x^2 - 1} - 1 \right| = \left| \frac{2x + 1}{x^2 - 1} \right|$$

$$< \left| \frac{2x + 2}{x^2 - 1} \right|$$

$$= \frac{2}{x - 1} \quad \text{if } x - 1 > 0 \qquad\qquad (2)$$

then (1) will follow if we find an $N > 1$ such that

$$\frac{2}{x - 1} < \epsilon \quad \text{whenever } x > N$$

or, equivalently,

$$\frac{x - 1}{2} > \frac{1}{\epsilon} \quad \text{whenever } x > N$$

or, equivalently,

$$x > 1 + \frac{2}{\epsilon} \quad \text{whenever} \quad x > N$$

We now write the proof.

PROOF: For any $\epsilon > 0$ let $N = 1 + \frac{2}{\epsilon}$. Then $N > 0$, and moreover if $x > N$, then $x - 1 > 0$, and it follows that

$$x > 1 + \frac{2}{\epsilon}$$

$$x - 1 > \frac{2}{\epsilon}$$

$$\frac{1}{x - 1} < \frac{\epsilon}{2}$$

$$\frac{2}{x - 1} < \epsilon \qquad\qquad (3)$$

By inequality (3) and inequality (2), inequality (1) follows. Thus,

$$\lim_{x \to +\infty} f(x) = 1$$

18. Prove that $\lim\limits_{x \to +\infty} (x^2 - 4) = +\infty$ by showing that for any $N > 0$ there exists an $M > 0$ such that $(x^2 - 4) > N$ whenever $x > M$.

SOLUTION: if $x > 0$, then

$$x^2 - 4 > N$$

if and only if

$$x^2 > 4 + N$$
$$x > \sqrt{4 + N}$$

Therefore, for any $N > 0$, choose $M = \sqrt{4 + N}$. Then $M > 0$, and furthermore if $x > M$, then

$$x > \sqrt{4 + N}$$
$$x^2 > 4 + N$$
$$x^2 - 4 > N$$

Therefore,

$$\lim_{x \to +\infty} (x^2 - 4) = +\infty$$

4.2 HORIZONTAL AND VERTICAL ASYMPTOTES

4.2.1 Definition   The line $x = a$ is said to be a vertical asymptote of the graph of the function f if at least one of the following statements is true:

(i) $\lim\limits_{x \to a^+} f(x) = +\infty$

(ii) $\lim\limits_{x \to a^+} f(x) = -\infty$

(iii) $\lim\limits_{x \to a^-} f(x) = +\infty$

(iv) $\lim\limits_{x \to a^-} f(x) = -\infty$

(i) $\lim\limits_{x \to a^+} f(x) = +\infty$ (ii) $\lim\limits_{x \to a^+} f(x) = -\infty$

(iii) $\lim\limits_{x \to a^-} f(x) = +\infty$ (iv) $\lim\limits_{x \to a^-} f(x) = -\infty$

Figure 4.2.1

In Fig. 4.2.1 we illustrate the graph of the function f for each of the four cases in Definition 4.2.1. In Fig. 4.2.1(i) and Fig. 4.2.1(ii) the curve "approaches" the asymptote from the right, whereas in Fig. 4.2.1(iii) and Fig. 4.2.1(iv) the curve approaches the asymptote from the left.

Definition 4.2.1 implies that the line $x = a$ is a vertical asymptote of the graph of the function f if the following three conditions are all satisfied:

(1) $f(x) = \dfrac{g(x)}{h(x)}$

(2) $\lim\limits_{x \to a} h(x) = 0$

(3) $\lim\limits_{x \to a} g(x) = c \neq 0$

That is, if $f(x)$ is a fraction such that as x approaches a the denominator has limit zero and the numerator has a finite nonzero limit, then the line $x = a$ is a vertical asymptote of the graph of f.

The following steps can be used to find a line $x = a$ that is a vertical asymptote of the graph of the function f.

1. Write $f(x)$ in the form $f(x) = \dfrac{g(x)}{h(x)}$.

2. Solve the equation $h(x) = 0$.
3. For each solution a of the equation in step 2, find $\lim\limits_{x \to a} f(x)$.

The line $x = a$ is a vertical asymptote of the graph of f if and only if the limit in step 3 is either $+\infty$ or $-\infty$.

4.2.2 Definition The line $y = b$ is said to be a horizontal asymptote of the graph of the function f if at least one of the following statements is true:

(i) $\lim\limits_{x \to +\infty} f(x) = b$

(ii) $\lim\limits_{x \to -\infty} f(x) = b$

If any of the tests for symmetry given in Theorem 1.3.6 are satisfied, we may use symmetry to shorten the process of finding all the horizontal and vertical asymptotes. This is illustrated in Exercises 8, 10, and 20.

Exercises 4.2

In Exercises 1-14, find the horizontal and vertical asymptotes of the graph of the function defined by the given equation and draw a sketch of the graph.

4. $F(x) = \dfrac{5}{x^2 + 8x + 16}$

SOLUTION: To find the horizontal asymptotes we find the limit of F as $x \to +\infty$ and as $x \to -\infty$.

$$\lim_{x \to +\infty} F(x) = \lim_{x \to +\infty} \frac{5}{x^2 + 8x + 16}$$

$$= \lim_{x \to +\infty} \frac{\dfrac{5}{x^2}}{1 + \dfrac{8}{x} + \dfrac{16}{x^2}} = 0$$

and

$$\lim_{x \to -\infty} F(x) = \lim_{x \to -\infty} \frac{5}{x^2 + 8x + 16}$$

$$= \lim_{x \to -\infty} \frac{\dfrac{5}{x^2}}{1 + \dfrac{8}{x} + \dfrac{16}{x^2}} = 0$$

Therefore, by Definition 4.2.2, we conclude that the line $y = 0$, (or the x-axis) is the only horizontal asymptote. Furthermore, since

$$F(x) = \frac{5}{(x + 4)^2} \tag{1}$$

we see that $F(x) > 0$ for all x, and thus the curve "approaches" the x-axis from above.

Because the denominator of the fraction in Eq. (1) is zero if $x = -4$, we test to see whether or not the line $x = -4$ is a vertical asymptote.

$$\lim_{x \to -4^+} f(x) = \lim_{x \to -4^+} \frac{5}{(x + 4)^2} = +\infty$$

Therefore by Definition 4.2.1 the line $x = -4$ is a vertical asymptote. Furthermore, the curve approaches the asymptote from the right in the manner illustrated by Fig. 4.2.1(i).

$$\lim_{x \to -4^-} f(x) = \lim_{x \to -4^-} \frac{5}{(x + 4)^2} = +\infty$$

Thus, the curve approaches the asymptote from the left in the manner illustrated by Fig. 4.2.1(iii). We use the asymptotes to draw a sketch of the graph of F, as illustrated in Fig. 4.2.4.

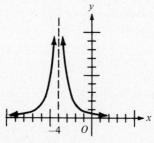

Figure 4.2.4

8. $f(x) = \dfrac{x^2}{4 - x^2}$

SOLUTION: Because $f(-x) = f(x)$, the graph of f is symmetric with respect to the y-axis. We draw a sketch of the graph for $x \geqslant 0$ and use symmetry to complete the sketch. Because,

$$\lim_{x \to +\infty} f(x) = \lim_{x \to +\infty} \frac{x^2}{4 - x^2}$$

$$= \lim_{x \to +\infty} \frac{1}{\dfrac{4}{x^2} - 1} = -1$$

we conclude that the line $y = -1$ is a horizontal asymptote.

Because, $4 - x^2 = 0$ if $x = \pm 2$, we test the line $x = 2$ as a possible vertical asymptote. (And by symmetry $x = -2$.) Since

$$\lim_{x \to 2^+} f(x) = \lim_{x \to 2^+} \frac{x^2}{4 - x^2} = -\infty$$

we conclude that the line $x = 2$ is a vertical asymptote, and furthermore the curve "approaches" the asymptote from the right in the manner illustrated in Fig. 4.2.1(ii). Since

$$\lim_{x \to 2^-} f(x) = \lim_{x \to 2^-} \frac{x^2}{4 - x^2} = +\infty$$

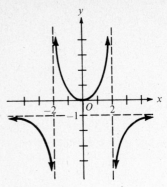

Figure 4.2.8

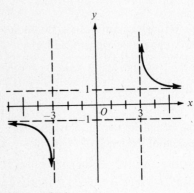

Figure 4.2.12

we conclude that the curve approaches the asymptote from the left in the manner illustrated by Fig. 4.2.1(iii). Because $f(0) = 0$, the curve contains the origin. We use the asymptotes, the origin, and symmetry to draw a sketch of the graph, as illustrated in Fig. 4.2.8.

12. $h(x) = \dfrac{x}{\sqrt{x^2 - 9}}$

SOLUTION: Because $h(-x) = -h(x)$, the graph of h is symmetric with respect to the origin. Because $x^2 - 9 > 0$ if $|x| > 3$, the domain of h is $\{x \mid |x| > 3\} = (-\infty, -3) \cup (3, +\infty)$.

$$\lim_{x \to +\infty} \frac{x}{\sqrt{x^2 - 9}} = \lim_{x \to +\infty} \frac{1}{\sqrt{1 - \dfrac{9}{x^2}}} = 1$$

Therefore, the line $y = 1$ is a horizontal asymptote. By symmetry, the line $y = -1$ is also a horizontal asymptote. Because $x^2 - 9 = 0$ if $x = \pm 3$, we test the line $x = 3$ as a possible vertical asymptote. (And by symmetry $x = -3$.)

$$\lim_{x \to 3^+} \frac{x}{\sqrt{x^2 - 9}} = +\infty$$

Therefore, the line $x = 3$ is a vertical asymptote and furthermore the curve approaches the asymptote from the right in the manner illustrated in Fig. 4.2.1(i). Because $h(x)$ is not defined for $-3 \leqslant x \leqslant 3$, the curve does not approach the asymptote from the left. We use the domain, symmetry, and the asymptotes to draw a sketch of the graph. See Fig. 4.2.12.

In Exercises 15-21, find the horizontal and vertical asymptotes of the graph of the given equation and draw a sketch of the graph.

16. $2xy + 4x - 3y + 6 = 0$

SOLUTION: Solving for y, we get

$$(2x - 3)y = -4x - 6$$

$$y = \frac{-4x - 6}{2x - 3}$$

We let $y = f(x)$.

$$\lim_{x \to +\infty} f(x) = \lim_{x \to +\infty} \frac{-4x - 6}{2x - 3}$$

$$= \lim_{x \to +\infty} \frac{-4 - \dfrac{6}{x}}{2 - \dfrac{3}{x}} = -2$$

And

$$\lim_{x \to -\infty} f(x) = \lim_{x \to -\infty} \frac{-4x - 6}{2x - 3}$$

$$= \lim_{x \to -\infty} \frac{-4 - \dfrac{6}{x}}{2 - \dfrac{3}{x}} = -2$$

Thus, the line $y = -2$ is the only horizontal asymptote. Because $2x - 3 = 0$ if $x = \frac{3}{2}$, we test to see whether or not the line $x = \frac{3}{2}$ is a vertical asymptote. By Limit Theorem 12(iii)

$$\lim_{x \to \frac{3}{2}^+} f(x) = \lim_{x \to \frac{3}{2}^+} \frac{-4x - 6}{2x - 3} = -\infty$$

Therefore, the line $x = \frac{3}{2}$ is a vertical asymptote, and furthermore the curve approaches the asymptote in the manner illustrated by Fig. 4.2.1(ii). Moreover, by Limit Theorem 12(iv)

$$\lim_{x \to \frac{3}{2}^-} f(x) = \lim_{x \to \frac{3}{2}^-} \frac{-4x - 6}{2x - 3} = +\infty$$

Thus, the curve approaches the line $x = \frac{3}{2}$ from the left in the manner illustrated by Fig. 4.2.1(iii). A sketch of the graph of the equation is shown in Fig. 4.2.16.

20. $xy^2 + 3y^2 - 9x = 0$

SOLUTION: Solving the given equation for y, we get

$$y^2 = \frac{9x}{x + 3}$$

$$y = \pm 3\sqrt{\frac{x}{x + 3}}$$

Therefore, the graph of the equation is symmetric with respect to the x-axis. Let f be the function defined by

$$f(x) = 3\sqrt{\frac{x}{x + 3}} \tag{1}$$

We draw a sketch of the graph of f and use symmetry to complete the sketch of the graph of the given equation. First, we determine the domain of f. Table 20 shows that the expression under the radical sign in Eq. (1) is nonnegative if $x \geq 0$ or $x < -3$. Therefore, the domain of f is $(-\infty, -3) \cup [0, +\infty)$.

Table 20

	$x < -3$	$x = -3$	$-3 < x < 0$	$x = 0$	$x > 0$
x	$-$	$-$	$-$	0	$+$
$x + 3$	$-$	0	$+$	$+$	$+$
$\dfrac{x}{x + 3}$	$+$	does not exist	$-$	0	$+$

$$\lim_{x \to +\infty} f(x) = \lim_{x \to +\infty} 3\sqrt{\frac{x}{x + 3}} = \lim_{x \to +\infty} 3\sqrt{\frac{1}{1 + \dfrac{3}{x}}} = 3$$

and

$$\lim_{x \to -\infty} f(x) = \lim_{x \to -\infty} 3\sqrt{\frac{x}{x + 3}} = \lim_{x \to -\infty} 3\sqrt{\frac{1}{1 + \dfrac{3}{x}}} = 3$$

Therefore, the line $y = 3$ is a horizontal asymptote. (And by symmetry, the line $y = -3$ is a horizontal asymptote.)

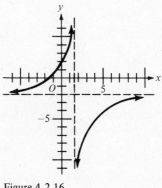

Figure 4.2.16

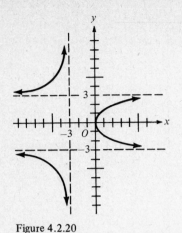

Figure 4.2.20

$$\lim_{x \to -3^-} f(x) = \lim_{x \to -3^-} 3\sqrt{\frac{x}{x+3}} = +\infty$$

Thus, the line $x = -3$ is a vertical asymptote, and the curve approaches the asymptote from the left in the manner illustrated by Fig. 4.2.1(iii). Because $f(0) = 0$, the curve contains the origin. We use the origin, the asymptotes, and symmetry to draw a sketch of the graph, which is shown in Fig. 4.2.20.

4.3 ADDITIONAL THEOREMS ON LIMITS OF FUNCTIONS

The theorems in this section are not often used for problem solving. However, they are used in later sections to prove other theorems. The most important theorem in this section is the following.

4.3.3 Theorem

(*Squeeze theorem*) Suppose that the functions $f, g,$ and h are defined on some open interval I containing a, except possibly at a itself, and that $f(x) \leqslant g(x) \leqslant h(x)$ for all x in I for which $x \neq a$. Also suppose that $\lim_{x \to a} f(x)$ and $\lim_{x \to a} h(x)$ both exist and are equal to L. Then $\lim_{x \to a} g(x)$ also exists and is equal to L.

Exercises 4.3

2. Prove Theorem 4.3.3.

SOLUTION: We are given that $f, g,$ and h are defined on some open interval I containing a, except possibly at a itself, and that

$$f(x) \leqslant g(x) \leqslant h(x) \quad \text{for all } x \text{ in } I \text{ with } x \neq a \tag{1}$$

We are also given that

$$\lim_{x \to a} f(x) = L \quad \text{and} \quad \lim_{x \to a} h(x) = L \tag{2}$$

We must show that for any $\epsilon > 0$ there is a $\delta > 0$ such that

$$|g(x) - L| < \epsilon \quad \text{whenever } 0 < |x - a| < \delta \tag{3}$$

Because of (2) we know that for any $\epsilon > 0$ there is some $\delta_1 > 0$ such that

$$|f(x) - L| < \epsilon \quad \text{whenever } 0 < |x - a| < \delta_1$$

or, equivalently,

$$L - \epsilon < f(x) < L + \epsilon \quad \text{whenever } 0 < |x - a| < \delta_1 \tag{4}$$

and some $\delta_2 > 0$ such that

$$|h(x) - L| < \epsilon \quad \text{whenever } 0 < |x - a| < \delta_2$$

or, equivalently,

$$L - \epsilon < h(x) < L + \epsilon \quad \text{whenever } 0 < |x - a| < \delta_2 \tag{5}$$

Hence, for any $\epsilon > 0$ choose $\delta = \min(\delta_1, \delta_2)$. Then $\delta \leqslant \delta_1$ and $\delta \leqslant \delta_2$. Whenever $0 < |x - a| < \delta$, it follows that $0 < |x - a| < \delta_1$. Therefore, by (4) we have, whenever $0 < |x - a| < \delta$, then

$$L - \epsilon < f(x) \tag{6}$$

And because $0 < |x - a| < \delta_2$, by (5) we have

$$h(x) < L + \epsilon \tag{7}$$

From (1), (6), and (7), we obtain:

$$L - \epsilon < f(x) \leqslant g(x) \leqslant h(x) < L + \epsilon$$

Thus,

$$L - \epsilon < g(x) < L + \epsilon$$

or, equivalently,

$$|g(x) - L| < \epsilon$$

Therefore,

$$\lim_{x \to a} g(x) = L$$

4. Let f be a function such that $|f(x)| \leqslant x^2$ for all x. Prove that f is differentiable at 0 and that $f'(0) = 0$.

SOLUTION: Because $|f(x)| \leqslant x^2$ for all x, we have

$$-x^2 \leqslant f(x) \leqslant x^2 \tag{1}$$

Replacing x by 0, in (1) gives

$$0 \leqslant f(0) \leqslant 0$$

Hence, $f(0) = 0$. By definition,

$$f'(0) = \lim_{x \to 0} \frac{f(x) - f(0)}{x - 0}$$

Because $f(0) = 0$, we must show that

$$f'(0) = \lim_{x \to 0} \frac{f(x)}{x} = 0 \tag{2}$$

If $x > 0$, we may divide each expression in (1) by x and obtain

$$-x \leqslant \frac{f(x)}{x} \leqslant x$$

Because both x and $-x$ have limit zero as $x \to 0^+$, it follows from Theorem 4.3.3 that

$$\lim_{x \to 0^+} \frac{f(x)}{x} = 0 \tag{3}$$

If $x < 0$, we may divide each expression in (1) by x, reverse the sense of the inequality, and obtain

$$x \leqslant \frac{f(x)}{x} \leqslant -x$$

Because both x and $-x$ have limit zero as $x \to 0^-$, it follows from Theorem 4.3.3, that

$$\lim_{x \to 0^-} \frac{f(x)}{x} = 0 \tag{4}$$

Thus, from (3) and (4)

$$\lim_{x \to 0} \frac{f(x)}{x} = 0$$

which is Eq. (2). Therefore, f is differentiable at 0 and $f'(0) = 0$.

4.4 CONTINUITY ON AN INTERVAL

4.4.1 Definition A function is said to be *continuous on an open interval* if and only if it is continuous at every number in the open interval.

4.4.2 Definition A function f is said to be *continuous from the right at the number* a if and only if the following three conditions are satisfied:

   (i) $f(a)$ exists
   (ii) $\lim\limits_{x \to a^+} f(x)$ exists
   (iii) $\lim\limits_{x \to a^+} f(x) = f(a)$

4.4.3 Definition The function f is said to be *continuous from the left at the number* a if and only if the following three conditions are satisfied:

   (i) $f(a)$ exists
   (ii) $\lim\limits_{x \to a^-} f(x)$ exists
   (iii) $\lim\limits_{x \to a^-} f(x) = f(a)$

4.4.4 Definition A function whose domain includes the closed interval $[a, b]$ is said to be continuous on $[a, b]$ if and only if it is continuous on the open interval (a, b), as well as continuous from the right at a and continuous from the left at b.

4.4.5 Definition

   (i) A function whose domain includes the interval half-open on the right $[a, b)$ is said to be continuous on $[a, b)$ if and only if it is continuous on the open interval (a, b) and continuous from the right at a.
   (ii) A function whose domain includes the interval half-open on the left $(a, b]$ is said to be continuous on $(a, b]$ if and only if it is continuous on the open interval (a, b) and continuous from the left at b.

Note that by Definition 4.4.5 we may conclude that a function f is continuous on $[a, b)$ even though f is discontinuous at a. To say that f is continuous at a means that f is continuous from both the left and the right at a. But f may be continuous on $[a, b)$ even though f is discontinuous from the left at a. For example, the function f defined by

$$f(x) = \begin{cases} x + 1 & \text{if } x < 0 \\ \dfrac{x + 1}{x - 1} & \text{if } 0 \leqslant x < 1 \end{cases}$$

is continuous on $[0, 1)$ because f is continuous at every number in $(0, 1)$ and f is continuous from the right at 0. However, f is discontinuous at 0 because $\lim\limits_{x \to 0^-} f(x) \neq \lim\limits_{x \to 0^+} f(x)$.

Similarly, f may be continuous on $(a, b]$ even though f is discontinuous at b, and f may be continuous on $[a, b]$ even though f is discontinuous at either a or b. Note, however, that if f is continuous on any interval I, then $f(x)$ must be defined at every number in the interval I.

Exercises 4.4

In Exercises 1-16, determine whether the function is continuous on each of the indicated intervals.

2. $f(r) = \dfrac{r+3}{r^2 - 4}$; $(0,4], (-2,2), (-\infty, -2], (2, +\infty), [-4,4], (-2,2)$.

SOLUTION: Because $r^2 - 4 = 0$ if $r = \pm 2$, f is not defined at $r = \pm 2$, and hence f is discontinuous on any interval that contains either 2 or -2. However, f is continuous at every number except 2 and -2. Therefore, f is continuous on $(-2,2)$ and $(2, +\infty)$, but f is discontinuous on $(0,4], (-\infty, -2], [-4,4]$ and $(-2,2)$.

6.
$$h(x) = \begin{cases} 2x - 3 & \text{if } x < -2 \\ x - 5 & \text{if } -2 \leq x \leq 1 \\ 3 - x & \text{if } 1 < x \end{cases}$$

$(-\infty, 1), (-2, +\infty), (-2, 1), [-2, 1), [-2, 1]$.

SOLUTION: The function h is continuous at all real numbers, except possibly at -2 and 1. Because $h(-2) = -7$ and

$$\lim_{x \to -2^-} h(x) = \lim_{x \to -2^-} (2x - 3) = -7$$

$$\lim_{x \to -2^+} h(x) = \lim_{x \to -2^+} (x - 5) = -7$$

then h is continuous at -2. Because $h(1) = -4$ and

$$\lim_{x \to 1^-} h(x) = \lim_{x \to 1^-} (x - 5) = -4$$

then h is continuous from the left at 1. Since

$$\lim_{x \to 1^+} h(x) = \lim_{x \to 1^+} (3 - x) = 2 \neq h(1)$$

then h is discontinuous from the right at 1. Therefore h is continuous on $(-\infty, 1)$, $(-2, 1), [-2, 1)$, and $[-2, 1]$. However, h is discontinuous on $(-2, +\infty)$.

10. $f(x) = \sqrt{3 + 2x - x^2}$; $(-1, 3), [-1, 3], [-1, 3), (-1, 3]$.

SOLUTION: The domain of f is $\{x \mid 3 + 2x - x^2 \geq 0\}$. We solve the inequality.

$$3 + 2x - x^2 \geq 0$$
$$x^2 - 2x - 3 \leq 0$$
$$(x + 1)(x - 3) \leq 0 \tag{1}$$

As Table 10 indicates, the factor $x + 1$ changes sign at $x = -1$; the factor $x - 3$ changes sign at $x = 3$; and the product $(x + 1)(x - 3)$ is either negative or zero when $-1 \leq x \leq 3$. Thus the domain of f is $[-1, 3]$. The function f is continuous at every number in $(-1, 3)$, but f is discontinuous at both -1 and 3 because the two-sided limits do not exist at these numbers.

Table 10

	$x < -1$	$x = -1$	$-1 < x < 3$	$x = 3$	$x > 3$
$x + 1$	$-$	0	$+$	$+$	$+$
$x - 3$	$-$	$-$	$-$	0	$+$
$(x+1)(x-3)$	$+$	0	$-$	0	$+$

However, $f(-1) = 0$ and

$$\lim_{x \to -1^+} f(x) = \lim_{x \to -1^+} \sqrt{3 + 2x - x^2} = 0 = f(-1)$$

Thus, f is continuous from the right at -1. Also $f(3) = 0$ and

$$\lim_{x \to 3^-} f(x) = \lim_{x \to 3^-} \sqrt{3 + 2x - x^2} = 0 = f(3)$$

and so f is continuous from the left at 3. Therefore, f is continuous on $(-1, 3)$, $[-1, 3], [-1, 3),$ and $(-1, 3]$.

12. $g(x) = \sqrt{\dfrac{2 + x}{25 - x^2}}; (-\infty, -5), (-\infty, -5], [-5, -2], [-2, 5], [-2, 5), (-2, 5],$

$(-2, 5), [5, +\infty), (5, +\infty).$

SOLUTION: The domain of g is the set of all replacements for x that give the expression under the radical sign a nonnegative value.

$$\frac{2 + x}{25 - x^2} \geqslant 0$$

$$\frac{x + 2}{x^2 - 25} \leqslant 0$$

$$\frac{x + 2}{(x + 5)(x - 5)} \leqslant 0 \qquad (1)$$

Table 12 indicates that the factor $x + 2$ changes sign at -2; the factor $x + 5$ changes sign at -5; and the factor $x - 5$ changes sign at 5. Because the fraction in (1) has a negative value for $x < -5$ or $-2 < x < 5$ and has value 0 when $x = -2$, the domain of g is $(-\infty, -5) \cup [-2, 5)$. The function g is continuous at every member of $(-\infty, -5)$ and at every member of $(-2, 5)$, and g is continuous from the right at -2, because

$$\lim_{x \to -2^+} \sqrt{\frac{2 + x}{25 - x^2}} = 0 = g(-2)$$

Because g is continuous on an interval I only if I is a subset of the domain of g, we conclude that g is continuous on $(-\infty, -5), [-2, 5), (-2, 5),$ and g is discontinuous on $(-\infty, -5], [-5, -2], [-2, 5], (-2, 5], [5, +\infty)$ and $(5, +\infty)$.

Table 12

	$x < -5$	$x = -5$	$-5 < x < -2$	$x = -2$	$-2 < x < 5$	$x = 5$	$x > 5$
$x + 2$	$-$	$-$	$-$	0	$+$	$+$	$+$
$x + 5$	$-$	0	$+$	$+$	$+$	$+$	$+$
$x - 5$	$-$	$-$	$-$	$-$	$-$	0	$+$
$\dfrac{x + 2}{(x + 5)(x - 5)}$	$-$	does not exist	$+$	0	$-$	does not exist	$+$

16. Determine the intervals on which the given function is continuous.

$$F(x) = \sqrt{\frac{x - 5}{x + 6}}$$

SOLUTION: The domain of F is the set of all replacements for x that give the expression under the radical sign a nonnegative value.

$$\frac{x-5}{x+6} \geq 0 \tag{1}$$

Table 16 indicates that the factor $x-5$ changes sign at 5; the factor $x+6$ changes sign at -6; and the fraction in (1) is positive for $x < -6$ or $x > 5$. Because the fraction has value 0 if $x = 5$, we conclude that the domain of F is $(-\infty, -6) \cup [5, +\infty)$. The function F is continuous at every number in $(-\infty, -6) \cup (5, +\infty)$, and furthermore F is continuous from the right at 5, because

$$\lim_{x \to 5^+} \sqrt{\frac{x-5}{x+6}} = 0 = f(5)$$

Therefore, F is continuous on $(-\infty, -6)$ and $[5, +\infty)$.

Table 16

	$x < -6$	$x = -6$	$-6 < x < 5$	$x = 5$	$x > 5$
$x - 5$	$-$	$-$	$-$	0	$+$
$x + 6$	$-$	0	$+$	$+$	$+$
$\dfrac{x-5}{x+6}$	$+$	does not exist	$-$	0	$+$

18. Determine the intervals on which the function of Exercise 26 in Exercises 1.8 is continuous.

SOLUTION: The function is defined by

$$f(x) = \text{sgn } x^2 - \text{sgn } x$$

In Section 1.8 we showed that

$$f(x) = \begin{cases} 2 & \text{if } x < 0 \\ 0 & \text{if } x \geq 0 \end{cases}$$

A sketch of the graph of f is shown in Fig. 1.8.26. The function f is continuous at all x except 0. Because

$$\lim_{x \to 0^+} f(x) = \lim_{x \to 0^+} 0 = 0 = f(0)$$

then f is continuous from the right at 0. Because

$$\lim_{x \to 0^-} f(x) = \lim_{x \to 0^-} 2 = 2 \neq f(0)$$

then f is discontinuous from the left at 0. Therefore, f is continuous on $(-\infty, 0)$ and $[0, +\infty)$.

24. Define $f \circ g$, and determine all values of x for which $f \circ g$ is continuous.

$$f(x) = \sqrt{x} \qquad g(x) = x + 1$$

SOLUTION: $f \circ g$ is defined by

$$(f \circ g)(x) = \sqrt{x + 1}$$

Because $x + 1 > 0$ if $x > -1$, then $f \circ g$ is continuous at all x in $(-1, +\infty)$. Note that because

$$\lim_{x \to -1^+} \sqrt{x + 1} = 0 = (f \circ g)(-1)$$

$f \circ g$ is also continuous from the right at -1. Thus $f \circ g$ is continuous on $[-1, +\infty)$. However, since $f \circ g$ is not defined for $x < -1$, $f \circ g$ is discontinuous at -1.

32. Find the values of the constants c and k that make the function continuous on $(-\infty, +\infty)$ and draw a sketch of the graph of the resulting function.

$$f(x) = \begin{cases} x + 2c & \text{if } x < -2 \\ 3cx + k & \text{if } -2 \leqslant x \leqslant 1 \\ 3x - 2k & \text{if } 1 < x \end{cases} \qquad (1)$$

SOLUTION: For all values of c and k the function f is continuous at all x, except possibly at $x = -2$ and $x = 1$. If f is continuous at -2, then

$$\lim_{x \to -2^-} f(x) = \lim_{x \to -2^+} f(x)$$

or, equivalently,

$$\lim_{x \to -2^-} (x + 2c) = \lim_{x \to -2^+} (3cx + k)$$

$$-2 + 2c = -6c + k \qquad (2)$$

Furthermore, if f is continuous at 1, then

$$\lim_{x \to 1^-} f(x) = \lim_{x \to 1^+} f(x)$$

or, equivalently,

$$\lim_{x \to 1^-} (3cx + k) = \lim_{x \to 1^+} (3x - 2k)$$

$$3c + k = 3 - 2k \qquad (3)$$

Solving Eq. (2) and Eq. (3) simultaneously, we get

$$c = \frac{1}{3} \quad \text{and} \quad k = \frac{2}{3}$$

Substituting the values for c and k into Eq. (1), we have

$$f(x) = \begin{cases} x + \dfrac{2}{3} & \text{if } x < -2 \\[2mm] x + \dfrac{2}{3} & \text{if } -2 \leqslant x \leqslant 1 \\[2mm] 3x - \dfrac{4}{3} & \text{if } 1 < x \end{cases}$$

or, equivalently,

$$f(x) = \begin{cases} x + \dfrac{2}{3} & \text{if } x \leqslant 1 \\[2mm] 3x - \dfrac{4}{3} & \text{if } x > 1 \end{cases}$$

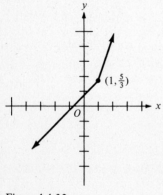

Figure 4.4.32

Now, $f(1) = \frac{5}{3}$. Furthermore,

$$\lim_{x \to 1^-} f(x) = \lim_{x \to 1^-} \left(x + \frac{2}{3} \right) = \frac{5}{3}$$

and

$$\lim_{x \to 1^+} f(x) = \lim_{x \to 1^+} \left(3x - \frac{4}{3} \right) = \frac{5}{3}$$

Thus, $\lim\limits_{x \to 1} f(x) = \frac{5}{3} = f(1)$. Therefore, f is continuous at 1. Hence, f is continuous on $(-\infty, +\infty)$. A sketch of the graph of f is shown in Fig. 4.4.32.

4.5 MAXIMUM AND MINIMUM VALUES OF A FUNCTION

4.5.1 Definition The function f is said to have a *relative maximum value* at c if there exists an open interval containing c, on which f is defined, such that $f(c) \geqslant f(x)$ for all x in this interval.

4.5.2 Definition The function f is said to have a *relative minimum value* at c if there exists an open interval containing c, on which f is defined, such that $f(c) \leqslant f(x)$ for all x in this interval.

If the function f has either a relative maximum or a relative minimum value at c, then f is said to have a *relative extremum* at c.

4.5.3 Theorem If $f(x)$ exists for all x in the open interval (a, b) and if f has a relative extremum at c, where $a < c < b$, then if $f'(c)$ exists, $f'(c) = 0$.

4.5.4 Definition If c is a number in the domain of the function f and if either $f'(c) = 0$ or $f'(c)$ does not exist, then c is called a *critical number* of f.

We conclude that if f has a relative extremum at the number c, then c must be a critical number of f. Therefore, if f does not have a critical number in some open interval I, then f does not have either a relative maximum value or a relative minimum value in the interval I. However, the converse does not hold. It is possible for c to be a critical number of f, and yet f may not have a relative extremum at c.

4.5.5 Definition The function f is said to have an *absolute maximum value on an interval* if there is some number c in the interval such that $f(c) \geqslant f(x)$ for all x in the interval. In such a case, $f(c)$ is the absolute maximum value of f on the interval.

4.5.6 Definition The function f is said to have an *absolute minimum value on an interval* if there is some number c in the interval such that $f(c) \leqslant f(x)$ for all x in the interval. In such a case, $f(c)$ is the absolute minimum value of f on the interval.

An absolute extremum of a function on an interval is either an absolute maximum value or an absolute minimum value of the function on the interval. If a function f has an absolute extremum on an interval, then the absolute extremum must occur either at a critical number of f or at an endpoint of the interval. If the interval is either an open interval or a half-open interval, then f may not have an absolute extremum on the interval. Furthermore, if f is discontinuous on the interval, then f may not have an absolute extremum on the interval. In particular, if the graph of f has a vertical asymptote $x = a$ where a is an interval I, then either f must fail to have an absolute maximum value or f must fail to have an absolute minimum value on I.

4.5.9 Theorem (Extreme-value theorem) If the function f is continuous on the closed interval $[a, b]$, then f has an absolute maximum value and an absolute minimum value on $[a, b]$.

The following steps can be used to find the absolute extrema of f on $[a, b]$ if f is continuous on $[a, b]$.

1. Find each critical number c of f in $[a, b]$.
2. Find the function value $f(c)$ for each critical number of step 1.
3. Find the function values $f(a)$ and $f(b)$.
4. The largest of the values from steps 2 and 3 is the absolute maximum value, and the smallest of the values from steps 2 and 3 is the absolute minimum value of f on the closed interval $[a, b]$.

Exercises 4.5

In Exercises 1-10, find the critical numbers of the given function.

4. $f(x) = x^{7/3} + x^{4/3} - 3x^{1/3}$

SOLUTION: The domain of f is $(-\infty, +\infty)$.

$$f'(x) = \frac{7}{3}x^{4/3} + \frac{4}{3}x^{1/3} - x^{-2/3}$$

$$= \frac{1}{3}x^{-2/3}(7x^2 + 4x - 3)$$

$$= \frac{(7x - 3)(x + 1)}{3x^{2/3}}$$

Because $f'(0)$ does not exist and 0 is in the domain of f, by Definition 4.5.4, 0 is a critical number of f. If $f'(x) = 0$, then

$$0 = (7x - 3)(x + 1)$$

$$x = \frac{3}{7} \quad \text{or} \quad x = -1$$

Because $f'(\frac{3}{7}) = 0$, $f'(-1) = 0$, and because $\frac{3}{7}$ and -1 are in the domain of f, by Definition 4.5.4, both $\frac{3}{7}$ and -1 are critical numbers of f.

8. $f(x) = (x^3 - 3x^2 + 4)^{1/3}$

SOLUTION: The domain of f is $(-\infty, +\infty)$.

$$f'(x) = \frac{1}{3}(x^3 - 3x^2 + 4)^{-2/3}(3x^2 - 6x)$$

$$= \frac{x(x - 2)}{x^3 - 3x^2 + 4}$$

$$= \frac{x(x - 2)}{(x + 1)(x - 2)^2}$$

$$= \frac{x}{(x + 1)(x - 2)}$$

The factored form of the denominator $x^3 - 3x^2 + 4$ was found by trial and error using synthetic division. Because $f'(x)$ is not defined at -1 and at 2, both -1 and 2 are critical numbers of f. Because $f'(0) = 0$, then 0 is also a critical number of f.

10. $f(x) = \dfrac{x + 1}{x^2 - 5x + 4}$

SOLUTION: Because the denominator of $f(x)$ is equivalent to $(x - 4)(x - 1)$, then neither $f(4)$ nor $f(1)$ is defined. The domain of f is $\{x \mid x \neq 4, x \neq 1\}$

$$f'(x) = \frac{(x^2 - 5x + 4) \cdot 1 - (x + 1)(2x - 5)}{(x^2 - 5x + 4)^2}$$

$$= \frac{-x^2 - 2x + 9}{(x^2 - 5x + 4)^2}$$

$f'(x) = 0$ if and only if

$$-x^2 - 2x + 9 = 0$$
$$x^2 + 2x + 1 = 10$$

$$(x + 1)^2 = 10$$
$$x + 1 = \pm\sqrt{10}$$
$$x = -1 \pm \sqrt{10}$$

Therefore $-1 + \sqrt{10}$ and $-1 - \sqrt{10}$ are critical numbers of f. We note that $f'(x)$ is undefined when $x^2 - 5x + 4 = 0$. However, the only solutions of this equation are 1 and 4, numbers which are not in the domain of f. Therefore, neither 1 nor 4 is a critical number of f.

In Exercises 11-24, find the absolute extrema of the given function on the given interval, if there are any, and find the values of x at which the absolute extrema occur. Draw a sketch of the graph of the function on the interval.

14. $f(x) = \dfrac{1}{x}$; $[2, 3)$

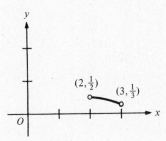

Figure 4.5.14

SOLUTION: Because

$$f'(x) = -\frac{1}{x^2}$$

$f'(x)$ exists for all x in the interval $[2, 3)$ and $f'(x) \neq 0$ for all x in $[2, 3)$. Hence, there are no critical numbers of f in $[2, 3)$. And thus, there are no relative extrema of f on $[2, 3)$. A sketch of the graph of f on $[2, 3)$ is shown in Fig. 4.5.14. The absolute maximum value of f is $\frac{1}{2}$, and this value occurs at $x = 2$, an end point of the graph. There is no absolute minimum value of f on $[2, 3)$.

18. $f(x) = \sqrt{4 - x^2}$; $(-2, 2)$

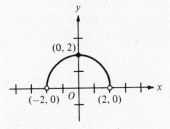

Figure 4.5.18

SOLUTION:

$$f'(x) = \frac{-x}{\sqrt{4 - x^2}}$$

The only critical number of f in the interval $(-2, 2)$ is 0. A sketch of the graph of f in the interval $(-2, 2)$ is shown in Fig. 4.5.18. The absolute maximum value of f on $(-2, 2)$ is 2, and this occurs at $x = 0$. There is no absolute minimum value of f on $(-2, 2)$.

22. $f(x) = \begin{cases} |x + 1| & \text{if } x \neq -1 \\ 3 & \text{if } x = -1 \end{cases}$; $[-2, 1]$

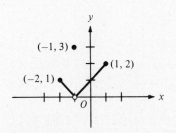

Figure 4.5.22

SOLUTION: If $x \neq -1$, then

$$f(x) = \sqrt{(x + 1)^2} = [(x + 1)^2]^{1/2}$$

$$f'(x) = \frac{1}{2}[(x + 1)^2]^{-1/2} \cdot 2(x + 1) = \frac{x + 1}{|x + 1|}$$

Therefore $f'(x)$ is not defined at $x = -1$, and thus -1 is a critical number of f. A sketch of the graph of f on the interval $[-2, 1]$ is shown in Fig. 4.5.22. The absolute maximum value of f on the interval is 3, and this maximum value occurs at $x = -1$. There is no absolute minimum value of f on the interval.

In Exercises 25-36, find the absolute maximum value and the absolute minimum value of the given function on the indicated interval by the method used in Examples 2 and 3 of this section (in the text). Draw a sketch of the graph of the function on the interval.

28. $f(x) = x^4 - 8x^2 + 16$; $[-1, 4]$

SOLUTION:

$$f(x) = (x^2 - 4)^2$$

Thus,

$$f'(x) = 2(x^2 - 4) \cdot 2x = 4x(x + 2)(x - 2)$$

The only critical numbers of f in the interval $[-1, 4]$ are 0 and 2. We find the value of $f(x)$ at each critical number and at each endpoint of $[-1, 4]$, and show the results in Table 28. From the table we see that the absolute maximum value of f on the interval $[-1, 4]$ is 144, and the absolute minimum value of f on the interval is 0. A sketch of the graph of f on the interval $[-1, 4]$ is shown in Fig. 4.5.28.

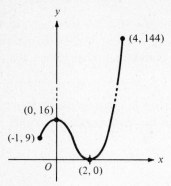

Figure 4.5.28

Table 28

x	-1	0	2	4
$f(x)$	9	16	0	144

32. $f(x) = \dfrac{x + 5}{x - 3}$; $[-5, 2]$

SOLUTION:

$$f'(x) = \frac{(x - 3) \cdot 1 - (x + 5) \cdot 1}{(x - 3)^2} = \frac{-8}{(x - 3)^2}$$

Because $f'(x)$ is defined and $f'(x) \neq 0$ for all x in the interval $[-5, 2]$, there are no critical numbers of f in this interval. The absolute extrema must occur at the endpoints of the interval. Because $f(-5) = 0$ and $f(2) = -7$, we conclude that 0 is the absolute maximum value of f on the interval $[-5, 2]$, and -7 is the absolute minimum value of f on the interval $[-5, 2]$. A sketch of the graph of f is shown in Fig. 4.5.32.

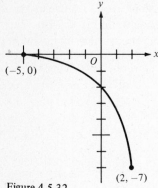

Figure 4.5.32

36. $f(x) = \begin{cases} 4 - (x + 5)^2 & \text{if } -6 \leqslant x \leqslant -4 \\ 12 - (x + 1)^2 & \text{if } -4 < x \leqslant 0 \end{cases}$; $[-6, 0]$

SOLUTION:

$$f'(x) = \begin{cases} -2(x + 5) & \text{if } -6 < x < -4 \\ -2(x + 1) & \text{if } -4 < x < 0 \end{cases}$$

Because $f'(-5) = 0$ and $f'(-1) = 0$, then -5 and -1 are critical numbers of f. Furthermore,

$$f'_-(-4) = \lim_{x \to -4^-} f'(x) = \lim_{x \to -4^-} -2(x + 5) = -2$$

and

$$f'_+(-4) = \lim_{x \to -4^+} f'(x) = \lim_{x \to -4^+} -2(x + 1) = 6$$

Therefore, $f'(-4)$ is not defined, and so -4 is a critical number of f. Table 36 shows the function value at each endpoint of $[-6, 0]$ and at each critical number of f. The absolute maximum value of f on the interval is 12, and the absolute minimum value of f on the interval is 3. A sketch of the graph of f on the interval $[-6, 0]$ is shown in Fig. 4.5.36.

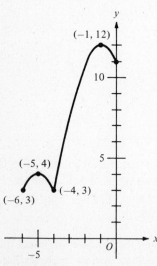

Figure 4.5.36

Table 36

x	−6	−5	−4	−1	0
$f(x)$	3	4	3	12	11

4.6 APPLICATIONS INVOLVING AN ABSOLUTE EXTREMUM ON A CLOSED INTERVAL

The procedure for solving the Exercises in this section is as follows.

1. Identify the variable for which you want to find an absolute extreme value. This is the dependent variable.
2. Express the dependent variable as a function of some other variable.
3. Verify that the function found in step 2 has a closed interval for its domain and that the function is continuous on that closed interval.
4. Follow the steps given in Section 4.5 for finding the absolute extrema of a function on a closed interval.

Often it is helpful to first express the dependent variable as a function of two other variables. In that case, you must find another equation involving those two variables and then use the equation to eliminate one of the variables. This is illustrated in Exercises 4, 10, and 12. Sometimes a figure is helpful, and sometimes the formulas from geometry given in Section 3.10 are helpful.

Exercises 4.6

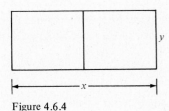

Figure 4.6.4

4. A rectangular plot of ground is to be enclosed by a fence and then divided down the middle by another fence. If the fence down the middle costs $1 per running foot and the other fence costs $2.50 per running foot, find the dimensions of the plot of largest possible area that can be enclosed with $480 worth of fence.

SOLUTION: See Fig. 4.6.4. Let

x = the number of feet in the length of the plot
y = the number of feet in the width of the plot
A = the number of square feet in the area of the plot

We want to determine x and y so that A has an absolute maximum value. We have

$$A = xy \tag{1}$$

To eliminate one of the variables which appear on the right side of Eq. (1), we find another equation in x and y. Because the fence down the middle costs $1 per running foot, then $1 \cdot y$ is the number of dollars in the cost of this fence. Because the other fence costs $2.50 per running foot, then $2.5(2x + 2y)$ is the number of dollars in the cost of this fence. Because the total cost of all the fence if $480, then

$$y + 2.5(2x + 2y) = 480$$

$$y = \frac{-5(x - 96)}{6} \tag{2}$$

Substituting the value of y from Eq. (2) into Eq. (1), we express A as a function of x as follows.

$$A(x) = \frac{-5x(x - 96)}{6} \tag{3}$$

From Eq. (3) we get $A(0) = 0$, $A(96) = 0$, and $A \geq 0$ if $0 \leq x \leq 96$. Therefore, A is defined on the closed interval $[0, 96]$, and because A is continuous on $[0, 96]$, then A must have an absolute maximum value on $[0, 96]$. From Eq. (3) we have

$$A(x) = \frac{-5x^2 + 480x}{6}$$

$$A'(x) = \frac{-10x + 480}{6}$$

If $A'(x) = 0$, then $x = 48$.

The absolute maximum value of A on $[0, 96]$ must occur at either $0, 48$, or 96. Because $A(0) = 0$ and $A(96) = 0$, while $A(48) = 1920$, we conclude that the absolute maximum value of A on $[0, 96]$ is 1920, occurring at 48. From Eq. (2), if $x = 48$, then $y = 40$. Therefore, the plot with the largest possible area is 48 ft. by 40 ft., and its area is 1920 sq. ft.

10. Find the dimensions of the right circular cylinder of greatest volume that can be inscribed in a sphere with a radius of 6 in.

SOLUTION: See Fig. 4.6.10. Let

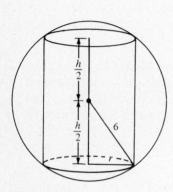

Figure 4.6.10

$r =$ the number of inches in the radius of the cylinder
$h =$ the number of inches in the altitude of the cylinder
$V =$ the number of cubic inches in the volume of the cylinder

We want to determine r and h so that V has an absolute maximum value. From geometry we have the formula

$$V = \pi r^2 h \tag{1}$$

To eliminate one of the two variables on the right side of Eq. (1), we find another equation involving r and h. Because the center of the sphere bisects the altitude of the cylinder and the altitude of the cylinder is perpendicular to the base, we have from the Pythagorean theorem

$$\left(\frac{1}{2}h\right)^2 + r^2 = 6^2$$

$$r^2 = 36 - \frac{1}{4}h^2 \tag{2}$$

Substituting the value for r^2 from Eq. (2) into Eq. (1), we express V as a function of h as follows.

$$V(h) = \pi h\left(36 - \frac{1}{4}h^2\right) \tag{3}$$

From Eq. (3) we have $V(0) = 0$, $V(12) = 0$, and $V \geqslant 0$ if $0 \leqslant h \leqslant 12$. Therefore, V is defined on the closed interval $[0, 12]$, and since V is continuous on $[0, 12]$, V must have an absolute maximum value on $[0, 12]$. From Eq. (3) we have

$$V(h) = \frac{\pi}{4}(144h - h^3)$$

$$V'(h) = \frac{\pi}{4}(144 - 3h^2)$$

If $V'(h) = 0$, then

$$144 - 3h^2 = 0$$
$$h = \pm 4\sqrt{3}$$

The only critical number in $[0, 12]$ is $4\sqrt{3}$. From Eq. (3) we get

$$V(4\sqrt{3}) = 4\sqrt{3}\,\pi \cdot 24 = 96\pi\sqrt{3}$$

Also

$$V(0) = 0 \quad \text{and} \quad V(12) = 0$$

Therefore, the greatest possible volume is $96\pi\sqrt{3}$ cu. in. Furthermore, from Eq. (2), if $h = \sqrt{48} = 4\sqrt{3}$, then $r = \sqrt{24} = 2\sqrt{6}$. Hence, for the cylinder of greatest volume the altitude is $4\sqrt{3}$ in., and the radius is $2\sqrt{6}$ in.

12.   A manufacturer can make a profit of $20 on each item if not more than 800 items are produced each week. The profit decreases 2 cents per item over 800. How many items should the manufacturer produce each week in order to have the greatest profit?

SOLUTION: Let

x = the number of items produced each week
y = the number of dollars in the profit per item if x items are produced each week
P = the number of dollars in the total weekly profit if x items are produced each week.

We want to determine x so that P is an absolute maximum. We have

$$P = x \cdot y \tag{1}$$

To eliminate y, we must express y as a function of x. We are given that

$$y = 20 \quad \text{if } x \leqslant 800 \tag{2}$$

If $x > 800$ then $x - 800$ is the number of items in excess of 800. Thus, we are also given that

$$y = 20 - 0.02(x - 800) \quad \text{if } x > 800$$

or, equivalently,

$$y = 36 - 0.02x \quad \text{if } x > 800 \tag{3}$$

Substituting from Eq. (2) and Eq. (3) into Eq. (1), we express P as a function of x as follows.

$$P(x) = \begin{cases} 20x & \text{if } x \leqslant 800 \\ 36x - 0.02x^2 & \text{if } x > 800 \end{cases} \tag{4}$$

Because $36x - 0.02x^2 = 0$ when $x = 1800$, from Eq. (4) we have $P(0) = 0$ and $P(1800) = 0$. Because $P \geqslant 0$ for $0 \leqslant x \leqslant 1800$, P is defined on the closed interval $[0, 1800]$. Furthermore, because

$$\lim_{x \to 800^-} P(x) = \lim_{x \to 800^-} 20x = 16,000$$

$$\lim_{x \to 800^+} P(x) = \lim_{x \to 800^+} (36x - .02x^2) = 16,000$$

Then $\lim_{x \to 800} P(x) = 16,000 = P(800)$. Hence, P is continuous on $[0, 1800]$, and therefore P has an absolute maximum value on $[0, 1800]$. From (4) we obtain

$$P'(x) = \begin{cases} 20 & \text{if } x < 800 \\ 36 - 0.04x & \text{if } x > 800 \end{cases}$$

Because $P'_-(800) = 20$ and $P'_+(800) = 36 - 0.04(800) = 4$, and $20 \neq 4$, then P is not differentiable at 800. Therefore, 800 is a critical number of P. Furthermore, because $36 - 0.04x = 0$ if $x = 900$, then $P'(900) = 0$, and 900 is also a critical number of P. From (4) we have $P(800) = 16,000$, $P(900) = 16,200$, $P(0) = 0$, and $P(1800) = 0$. Thus, the greatest weekly profit is $16,200, and 900 items per week should be produced in order to make this profit.

16. Suppose a weight is to be held 10 ft. below a horizontal line AB by a wire in the shape of a Y. If the points A and B are 8 ft. apart, what is the shortest total length of wire that can be used?

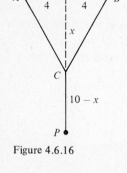

Figure 4.6.16

SOLUTION: See Fig. 4.6.16. The weight is at point P and the line PD is perpendicular to AB. We are given that $|\overline{AB}| = 8$ and $|\overline{PD}| = 10$. Because triangle ABC is isosceles, then $|\overline{AD}| = |\overline{DB}| = 4$. Let

x = the number of feet in $|\overline{CD}|$
z = the number of feet in the total length of the wire

We want to find the absolute minimum value of z. We first express z as a function of x. We note that

$$z = |\overline{AC}| + |\overline{BC}| + |\overline{PC}| \tag{1}$$

Because triangles ADC and BDC are right triangles, we have $|\overline{AC}| = \sqrt{x^2 + 16}$ and $|\overline{BC}| = \sqrt{x^2 + 16}$. And because $|\overline{PC}| = 10 - x$, from Eq. (1) we have

$$z = 2\sqrt{x^2 + 16} + 10 - x \tag{2}$$

We note that z is defined on $[0, 10]$ because $|\overline{PC}| \geqslant 0$, and that z is continuous on $[0, 10]$. Thus, z must have an absolute minimum value on $[0, 10]$. From Eq. (2) we have

$$D_x z = \frac{2x}{\sqrt{x^2 + 16}} - 1$$

If $D_x z = 0$, then

$$\frac{2x}{\sqrt{x^2 + 16}} = 1$$

$$4x^2 = x^2 + 16$$

$$x = \frac{4}{3}\sqrt{3}$$

Therefore, $\frac{4}{3}\sqrt{3}$ is a critical number of z. We test the critical number and each endpoint of the interval $[0, 10]$. From Eq. (2), if $x = 0$, then $z = 18$; if $x = \frac{4}{3}\sqrt{3}$, then $z = 10 + 4\sqrt{3} \approx 16.9$; if $x = 10$, then $z = 4\sqrt{29} \approx 21.5$. Therefore, the shortest total length of wire that can be used is $(10 + 4\sqrt{3})$ ft.

4.7 ROLLE'S THEOREM AND THE MEAN-VALUE THEOREM

4.7.1 Theorem

(*Rolle's theorem*) Let f be a function such that

(i) it is continuous on the closed interval $[a, b]$
(ii) it is differentiable on the open interval (a, b)
(iii) $f(a) = f(b) = 0$

Then there is a number c in the open interval (a, b) such that $f'(c) = 0$.

4.7.2 Theorem

(*Mean-value theorem*) Let f be a function such that

(i) it is continuous on the closed interval $[a, b]$
(ii) it is differentiable on the open interval (a, b)

Then there is a number c in the open interval (a, b) such that

$$f'(c) = \frac{f(b) - f(a)}{b - a}$$

The conclusion of Rolle's theorem states that there is a point on the graph of f

between the points where $x = a$ and $x = b$ at which the line tangent to the curve is a horizontal line. The conclusion of the mean-value theorem states that there is a point on the graph of f between the points where $x = a$ and $x = b$ at which the line tangent to the curve is parallel to the line that contains the points on the curve where $x = a$ and $x = b$. Note that Rolle's theorem is actually a special case of the mean-value theorem. The theorems in this section are not often used for problem solving. However, they are among the most important theorems of the calculus because they are used in the proof of many later theorems.

Exercises 4.7

4. Verify that conditions (i), (ii), and (iii) of the hypothesis of Rolle's theorem are satisfied by the given function on the indicated interval. Then find a suitable value for c that satisfies the conclusion of Rolle's theorem.

$$f(x) = x^3 - 16x; \ [-4, 0]$$

SOLUTION: Because f is continuous and differentiable for all real x, conditions (i) and (ii) are satisfied. Because $f(-4) = 0$ and $f(0) = 0$, condition (iii) is satisfied.

$$f'(x) = 3x^2 - 16$$

To find c, we let $f'(c) = 0$. Thus,

$$3c^2 - 16 = 0$$

$$c = \pm \frac{4}{3}\sqrt{3}$$

Because only $-\frac{4}{3}\sqrt{3}$ is in the open interval $(-4, 0)$, the only suitable choice for c is $-\frac{4}{3}\sqrt{3}$.

8. Verify that the hypothesis of the mean-value theorem is satisfied for the given function on the indicated interval. Then find a suitable choice for c that satisfies the conclusion of the mean-value theorem.

$$f(x) = x - 1 + \frac{1}{x - 1}; \ \left[\frac{3}{2}, 3\right] \tag{1}$$

SOLUTION: Because f is continuous at every number except 1, then f is continuous on $[\frac{3}{2}, 3]$, and thus condition (i) of the hypothesis of the mean-value theorem is satisfied.

$$f'(x) = 1 - \frac{1}{(x - 1)^2} \tag{2}$$

Because f' is defined for all $x \neq 1$, then f is differentiable on $(\frac{3}{2}, 3)$, and thus condition (ii) of the hypothesis of the mean-value theorem is satisfied. To find c we solve the equation

$$f'(c) = \frac{f(b) - f(a)}{b - a}$$

which, by Eq. (2), is equivalent to

$$1 - \frac{1}{(c - 1)^2} = \frac{f(b) - f(a)}{b - a} \tag{3}$$

Because $a = \frac{3}{2}$ and $b = 3$, then

$$\frac{f(b) - f(a)}{b - a} = \frac{f(3) - f\left(\frac{3}{2}\right)}{3 - \frac{3}{2}}$$

$$= \frac{\frac{5}{2} - \frac{5}{2}}{\frac{3}{2}}$$

$$= 0 \qquad\qquad\qquad (4)$$

Substituting from (4) into (3), we get

$$1 - \frac{1}{(c - 1)^2} = 0$$

$$c = 2 \quad\text{or}\quad c = 0$$

Because 0 is not in the interval $(\frac{3}{2}, 3)$, the only suitable choice for c is 2.

In Exercises 14-23, (a) draw a sketch of the graph of the given function on the indicated interval; (b) test the three conditions (i), (ii), and (iii) of the hypothesis of Rolle's theorem and determine which conditions are satisfied and which, if any, are not satisfied; and (c) if the three conditions in part (b) are satisfied, determine a point at which there is a horizontal tangent line.

14. $f(x) = x^{3/4} - 2x^{1/4}$; $[0, 4]$

SOLUTION:

(a) A sketch of the graph of f on $[0, 4]$ is shown in Fig. 4.7.14.

(b) Because f is continuous at every positive number and continuous from the right at 0, then f is continuous on $[0, 4]$. Thus condition (i) of the hypothesis of Rolle's theorem is satisfied.

$$f'(x) = \frac{3}{4}x^{-1/4} - \frac{2}{4}x^{-3/4}$$

$$= \frac{1}{4}x^{-3/4}(3x^{1/2} - 2)$$

$$= \frac{3x^{1/2} - 2}{4x^{3/4}}$$

Because $f'(x)$ exists for all x in the open interval $(0, 4)$, then condition (ii) is satisfied. Note that $f'(0)$ does not exist, but this is not required for condition (ii). Because $f(x) = x^{1/4}(x^{1/2} - 2)$, then $f(0) = 0$ and $f(4) = 4^{1/4}(4^{1/2} - 2) = 0$. Hence, condition (iii) is also satisfied.

(c) To find a point at which there is a horizontal tangent line, we set $f'(x) = 0$. Thus,

$$\frac{3x^{1/2} - 2}{4x^{3/4}} = 0$$

$$3x^{1/2} - 2 = 0$$

$$x = \frac{4}{9}$$

There is a horizontal tangent line at the point where $x = \frac{4}{9}$.

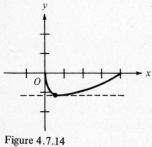

Figure 4.7.14

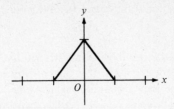

Figure 4.7.22

22. $f(x) = 1 - |x|; [-1, 1]$

SOLUTION:

(a) A sketch of the graph of f on $[-1, 1]$ is shown in Fig. 4.7.22.

(b) Because f is continuous in $[-1, 1]$, condition (i) of the hypothesis of Rolle's theorem is satisfied.

$$f(x) = 1 - \sqrt{x^2}$$

$$f'(x) = \frac{-x}{\sqrt{x^2}} = \frac{-x}{|x|}$$

Because $f'(0)$ is not defined, f' is not differentiable on $(-1, 1)$, and thus condition (ii) is not satisfied. Because $f(1) = 0$ and $f(-1) = 0$, condition (iii) is satisfied.

(c) Because one of the conditions of the hypothesis of Rolle's theorem is not satisfied, we cannot conclude that there is a point at which there is a horizontal tangent. Indeed, as Fig. 4.7.22 illustrates, there is no such point.

34. There is no number in the open interval (a, b) that satisfies the conclusion of the mean-value theorem. Determine which part of the hypothesis of the mean-value theorem fails to hold. Draw a sketch of the graph of $y = f(x)$ and the line through the points $(a, f(a))$ and $(b, f(b))$.

$$f(x) = \begin{cases} 2x + 3 & \text{if } x < 3 \\ 15 - 2x & \text{if } 3 \leqslant x \end{cases}; \ a = -1, b = 5$$

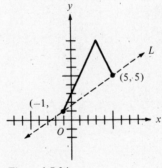

Figure 4.7.34

SOLUTION:

$$f'(x) = \begin{cases} 2 & \text{if } x < 3 \\ -2 & \text{if } x > 3 \end{cases}$$

$f'_-(3) = 2$ and $f'_+(x) = -2$. Thus $f'(3)$ does not exist. Therefore, f is not differentiable on the open interval $(-1, 5)$. Hence, condition (ii) of the hypothesis of the mean-value theorem is not satisfied. A sketch of the graph of f on $[-1, 5]$ is shown in Fig. 4.7.34. Note that there is no point on the graph of f at which the tangent line to the graph is parallel to L, the line through the end points of the interval $[-1, 5]$.

36. Prove by Rolle's theorem that the equation $x^3 + 2x + c = 0$, where c is any constant, cannot have more than one real root.

SOLUTION: Suppose that x_1 and x_2 are two distinct real roots of the given equation with $x_1 < x_2$. Let f be the function defined by $f(x) = x^3 + 2x + c$. Then

(i) f is continuous on $[x_1, x_2]$
(ii) f is differentiable on (x_1, x_2)
(iii) Because x_1 and x_2 are roots of the equation $x^3 + 2x + c = 0$, then
   $f(x_1) = 0$ and $f(x_2) = 0$.

Because the hypothesis of Rolle's theorem is satisfied, there must be a number $\bar{x}$ with $\bar{x}$ in (x_1, x_2) such that

$$f'(\bar{x}) = 0 \tag{1}$$

But

$$f'(x) = 3x^2 + 2$$

Because $f'(x) > 0$ for all real x, then $f'(\bar{x}) \neq 0$. Because this contradicts Eq. (1) (the conclusion of Rolle's theorem), we conclude that there cannot be two distinct real roots of the given equation. Therefore, this equation cannot have more than one real root.

38. Suppose $s = f(t)$ is an equation of motion of a particle moving in a straight line where f satisfies the hypothesis of the mean-value theorem. Show that the conclusion of the mean-value theorem assures us that there will be some instant during any time interval when the instantaneous velocity will equal the average velocity during that time interval.

SOLUTION: Let t_1 and t_2 be any two instants of time with $t_1 < t_2$. We are given that f satisfies the hypothesis of the mean-value theorem on any interval, and in particular on $[t_1, t_2]$. Therefore, there is some $\bar{t}$ with $t_1 < \bar{t} < t_2$ such that

$$f'(\bar{t}) = \frac{f(t_2) - f(t_1)}{t_2 - t_1} \tag{1}$$

The derivative on the left of Eq. (1) is the instantaneous velocity of the particle at the instant when $t = \bar{t}$. The fraction on the right of Eq. (1) is the average velocity of the particle during the time interval $[t_1, t_2]$. Therefore, there is some instant during any time interval when the instantaneous velocity equals the average velocity during that time interval.

Review Exercises

4. Evaluate the limit:

$$\lim_{t \to +\infty} \sqrt{t^2 + t} - \sqrt{t^2 + 4}$$

SOLUTION:

$$\lim_{t \to +\infty} (\sqrt{t^2 + t} - \sqrt{t^2 + 4}) = \lim_{t \to +\infty} \frac{\sqrt{t^2 + t} - \sqrt{t^2 + 4}}{1} \cdot \frac{\sqrt{t^2 + t} + \sqrt{t^2 + 4}}{\sqrt{t^2 + t} + \sqrt{t^2 + 4}}$$

$$= \lim_{t \to +\infty} \frac{(t^2 + t) - (t^2 + 4)}{\sqrt{t^2 + t} + \sqrt{t^2 + 4}}$$

$$= \lim_{t \to +\infty} \frac{t - 4}{\sqrt{t^2 + t} + \sqrt{t^2 + 4}} \cdot \frac{\frac{1}{t}}{\frac{1}{\sqrt{t^2}}}$$

$$= \lim_{t \to +\infty} \frac{1 - \frac{4}{t}}{\sqrt{1 + \frac{1}{t}} + \sqrt{1 + \frac{4}{t^2}}}$$

$$= \frac{1}{2}$$

6. Find the horizontal and vertical asymptotes of the graph of the function defined by the given equation and draw a sketch of the graph.

$$g(x) = \frac{-2x}{\sqrt{x^2 - 5x + 6}}$$

SOLUTION: Because $x^2 - 5x + 6 > 0$ if and only if $x > 3$ or $x < 2$, then the domain of g if $(-\infty, 2) \cup (3, +\infty)$.

$$\lim_{x \to +\infty} g(x) = \lim_{x \to +\infty} \frac{-2x}{\sqrt{x^2 - 5x + 6}} \cdot \frac{\dfrac{1}{x}}{\dfrac{1}{\sqrt{x^2}}}$$

$$\lim_{x \to +\infty} \frac{-2}{\sqrt{1 - \dfrac{5}{x} + \dfrac{6}{x^2}}} = -2$$

Therefore, the line $y = -2$ is a horizontal asymptote. If $x < 0$, then $-\sqrt{x^2} = x$. Hence,

$$\lim_{x \to -\infty} g(x) = \lim_{x \to -\infty} \frac{-2x}{\sqrt{x^2 - 5x + 6}} \cdot \frac{\dfrac{1}{x}}{\dfrac{-1}{\sqrt{x^2}}}$$

$$\lim_{x \to -\infty} \frac{-2}{-\sqrt{1 - \dfrac{5}{x} + \dfrac{6}{x^2}}} = 2$$

Therefore, the line $y = 2$ is a horizontal asymptote. Because $x^2 - 5x + 6 = 0$ if $x = 2$ or $x = 3$, we test to see whether the lines $x = 2$ and $x = 3$ are vertical asymptotes.

$$\lim_{x \to 2^-} g(x) = \lim_{x \to 2^-} \frac{-2x}{\sqrt{(x - 2)(x - 3)}} = -\infty$$

Therefore, the line $x = 2$ is a vertical asymptote. The curve approaches this asymptote from the left only because g is not defined for $2 < x < 3$. Similarly,

$$\lim_{x \to 3^+} g(x) = -\infty$$

and thus the curve approaches the asymptote $x = 3$ from the right only. Because $g(0) = 0$, the curve contains the origin. We use the origin, the domain, and the asymptotes to draw a sketch of the graph. See Fig. 4.6R

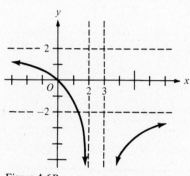

Figure 4.6R

8. Define $f \circ g$ for the given functions f and g, and determine all values of x for which $f \circ g$ is continuous.

$$f(x) = \frac{4}{3x - 5}; \ g(x) = \sqrt{x}$$

SOLUTION: $f \circ g$ is defined by

$$(f \circ g)(x) = \frac{4}{3\sqrt{x} - 5}$$

$f \circ g$ is continuous for all x for which $x > 0$ and $3\sqrt{x} - 5 \neq 0$. If $3\sqrt{x} - 5 \neq 0$, then $x \neq \frac{25}{9}$. Therefore, $f \circ g$ is continuous at all x in $(0, \frac{25}{9})$ and at all x in $(\frac{25}{9}, +\infty)$.

14. Find the absolute extrema of the given function on the given interval.

$$f(x) = x^4 - 12x^2 + 36; \ [-2, 6]$$

SOLUTION:

$$f(x) = (x^2 - 6)^2$$
$$f'(x) = 2(x^2 - 6)(2x) = 4x(x^2 - 6)$$

Because $f'(x)$ is defined for all x and $f'(x) = 0$ only if $x = 0$ or $x = \pm\sqrt{6}$, the only critical numbers of f in $[-2, 6]$ are 0 and $\sqrt{6}$. Table 14 shows the function values at each critical number and at each end point of $[-2, 6]$. We conclude that the absolute maximum value is 900, which occurs at $x = 6$, and the absolute minimum value is 0, which occurs at $x = \sqrt{6}$.

Table 14

x	-2	0	$\sqrt{6}$	6
$f(x)$	4	36	0	900

18. A piece of wire 80 in. long is bent to form a rectangle. Find the dimensions of the rectangle so that its area is as large as possible.

SOLUTION: Let

x = the number of inches in the length of the rectangle
y = the number of inches in the width of the rectangle
A = the number of square inches in the area of the rectangle

We want to determine x and y so that A has an absolute maximum value. Now

$$A = xy \qquad (1)$$

We are given that

$$2x + 2y = 80$$
$$y = 40 - x \qquad (2)$$

From (1) and (2) we get

$$A(x) = x(40 - x)$$

Because $A \geqslant 0$, then $0 \leqslant x \leqslant 40$, and thus $A(x)$ is defined and continuous on the closed interval $[0, 40]$.

$$A'(x) = 40 - 2x$$

If $A'(x) = 0$, then $x = 20$. Because $A(20) = 400, A(0) = A(40) = 0$, we conclude that 400 is the absolute maximum value of the function A. If $x = 20$, then $y = 20$. Therefore, if the area of the rectangle is as large as possible, its dimensions are 20 in. by 20 in. It is a square.

22. If f is a polynomial function, show that between any two consecutive roots of the equation $f'(x) = 0$ there is at most one root of the equation $f(x) = 0$.

SOLUTION: If x_1 is a root of the equation $f'(x) = 0$, then $f'(x_1) = 0$, and thus the tangent line to the graph of f at the point where $x = x_1$ is a horizontal line. If a is a root of the equation $f(x) = 0$, then $f(a) = 0$, and thus the graph of f intersects the x-axis at the point where $x = a$. See Fig. 4.22R. Let x_1 and x_2 be any two consecutive roots of the equation $f'(x) = 0$ with $x_1 < x_2$. Then because there are no other roots between x_1 and x_2, we have

$$f'(x) \neq 0 \quad \text{if} \quad x_1 < x < x_2 \qquad (1)$$

Suppose that a and b are two roots of the equation $f(x) = 0$ with

$$x_1 < a < b < x_2 \qquad (2)$$

Then $f(a) = 0$ and $f(b) = 0$. Moreover, because f is a polynomial function, then f is continuous and differentiable at all real numbers. Hence, the three conditions of the hypothesis of Rolle's theorem are satisfied, and thus we may conclude that there

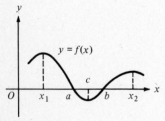

Figure 4.22R

is some c with $a < c < b$ such that

$$f'(c) = 0 \tag{3}$$

But because $a < c < b$, by (2) we have $x_1 < c < x_2$, and thus Eq. (3) contradicts (1). Therefore, there cannot be two roots of the equation $f(x) = 0$ that are between x_1 and x_2. Thus, there is at most one such root.

28. A manufacturer offers to deliver to a dealer 300 tables at $90 per table and to reduce the price per table on the entire order by 25 cents for each additional table over 300. Find the dollar total involved in the largest possible transaction between the manufacturer and the dealer under these circumstances.

SOLUTION: Let

 $x =$ the number of tables ordered
 $y =$ the number of dollars in the price per table if x tables are ordered
 $T =$ the number of dollars in the total price of x tables

We want to find the absolute maximum value of T. we have

$$T = xy \quad \text{and} \quad x \geqslant 300 \tag{1}$$

Moreover, we are given that

$$y = 90 - 0.25(x - 300) \quad \text{if } x \geqslant 300 \tag{2}$$

From (1) and (2) we get

$$T(x) = x\left(165 - \frac{1}{4}x\right) \quad \text{if } x \geqslant 300 \tag{3}$$

Because $x(165 - \frac{1}{4}x) \geqslant 0$ if $0 \leqslant x \leqslant 660$, by (3) we know that the function T is defined on the closed interval $[300, 660]$. From (3) we have

$$T(x) = 165x - \frac{1}{4}x^2$$

$$T'(x) = 165 - \frac{1}{2}x$$

If $T'(x) = 0$, then $x = 330$. Because $T(330) = 27,225$, $T(300) = 27,000$, and $T(660) = 0$, we conclude that $27,225 is the largest possible transaction, and this occurs when 330 tables are ordered.

5
Additional applications of the derivative

5.1 INCREASING AND DECREASING FUNCTIONS AND THE FIRST DERIVATIVE TEST

5.1.1 Definition A function f defined on an interval is said to be *increasing* on that interval if and only if $f(x_1) < f(x_2)$ whenever $x_1 < x_2$ where x_1 and x_2 are any numbers in the interval.

5.1.2 Definition A function f defined on an interval is said to be *decreasing* on that interval if and only if $f(x_1) > f(x_2)$ whenever $x_1 < x_2$ where x_1 and x_2 are any numbers in the interval.

5.1.3 Theorem Let the function f be continuous on the closed interval $[a, b]$ and differentiable on the open interval (a, b):

(i) if $f'(x) > 0$ for all x in (a, b), then f is increasing on $[a, b]$;
(ii) if $f'(x) < 0$ for all x in (a, b), then f is decreasing on $[a, b]$.

Note that the hypothesis of Theorem 5.1.3 may be satisfied even when $f'(a) = 0$ and $f'(b) = 0$ or when f is not differentiable at a or at b.

5.1.4 Theorem (*First derivative test for relative extrema*) Let the function f be continuous at all points of the open interval (a, b) containing the number c, and suppose that f' exists at all points of (a, b) except possibly at c:

(i) if $f'(x) > 0$ for all values of x in some open interval having c as its right endpoint, and if $f'(x) < 0$ for all values of x in some open interval having c as its left endpoint, then f has a relative maximum value at c;

(ii) if $f'(x) < 0$ for all values of x in some open interval having c as its right endpoint, and if $f'(x) > 0$ for all values of x in some open interval containing c as its left endpoint, then f has a relative minimum value at c.

From the first derivative test we conclude the following.

1. If f is continuous at c and $f'(x)$ changes sign from positive to negative as x increases through the number c, then f has a relative maximum value at c.
2. If f is continuous at c and $f'(x)$ changes sign from negative to positive as x increases through the number c, then f has a relative minimum value at c.

If f has a relative extremum at c, then either $f'(c) = 0$ or $f'(c)$ is not defined. When $f'(c) = 0$, we conclude that the tangent line to the graph of f at the point where $x = c$ is a horizontal line. If $f'(c)$ does not exist, there are two possibilities:

1. If $\lim\limits_{x \to c} f'(x) = \pm\infty$, then the tangent line to the graph of f at the point where $x = c$ is a vertical line.
2. If $\lim\limits_{x \to c^+} f'(x) \neq \lim\limits_{x \to c^-} f'(x)$, then a single tangent line to the graph of f does not exist at the point where $x = c$. The graph has a "corner" at this point, and there may be a right-hand tangent line that is distinct from a left-hand tangent line.

The following steps make use of the first derivative test to draw a sketch of the graph of a function f.

1. Find each number at which $f'(x)$ is either zero or undefined, and arrange the numbers in increasing order. (This includes each critical number of f and each number at which f is discontinuous.)
2. Use the numbers found in step 1 to partition the number line into open intervals.
3. For each interval determine whether $f'(x)$ is positive or negative (or zero).
4. Use the result of step 3 and the first derivative test to identify the values of x at which the relative extrema occur.
5. For each x at which a relative extremum occurs calculate the function value and determine whether the tangent line to the graph is horizontal, vertical, or not defined.
6. Use the information found in step 5, the domain of f, the continuity of f, symmetry, and asymptotes to draw a sketch of the graph.

Exercises 5.1

In Exercises 1-30 do each of the following:

(a) find the relative extrema of f by applying the first derivative test
(b) determine the values of x at which the relative extrema occur
(c) determine the intervals on which f is increasing
(d) determine the intervals on which f is decreasing
(e) draw a sketch of the graph.

6. $f(x) = 2x + \dfrac{1}{2x}$

SOLUTION: First, we note that f is continuous at all $x \neq 0$, but f is discontinuous at 0. Because

$$f'(x) = 2 - \frac{1}{2x^2}$$

$$= \frac{(2x + 1)(2x - 1)}{2x^2}$$

then $f'(x) = 0$ if $x = \frac{1}{2}$ or $x = -\frac{1}{2}$, and $f'(x)$ is undefined at $x = 0$. We note that $f'(x) > 0$ if $x > \frac{1}{2}$ and that $f'(x)$ changes sign at $x = \frac{1}{2}$ and $x = -\frac{1}{2}$. However, $f'(x)$ does not change sign at $x = 0$, because the factor $2x^2$, which appears in the denominator of $f'(x)$, is always nonnegative. This discussion is summarized in Table 6a. By Theorem 5.1.3 f is increasing on each interval in which $f'(x) > 0$, and f is decreasing on each interval in which $f'(x) < 0$. Thus f is increasing on $(-\infty, -\frac{1}{2}]$, increasing on $[\frac{1}{2}, +\infty)$, decreasing on $[-\frac{1}{2}, 0)$, and decreasing on $(0, \frac{1}{2}]$. (We exclude $x = 0$ because $f(0)$ is not defined.)

By the first derivative test we conclude that f has a relative maximum value at $x = -\frac{1}{2}$ and a relative minimum value at $x = \frac{1}{2}$. Furthermore, because $f'(-\frac{1}{2}) = 0$ we conclude that the tangent line to the graph of f at the point where $x = -\frac{1}{2}$ is a horizontal line. Similarly, the tangent line to the graph of f at the point where $x = \frac{1}{2}$ is horizontal because $f'(\frac{1}{2}) = 0$. This discussion is summarized in Table 6b, which also gives the function value at each relative extremum. Because

$$\lim_{x \to 0^+} f(x) = \lim_{x \to 0^+} \left(2x + \frac{1}{2x}\right) = +\infty$$

and

$$\lim_{x \to 0^-} f(x) = \lim_{x \to 0^-} \left(2x + \frac{1}{2x}\right) = -\infty$$

we conclude that the line $x = 0$ (that is, the y-axis) is a vertical asymptote. We use the information in Table 6a, Table 6b, and the asymptote to draw a sketch of the graph. See Fig. 5.1.6.

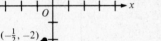

Figure 5.1.6

Table 6a

	$-\infty$		$-\frac{1}{2}$		0		$\frac{1}{2}$		$+\infty$
$f'(x)$		$+$		$-$		$-$		$+$	
f is		increasing		decreasing		decreasing		increasing	

Table 6b

x	$f(x)$	$f'(x)$	Description of Graph
$-\frac{1}{2}$	-2	0	f has a relative maximum value; graph has a horizontal tangent line.
$\frac{1}{2}$	2	0	f has a relative minimum value; graph has a horizontal tangent line.

10. $f(x) = x\sqrt{5 - x^2}$

SOLUTION: Because $5 - x^2 \geq 0$ if $|x| \leq \sqrt{5}$, the domain of f is $[-\sqrt{5}, \sqrt{5}]$, and f is continuous on $(-\sqrt{5}, \sqrt{5})$.

$$f'(x) = \frac{5 - 2x^2}{\sqrt{5 - x^2}}$$

Then $f'(x) = 0$ if $x = \pm\sqrt{\frac{5}{2}}$ and $f'(x)$ is not defined if $x = \pm\sqrt{5}$. We note that $f'(x) < 0$ if $\sqrt{\frac{5}{2}} < x < \sqrt{5}$ and that $f'(x)$ changes sign at $x = \sqrt{\frac{5}{2}}$ and $x = -\sqrt{\frac{5}{2}}$. As in Exercise 6, we summarize this discussion in Table 10a, which also indicates that f is increasing on $[-\sqrt{\frac{5}{2}}, \sqrt{\frac{5}{2}}]$, decreasing on $[-\sqrt{5}, -\sqrt{\frac{5}{2}}]$, and decreasing on $[\sqrt{\frac{5}{2}}, \sqrt{5}]$. Therefore, by the first derivative test, f has a relative minimum value at $x = -\sqrt{\frac{5}{2}}$ and a relative maximum value at $x = \sqrt{\frac{5}{2}}$. Because $f'(\sqrt{\frac{5}{2}}) = f'(-\sqrt{\frac{5}{2}}) = 0$, the tangent line to the graph of f at each relative extremum is a horizontal line. Because

$$\lim_{x \to \sqrt{5^-}} f'(x) = -\infty \quad \text{and} \quad \lim_{x \to -\sqrt{5}^+} f'(x) = -\infty$$

we conclude that the tangent lines to the graph of f "approach" a vertical line as the point of tangency approaches the points where $x = \sqrt{5}$ and $x = -\sqrt{5}$. This discussion is summarized in Table 10b which also gives the function value at each critical number. We use the continuity of f and the information from Table 10a and Table 10b to draw a sketch of the graph of f. See Fig. 5.1.10.

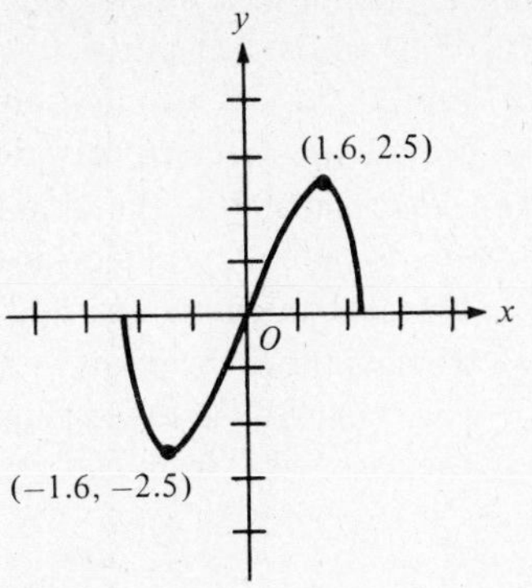

Table 10a

	$-\sqrt{5}$	$-\sqrt{\frac{5}{2}}$	$\sqrt{\frac{5}{2}}$	$\sqrt{5}$
$f'(x)$		$-$	$+$	$-$
f is		decreasing	increasing	decreasing

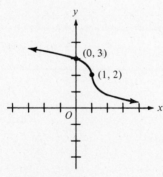

Figure 5.1.10

Table 10b

x	$f(x)$	$f'(x)$	Description of Graph
$-\sqrt{5}$	0	$-\infty$	Endpoint; graph has a vertical tangent line.
$-\sqrt{\frac{5}{2}}$	$-\frac{5}{2}$	0	f has a relative minimum value; graph has a horizontal tangent line.
$\sqrt{\frac{5}{2}}$	$\frac{5}{2}$	0	f has a relative maximum value; graph has a horizontal tangent line.
$\sqrt{5}$	0	$-\infty$	Endpoint; graph has a vertical tangent line.

14. $f(x) = 2 - (x - 1)^{1/3}$

SOLUTION: f is continuous at all real numbers. Because

$$f'(x) = \frac{-1}{3(x - 1)^{2/3}}$$

then $f'(x)$ is not defined at $x = 1$. Because $f'(x) < 0$ if $x > 1$ and $f'(x) < 0$ if $x < 1$, we conclude that f is decreasing on $[1, +\infty)$ and decreasing on $(-\infty, 1]$. By the first derivative test f does not have a relative extremum. Because

$$\lim_{x \to 1} f'(x) = -\infty$$

the tangent line to the graph of f is vertical at the point where $x = 1$ [namely at the point $(1, 2)$]. A sketch of the graph of f is shown in Fig. 5.1.14.

Figure 5.1.14

20. $f(x) = \begin{cases} \sqrt{25 - (x + 7)^2} & \text{if } x \leqslant -3 \\ 12 - x^2 & \text{if } -3 < x \end{cases}$

SOLUTION: Because $25 - (x + 7)^2 \geqslant 0$ if $-12 \leqslant x \leqslant -2$, then f is defined on $[-12, +\infty)$. Because the left-hand limit of f and the right-hand limit of f as x approaches -3 are both 3, we conclude that f is continuous on $[-12, +\infty)$. Because

$$f'(x) = \begin{cases} \dfrac{-(x + 7)}{\sqrt{25 - (x + 7)^2}} & \text{if } -12 < x < -3 \\ -2x & \text{if } x > -3 \end{cases}$$

then $f'(x) = 0$ if $x = -7$ or $x = 0$, and $f'(x)$ is not defined if $x = -12$. Furthermore, because $f'_-(-3) = -\frac{4}{3}$ and $f'_+(-3) = 6$, we conclude that $f'(-3)$ does not exist. As in Exercise 6, we determine whether $f'(x)$ is positive or negative in each interval between consecutive critical numbers. For this function, this is most easily done by choosing a replacement for x from each of the intervals and substituting in the formula for $f'(x)$. The results are summarized in Table 20a, which also indicates for each interval whether f is increasing or decreasing. We use the first derivative test to identify each x at which f has a relative extremum. We calculate the function value at each critical number, and use $f'(x)$ at this number to determine the behavior of the tangent line, if it exists. See Table 20b. Because the left-hand derivative and the right-hand derivative of f at -3 are not equal, we conclude that the point $(-3, 3)$ is a "corner" point on the graph. Because $f'_+(12) = +\infty$, we conclude that the tangent lines to the graph of f approach a vertical line as x approaches -12 from the right. A sketch of the graph of f is shown in Fig. 5.1.20.

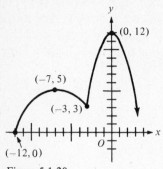

Figure 5.1.20

Table 20a

	-12	-7	-3	0	$+\infty$
$f'(x)$		+	−	+	−
f is		increasing	decreasing	increasing	decreasing

Table 20b

x	$f(x)$	$f'(x)$	Description of Graph
-12	0	$+\infty$	Endpoint; graph has a vertical tangent line.
-7	5	0	f has a relative maximum value; graph has a horizontal tangent line.
-3	3	does not exist	f has a relative minimum value; graph has a corner point.
0	12	0	f has a relative maximum value; graph has a horizontal tangent line.

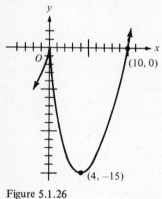

Figure 5.1.26

26. $f(x) = x^{5/3} - 10x^{2/3}$

SOLUTION: f is continuous at all real numbers. Because

$$f'(x) = \frac{5(x - 4)}{3x^{1/3}}$$

then 0 and 4 are critical numbers of f. Table 26a indicates for each interval determined by these critical numbers whether $f'(x)$ is positive or negative and whether f is increasing or decreasing on the interval. Table 26b gives the function value at each critical number, the behavior of $f'(x)$ at that critical number, and a description of the graph of f. A sketch of the graph is shown in Fig. 5.1.26.

Table 26a

	$-\infty$	0	4	$+\infty$
$f'(x)$		+	−	+
f is		increasing	decreasing	increasing

Table 26b

x	$f(x)$	$f'(x)$	Description of Graph
0	0	$\pm\infty$	f has a relative maximum value; graph has a vertical tangent line.
4	$-12\sqrt[3]{2}$	0	f has a relative minimum value; graph has a horizontal tangent line.

32. Find a, b, and c so that the function defined by $f(x) = ax^2 + bx + c$ will have a relative maximum value of 7 at 1 and the graph of $y = f(x)$ will go through the point $(2, -2)$.

SOLUTION: Because f is a polynomial function, then f' is defined for all x. Therefore, a relative extremum of f must occur at a point where $f'(x) = 0$. Because $f'(x) = 2ax + b$, and we are given that a relative maximum value occurs at $x = 1$, then $f'(1) = 0$. Since $f'(1) = 2a + b$, we have

$$2a + b = 0 \tag{1}$$

Because 7 is the value of the function at 1, we know that $f(1) = 7$. Since $f(1) = a + b + c$, we have

$$a + b + c = 7 \tag{2}$$

Because the graph of $y = f(x)$ contains the point $(2, -2)$, then $f(2) = -2$. Since $f(2) = 4a + 2b + c$, we have

$$4a + 2b + c = -2 \tag{3}$$

We solve Eqs. (1), (2), and (3) simultaneously and obtain $a = -9$, $b = 18$, $c = -2$. Therefore, the function is defined by $f(x) = -9x^2 + 18x - 2$.

40. The function f is differentiable at each number in the closed interval $[a, b]$. Prove that if $f'(a) \cdot f'(b) < 0$, there is a number c in the open interval (a, b) such that $f'(c) = 0$.

SOLUTION: Because f is differentiable at each number in $[a, b]$, then f is continuous on $[a, b]$ and by Theorem 4.5.9 f has an absolute maximum value and an absolute minimum value on $[a, b]$. If either of these absolute extrema occurs at some number c with $a < c < b$, then by Theorem 4.5.3 $f'(c) = 0$, which is the desired result. Otherwise, both of the absolute extrema occur at endpoints of $[a, b]$. We show that this is impossible. We consider two cases.

Case 1: The absolute maximum value of f is at a, and the absolute minimum value of f is at b. Because f is differentiable at a, then the derivative from the right of f at a exists. Furthermore, since the absolute maximum value of f is at a, then $f(a) \geqslant f(x)$ for all x in $[a, b]$, and thus

$$\frac{f(x) - f(a)}{x - a} \leqslant 0 \quad \text{if } a < x < b$$

Hence

$$f'_+(a) = \lim_{x \to a^+} \frac{f(x) - f(a)}{x - a} \leqslant 0 \tag{1}$$

Because we are given that $f'(a) \cdot f'(b) < 0$, then $f'(a) \neq 0$, and thus (1) implies that

$$f'(a) < 0 \tag{2}$$

Because f is differentiable at b, the derivative from the left of f at b exists. Since the absolute minimum value of f is at b, then $f(b) \leqslant f(x)$ for all x in $[a, b]$, and thus

$$\frac{f(x) - f(b)}{x - b} \leqslant 0 \quad \text{if} \quad a < x < b$$

Hence,

$$f'_-(b) = \lim_{x \to b^-} \frac{f(x) - f(b)}{x - b} \leqslant 0 \tag{3}$$

Because $f'(a) \cdot f'(b) < 0$, then $f'(b) \neq 0$, and thus (3) implies

$$f'(b) < 0 \tag{4}$$

But (2) and (4) contradict the hypothesis that $f'(a) \cdot f'(b) < 0$. Therefore, the Case 1 assumption is false. We consider the remaining possibility.

Case 2: The absolute maximum value of f is at b, and the absolute minimum value of f is at a. Then inequalities (1), (2), (3), and (4) of Case 1 are reversed, and we again contradict the hypothesis that $f'(a) \cdot f'(b) < 0$. Therefore both Case 1 and Case 2 are impossible, and we have proved that $f'(c) = 0$ for some c in (a, b).

5.2 THE SECOND DERIVATIVE TEST FOR RELATIVE EXTREMA

5.2.1 Theorem

(*Second derivative test for relative extrema*) Let c be a critical number of a function f at which $f'(c) = 0$, and let f' exist for all values of x in some open interval containing c. Then if $f''(c)$ exists and

(i) if $f''(c) < 0$, then f has a relative maximum value at c;
(ii) if $f''(c) > 0$, then f has a relative minimum value at c.

If c is a critical number of a function f at which $f'(c)$ does not exist or at which both $f'(c)$ and $f''(c)$ are zero, then we cannot use the second-derivative test to determine whether f has a relative extremum at c. For each of these cases, however, we may use the first derivative test to determine the relative extrema.

Exercises 5.2

In Exercises 1-14, find the relative extrema of the given function by using the second derivative test, if it can be applied. If the second derivative test cannot be applied, use the first derivative test.

4. $h(x) = 2x^3 - 9x^2 + 27$

SOLUTION:

$$h'(x) = 6x^2 - 18x = 6x(x - 3)$$

The critical numbers of h are 0 and 3.

$$h''(x) = 12x - 18$$

Because $h'(0) = 0$ and $h''(0) = -18 < 0$, by the second derivative test we conclude that h has a relative maximum value at $x = 0$. Because $h(0) = 27$, this relative maximum value is 27. Because $h'(3) = 0$, $h''(3) = 18 > 0$, then by the second derivative test h has a relative minimum value at $x = 3$. Because $h(3) = 0$, this relative minimum value is 0.

8. $f(x) = x(x - 1)^3$

SOLUTION:

$$f'(x) = 3x(x-1)^2 + (x-1)^3 \cdot 1$$
$$= (x-1)^2(4x-1)$$

The critical numbers of f are $\frac{1}{4}$ and 1.

$$f''(x) = (x-1)^2 \cdot 4 + (4x-1) \cdot 2(x-1)$$
$$= 6(x-1)(2x-1)$$

Because $f'(\frac{1}{4}) = 0$ and $f''(\frac{1}{4}) > 0$, by the second derivative test f has a relative minimum value at $\frac{1}{4}$. This relative minimum value is $f(\frac{1}{4}) = -\frac{27}{256}$. Because $f'(1) = 0$ and $f''(1) = 0$, the second derivative test fails at $x = 1$. We must use the first derivative test. Table 8 gives the sign of $f'(x)$ for each of the intervals determined by the critical numbers of f. Note that $f'(x)$ does not change sign at $x = 1$, because the factor $(x-1)^2$, which appears in $f'(x)$, is always nonnegative.

From Table 8 and the first derivative test we conclude that f does not have a relative extremum at $x = 1$. Note that the first derivative test indicates that f has a relative minimum value at $x = \frac{1}{4}$, which agrees with the result of applying the second derivative test at $x = \frac{1}{4}$.

Table 8

	$-\infty$	$\frac{1}{4}$	1	$+\infty$
$f'(x)$		$-$	$+$	$+$

12. $g(x) = \dfrac{9}{x} + \dfrac{x^2}{9}$

SOLUTION:

$$g'(x) = \frac{-9}{x^2} + \frac{2x}{9} = \frac{2x^3 - 81}{9x^2}$$

If $g'(x) = 0$, then $2x^3 - 81 = 0$ and $x = \frac{3}{2}\sqrt[3]{12}$, which is the only critical number of g.

$$g''(x) = \frac{18}{x^3} + \frac{2}{9}$$

We note that $g''(x) > 0$ for all $x > 0$. Because $g'(\frac{3}{2}\sqrt[3]{12}) = 0$ and $g''(\frac{3}{2}\sqrt[3]{12}) > 0$, by the second derivative test, g has a relative minimum value at $x = \frac{3}{2}\sqrt[3]{12}$. The relative minimum value of g is $g(\frac{3}{2}\sqrt[3]{12}) = \frac{3}{2}\sqrt[3]{18}$.

14. $G(x) = x^{2/3}(x-4)^2$

SOLUTION:

$$G'(x) = x^{2/3} \cdot 2(x-4) + (x-4)^2 \cdot \frac{2}{3}x^{-1/3}$$

$$= \frac{8}{3}x^{-1/3}(x-4)(x-1)$$

The critical numbers of G are 0, 1, and 4. Because $G'(0)$ is not defined, we cannot use the second derivative test to determine whether G has a relative extremum at $x = 0$. However, by Table 14 and the first derivative test, we conclude that G has a relative minimum value at $x = 0$, a relative maximum value at $x = 1$, and a relative minimum value at $x = 4$. Because $G(0) = 0$, $G(1) = 9$, and $G(4) = 0$, these relative extrema are 0, 9 and 0, respectively.

Table 14

	$-\infty$	0	1	4	$+\infty$
$G'(x)$		$-$	$+$	$-$	$+$

16. Given $f(x) = x^r - rx + k$, where $r > 0$ and $r \neq 1$, prove that (a) if $0 < r < 1$, f has a relative maximum value at 1; (b) if $r > 1$, f has a relative minimum value at 1.

SOLUTION:

$$f'(x) = rx^{r-1} - r$$

We note that $f'(1) = r \cdot 1^{r-1} - r = 0$ for all r.

$$f''(x) = r(r-1)x^{r-2}$$
$$f''(1) = r(r-1) \cdot 1^{r-2} = r(r-1)$$

(a) If $0 < r < 1$, then $r(r-1) < 0$. Because $f'(1) = 0$ and $f''(1) = r(r-1) < 0$, by the second derivative test f has a relative maximum value at 1.

(b) If $r > 1$, then $r(r-1) > 0$. Because $f'(1) = 0$ and $f''(1) = r(r-1) > 0$, by the second derivative test f has a relative minimum value at 1.

5.3 ADDITIONAL PROBLEMS INVOLVING ABSOLUTE EXTREMA

A function f may not have an absolute maximum value or an absolute minimum value on an interval I if I is not a closed interval. However, we may sometimes use the following theorem to show that f has an absolute extremum on I.

5.3.1 Theorem

Let the function f be continuous on the interval I containing the number c. If $f(c)$ is a relative extremum of f on I and c is the only number in I for which f has a relative extremum, then $f(c)$ is an absolute extremum of f on I. Furthermore,

(i) if $f(c)$ is a relative maximum value of f on I, then $f(c)$ is an absolute maximum value of f on I;

(ii) if $f(c)$ is a relative minimum value of f on I, then $f(c)$ is an absolute minimum value of f on I.

The following steps may be used to solve the exercises in this section. We illustrate the steps in Exercises 12, 16, and 18.

1. Identify the variable for which you want to find the absolute maximum value of the absolute minimum value. This is the dependent variable.
2. Express the dependent variable as a function of some other variable and show that the function is continuous on I.
3. Show that the dependent variable has only one critical number in I.
4. Use either the first derivative test or the second derivative test to show that the dependent variable has a relative extremum at this critical number.
5. The absolute extremum of the dependent variable is the relative extremum found in step 4.

When it is not convenient to express the dependent variable as a function of only one variable, the steps that follow may be used to find the critical numbers. We illustrate the steps in Exercise 14.

1. Express the dependent variable, say z, as a function of two other variables, say x and y.
2. Find another equation involving x and y.
3. Differentiate on both sides of the equations found in steps 1 and 2 with respect to x.
4. Eliminate $D_x y$ from the pair of equations found in step 3.
5. Set $D_x z = 0$ in the equation found in step 4 and solve for z. This gives the critical numbers for z.

Exercises 5.3

In Exercises 1-8, find the absolute extrema of the given function on the given interval if there are any.

4. $f(x) = \dfrac{x^2}{x+3}; [-4, -1]$

SOLUTION: f is discontinuous at -3. Because

$$\lim_{x \to -3^-} \frac{x^2}{x+3} = -\infty$$

f does not have an absolute minimum value on $[-4, -1]$. Because

$$\lim_{x \to -3^+} \frac{x^2}{x+3} = +\infty$$

f does not have an absolute maximum value on $[-4, -1]$.

6. $G(x) = (x-5)^{2/3}; (-\infty, +\infty)$

SOLUTION: G is continuous on $(-\infty, +\infty)$.

$$G'(x) = \frac{2}{3(x-5)^{1/3}}$$

The only critical number of G is 5. Because $G'(x) < 0$ if $x < 5$, then G is decreasing on $(-\infty, 5]$. Because $G'(x) > 0$ if $x > 5$, then G is increasing on $[5, +\infty)$. Hence, G has a relative minimum value at $x = 5$. Because the only relative extremum of G occurs at $x = 5$, by Theorem 5.3.1 we conclude that $G(5) = 0$ is the absolute minimum value of G on $(-\infty, +\infty)$. Because

$$\lim_{x \to +\infty} G(x) = +\infty$$

G does not have an absolute maximum value.

12. A box manufacturer is to produce a box with an open top of specific volume whose base is a rectangle having a length that is three times its width. Find the most economical dimensions.

SOLUTION: Because the cost of the box is proportional to its total surface area, we want to determine the dimensions so that the surface area is an absolute minimum. Let

$x =$ the number of units in the width of the base
$y =$ the number of units in the length of the base
$z =$ the number of units in the depth of the box
$A =$ the number of square units in the total surface area of the box

Because the box has an open top, we have

$$A = xy + 2xz + 2yz \tag{1}$$

Because we are given that $y = 3x$, we may replace y by $3x$ in Eq. (1) and obtain

$$A = 3x^2 + 8xz \tag{2}$$

If V is the number of cubic units in the volume of the box, then

$$V = 3x^2 z$$

$$z = \frac{V}{3x^2} \tag{3}$$

Substituting the value for z from Eq. (3) into Eq. (2), we express A as a function of x as follows.

$$A(x) = 3x^2 + \frac{8V}{3x} \tag{4}$$

Because V is a constant, we have

$$A'(x) = 6x - \frac{8V}{3x^2} \tag{5}$$

Setting $A'(x) = 0$, we get

$$0 = 6x - \frac{8V}{3x^2}$$

$$x = \frac{1}{3}\sqrt[3]{12V}$$

We use the second derivative test to see whether A has a relative extremum at the critical number $\frac{1}{3}\sqrt[3]{12V}$. From Eq. (5) we get

$$A''(x) = 6 + \frac{16V}{3x^3}$$

Because $A''(x) > 0$ for all $x > 0$, then A has a relative minimum value at the critical number. From Eq. (4), A is continuous for all $x > 0$. Because A has only one relative extremum, then the relative minimum value must be the absolute minimum value. If $x = \frac{1}{3}\sqrt[3]{12V}$, then $y = 3x = \sqrt[3]{12V}$, and from Eq. (3) we get $z = \frac{1}{4}\sqrt[3]{12V}$. Therefore, the most economical dimensions are $\frac{1}{3}\sqrt[3]{12V}$ units by $\sqrt[3]{12V}$ units for the base and $\frac{1}{4}\sqrt[3]{12V}$ units for the altitude.

14. A Norman window consists of a rectangle surmounted by a semicircle. Find the shape of such a window that will admit the most light for a given perimeter.

SOLUTION: Because the amount of light admitted is proportional to the area of the window, we want to determine the dimensions of the window so that the area is an absolute maximum. Refer to Fig. 5.3.14. Let

x = the number of units in the width of the rectangle
r = the number of units in the radius of the semicircle
A = the number of square units in the area of the window

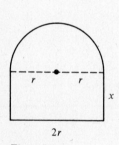

Figure 5.3.14

Because $2r$ units is the length of the rectangle, then the area of the rectangle is $2rx$ square units. Because the area of the semicircle is $\frac{1}{2}\pi r^2$ square units, then

$$A = 2rx + \frac{1}{2}\pi r^2 \tag{1}$$

If P is the number of units in the perimeter of the window, then

$$P = 2x + 2r + \pi r \tag{2}$$

Differentiating on both sides of Eq. (1) and Eq. (2) with respect to r, we get

$$D_r A = 2r\,D_r x + 2x + \pi r \tag{3}$$
$$D_r P = 2\,D_r x + 2 + \pi \tag{4}$$

Because P is a constant, then $D_r P = 0$, and thus Eq. (4) may be written as

$$0 = 2\,D_r x + 2 + \pi$$

$$D_r x = \frac{-2 - \pi}{2} \tag{5}$$

Substituting this value for $D_r x$ into Eq. (3), we obtain

$$D_r A = 2r\,\frac{-2 - \pi}{2} + 2x + \pi r$$

$$= -2r + 2x \tag{6}$$

To find the critical numbers of A, we set $D_r A = 0$. Then Eq. (6) gives

$$0 = -2r + 2x$$
$$r = x \tag{7}$$

Hence, A has a critical number when $r = x$. We use the second derivative test to see whether A has a relative extremum at this critical number. Differentiating on both sides of Eq. (6) with respect to r, we have

$$D_r^2 A = -2 + 2\,D_r x \tag{8}$$

Substituting the value of $D_r x$ given in Eq. (5) into Eq. (8), we have

$$D_r^2 A = -2 + 2\,\frac{-2 - \pi}{2} = -4 - \pi$$

Because $D_r^2 A < 0$, then A has a relative maximum value when $r = x$. Because there is only one critical number for A and because A is a continuous function, then the relative maximum value of A is the absolute maximum value of A. Thus, the window will admit the most light when $x = r$; that is, when the length of the rectangle equals the radius of the semicircle.

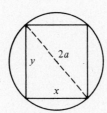

Figure 5.3.16

16. The strength of a rectangular beam is proportional to the breadth and the square of its depth. Find the dimensions of the strongest beam that can be cut from a log in the shape of a right circular cylinder of radius a in.

SOLUTION: Refer to Fig. 5.3.16, which illustrates a cross section of the rectangular beam cut from the log. Let

$x =$ the number of inches in the breadth of the beam
$y =$ the number of inches in the depth of the beam
$S =$ the number of units in the strength of the beam

We want to determine x and y so that S has an absolute maximum value. Because the strength of the beam is proportional to the breadth and the square of the depth, we are given that

$$S = kxy^2 \tag{1}$$

where k is a constant. Because the radius of the log is a inches, then its diameter is $2a$ inches, and we have

$$x^2 + y^2 = (2a)^2 \tag{2}$$

Solving Eq. (2) for y^2 and substituting this value for y^2 into Eq. (1), we express S as a function of x as follows.

$$S(x) = kx(4a^2 - x^2)$$
$$= k(4a^2 x - x^3) \tag{3}$$

Because k and a are constants, from Eq. (3) we have

$$S'(x) = k(4a^2 - 3x^2) \tag{4}$$

If $S'(x) = 0$, then

$$4a^2 - 3x^2 = 0$$

$$x = \frac{2}{3}a\sqrt{3} \qquad \qquad \textbf{(5)}$$

Hence, S has a critical number when $x = \frac{2}{3}a\sqrt{3}$. From Eq. (4) we get $S''(x) = -6kx$. Because $S''(x) < 0$ for all positive x, then S has a relative maximum value at the critical number. Because S is a continuous function and S has only one critical number, then the relative maximum value of S is the absolute maximum value of S. When $x = \frac{2}{3}a\sqrt{3}$, from Eq. (2) we get $y = \frac{2}{3}a\sqrt{6}$. Therefore, for the strongest beam the breadth is $\frac{2}{3}a\sqrt{3}$ in and the depth is $\frac{2}{3}a\sqrt{6}$ in.

18. A page of print is to contain 24 in.^{2} of printed area, a margin of $1\frac{1}{2}$ in. at the top and bottom, and a margin of 1 in. at the sides. What are the dimensions of the smallest page that would fill these requirements?

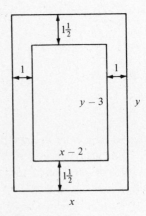

Figure 5.3.18

SOLUTION: Refer to Fig. 5.3.18. Let

    x = the number of inches in the width of the page
    y = the number of inches in the length of the page
    A = the number of square inches in the area of the page

We want to determine x and y so that A has an absolute minimum value. We have

$$A = xy \qquad \qquad \textbf{(1)}$$

Because the margin at the top and bottom is $1\frac{1}{2}$ in., the number of inches in the length of the printed part of the page is $y - 3$. Because the side margins are 1 in. each, the number of inches in the width of the printed part of the page is $x - 2$. Because the printed part of the page is to contain 24 in.^{2}, we have

$$(x - 2)(y - 3) = 24 \qquad \qquad \textbf{(2)}$$

Solving Eq. (2) for y and substituting this value for y into Eq. (1), we express A as a function of x. Thus,

$$A(x) = \frac{3(x^2 + 6x)}{x - 2} \qquad \qquad \textbf{(3)}$$

Because $A > 0$, we have $x > 2$, and A is continuous on $(2, +\infty)$.

$$A'(x) = \frac{3[(x - 2)(2x + 6) - (x^2 + 6x) \cdot 1]}{(x - 2)^2}$$

$$= \frac{3(x^2 - 4x - 12)}{(x - 2)^2} = \frac{3(x - 6)(x + 2)}{(x - 2)^2} \qquad \qquad \textbf{(4)}$$

From Eq. (4) we see that $A'(6) = 0$. Furthermore, 6 is the only critical number for A in the interval $(2, +\infty)$. We use the first derivative test to see if A has a relative extremum at 6. Because $A'(x) < 0$ if $0 < x < 6$, then A is decreasing on $(0, 6]$. Because $A'(x) > 0$ if $x > 6$, then A is increasing on $[6, +\infty)$. Thus, A has a relative minimum value at $x = 6$. Because 6 is the only critical number of A in $(2, +\infty)$, we conclude that A has an absolute minimum value at $x = 6$. When $x = 6$, from Eq. (2) we get $y = 9$. Hence, the smallest page that fills the requirements is 6 in. $\times$ 9 in.

22. A right circular cone is to be inscribed in a sphere of given radius. Find the ratio of the altitude to the base radius of the cone of largest possible volume.

SOLUTION: Refer to Fig. 5.3.22. Let

    r = the number of units in the radius of the cone
    h = the number of units in the altitude of the cone
    V = the number of cubic units in the volume of the cone

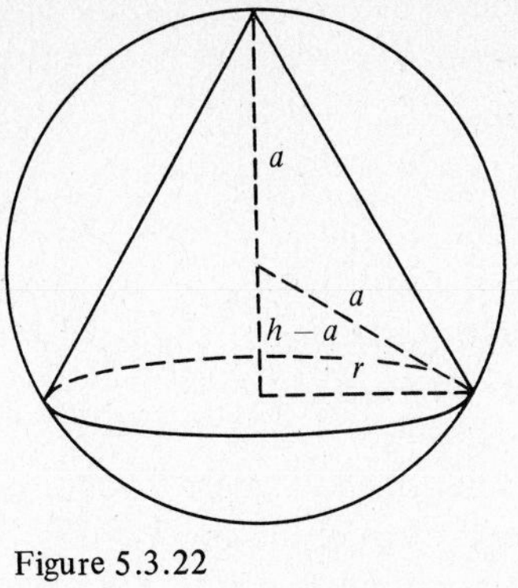

Figure 5.3.22

We want to find h/r so that V has an absolute maximum value. We have

$$V = \frac{1}{3}\pi r^2 h \tag{1}$$

If a is the number of units in the radius of the sphere, then

$$(h-a)^2 + r^2 = a^2 \tag{2}$$

Solving Eq. (2) for r^2 and substituting this value for r^2 into Eq. (1), we express V as a function of h as follows.

$$V(h) = \frac{1}{3}\pi(-h^3 + 2ah^2)$$

We note that V is continuous and defined on $(0, 2a)$.

$$V'(h) = \frac{1}{3}\pi(-3h^2 + 4ah)$$

If $V'(h) = 0$, then

$$-3h^2 + 4ah = 0$$

$$h = \frac{4}{3}a$$

Thus, V has a critical number when $h = \frac{4}{3}a$. Because

$$V''(h) = \frac{1}{3}\pi(-6h + 4a)$$

then

$$V''\left(\frac{4}{3}a\right) = \frac{1}{3}\pi\left(-6 \cdot \frac{4}{3}a + 4a\right) = -\frac{4}{3}\pi a$$

Because $V'(\frac{4}{3}a) = 0$ and $V''(\frac{4}{3}a) < 0$, V has a relative maximum value at $\frac{4}{3}a$. Since $\frac{4}{3}a$ is the only critical number for V in $(0, 2a)$, then V has an absolute maximum value at $\frac{4}{3}a$. When $h = \frac{4}{3}a$ by Eq. (2) we get

$$r = \frac{2}{3}a\sqrt{2}$$

Therefore,

$$\frac{h}{r} = \frac{\frac{4}{3}a}{\frac{2}{3}\sqrt{2}\,a} = \sqrt{2}$$

For the cone of largest possible volume, the ratio of altitude to base radius is $\sqrt{2}$.

5.4 CONCAVITY AND POINTS OF INFLECTION

5.4.1 Definition   The graph of a function f is said to be *concave upward* at the point $(c, f(c))$ if $f'(c)$ exists and if there is an open interval I containing c such that for all values of $x \neq c$ in I the point $(x, f(x))$ on the graph is above the tangent line to the graph at $(c, f(c))$.

5.4.2 Definition   The graph of a function f is said to be *concave downward* at the point $(c, f(c))$ if $f'(c)$ exists and if there is an open interval I containing c such that for all values of $x \neq c$ in I the point $(x, f(x))$ on the graph is below the tangent line to the graph at $(c, f(c))$.

5.4.3 Theorem   Let f be a function which is differentiable on some open interval containing c. Then

(i) if $f''(c) > 0$, the graph of f is concave upward at $(c, f(c))$;
(ii) if $f''(c) < 0$, the graph of f is concave downward at $(c, f(c))$.

5.4.4 Definition The point $(c, f(c))$ is a *point of inflection* of the graph of the function f if the graph has a tangent line there and if there exists an open interval I containing c such that if x is in I, then either

(i) $f''(x) < 0$ if $x < c$ and $f''(x) > 0$ if $x > c$; or
(ii) $f''(x) > 0$ if $x < c$ and $f''(x) < 0$ if $x > c$.

5.4.5 Theorem If the function f is differentiable on some open interval containing c and if $(c, f(c))$ is a point of inflection of the graph of f, then if $f''(c)$ exists, $f''(c) = 0$.

If the graph of f has a tangent line at the point where $x = c$ and if $f''(x)$ changes sign as x increases through the value c, then the graph of f has a point of inflection at the point where $x = c$. Every point of inflection of the graph of the function f occurs at a point where $f''(x)$ is either zero or undefined. If f is a polynomial function, then $f''(x)$ exists for all x, and thus every point of inflection of the graph of f occurs at a point where $f''(x) = 0$. However, if $f''(c) = 0$, we cannot conclude that the graph of f has a point of inflection at c.

The following steps may be used to locate the points of inflection of the graph of f.

1. Find $f'(x)$ and $f''(x)$.
2. Find all numbers c for which $f''(c) = 0$ or $f''(c)$ is not defined.
3. If the graph of f has a tangent line at the point where $x = c$ and if $f''(x)$ changes sign as x increases through the value c, then the graph of f has a point of inflection at the point where $x = c$.

Exercises 5.4

In Exercises 1-10, determine where the graph of the given function is concave upward, where it is concave downward, and find the points of inflection if there are any.

2. $g(x) = x^3 + 3x^2 - 3x - 3$

SOLUTION:

$$g'(x) = 3x^2 + 6x - 3$$
$$g''(x) = 6x + 6 = 6(x + 1)$$

We note that $g''(x) = 0$ if $x = -1$. Because $g''(x) > 0$ if $x > -1$, the graph of g is concave upward for all x in $(-1, +\infty)$. Because $g''(x) < 0$ if $x < -1$, the graph of g is concave downward for all x in $(-\infty, -1)$. Because $g'(-1)$ exists, the graph of g has a tangent line at the point where $x = -1$. Furthermore, $g''(x)$ changes sign as x increases through the value -1. Therefore, the graph of g has a point of inflection at the point $(-1, 2)$.

6. $G(x) = \dfrac{2x}{(x^2 + 4)^{3/2}}$

SOLUTION:

$$G(x) = 2x(x^2 + 4)^{-3/2}$$

$$G'(x) = 2x\left(-\frac{3}{2}\right)(x^2 + 4)^{-5/2} \cdot 2x + (x^2 + 4)^{-3/2} \cdot 2$$

$$= -4(x^2 + 4)^{-5/2}(x^2 - 2)$$

$$G''(x) = -4 \left[(x^2 + 4)^{-5/2} \cdot 2x + (x^2 - 2)\left(-\frac{5}{2}\right)(x^2 + 4)^{-7/2} \cdot 2x \right]$$

$$= \frac{12x(x^2 - 6)}{(x^2 + 4)^{7/2}}$$

$G''(x) = 0$ if $x = 0$ or $x = \pm\sqrt{6}$. $G''(x)$ is defined for all x. We note that $G''(x) > 0$ if $x > \sqrt{6}$ and that $G''(x)$ changes sign at $x = \sqrt{6}$, $x = 0$, and $x = -\sqrt{6}$. This discussion is summarized in Table 6, which indicates that the graph of G is concave upward if $x > \sqrt{6}$ or $-\sqrt{6} < x < 0$, and the graph of G is concave downward if $x < -\sqrt{6}$ or $0 < x < \sqrt{6}$. Because $G'(x)$ exists at each of the points where $G''(x)$ changes sign, the graph of G has a point of inflection at each of these points, namely $(\sqrt{6}, \frac{1}{25}\sqrt{15})$, $(0, 0)$, and $(-\sqrt{6}, -\frac{1}{25}\sqrt{15})$.

Table 6

	$-\infty$		$-\sqrt{6}$		0		$\sqrt{6}$		$+\infty$
$G''(x)$		$-$		$+$		$-$		$+$	
Graph is		concave downward		concave upward		concave downward		concave upward	

8. $F(x) = (2x - 6)^{3/2} + 1$

SOLUTION: Because $2x - 6 \geq 0$ if $x \geq 3$, the domain of F is $[3, +\infty)$.

$$F'(x) = 3(2x - 6)^{1/2}$$

$$F''(x) = \frac{3}{(2x - 6)^{1/2}}$$

Because $F''(x) > 0$ for all $x > 3$, the graph of F is concave upward for all $x > 3$. There are no points at which the graph of F is concave downward, and there are no points of inflection.

12. If $f(x) = ax^3 + bx^2 + cx$, determine a, b, and c so that the graph of f will have a point of inflection at $(1, 2)$ and so that the slope of the inflectional tangent there will be -2.

SOLUTION:

$$f'(x) = 3ax^2 + 2bx + c$$
$$f''(x) = 6ax + 2b$$

Because f is a polynomial function, $f''(x) = 0$ at each point of inflection. We let $f''(1) = 0$ because $(1, 2)$ is to be a point of inflection. Thus

$$6a + 2b = 0 \tag{1}$$

Because the slope of the tangent line at the point of inflection is -2, we must have $f'(1) = -2$. Thus

$$3a + 2b + c = -2 \tag{2}$$

Because $(1, 2)$ is a point on the graph of f, then $f(1) = 2$. Thus,

$$a + b + c = 2 \tag{3}$$

We solve Eqs. (1), (2), and (3) simultaneously and find $a = 4$, $b = -12$, and $c = 10$. Thus, $f(x) = 4x^3 - 12x^2 + 10x$.

In Exercises 15-24, draw a sketch of a portion of the graph of a function f through the point where $x = c$ if the given conditions are satisfied. If the conditions are incomplete or inconsistent, explain. It is assumed that f is continuous on some open interval containing c.

16. $f'(x) > 0$ if $x < c$; $f'(x) > 0$ if $x > c$;
      $f''(x) > 0$ if $x < c$; $f''(x) < 0$ if $x > c$.

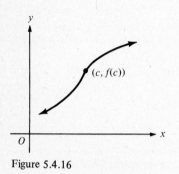

Figure 5.4.16

SOLUTION: Because $f'(x) > 0$ if $x < c$, then f is increasing on $(-\infty, c]$. Because $f'(x) > 0$ if $x > c$, then f is increasing on $[c, +\infty)$. Because $f''(x) > 0$ if $x < c$, then the graph of f is concave upward at every x in $(-\infty, c)$. Because $f''(x) < 0$ if $x > c$, then the graph of f is concave downward at every x in $(c, +\infty)$. This discussion is summarized in Table 16. If there is a tangent line to the graph of f at the point where $x = c$, then this point must be a point of inflection. However, we are not given information that is sufficient to conclude that there is a tangent line at this point. Fig. 5.4.16 shows a sketch of the graph if $f'_-(c) \neq f'_+(c)$. Note there is a corner point at $x = c$.

Table 16

	$x < c$	$x > c$
Function is	increasing	increasing
Graph is	concave upward	concave downward

20. $f''(c) = 0$; $f'(c) = \dfrac{1}{2}$; $f''(x) > 0$ if $x < c$; $f''(x) < 0$ if $x > c$.

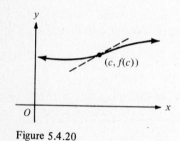

Figure 5.4.20

SOLUTION: Because $f''(x)$ changes sign at $x = c$ and $f'(c)$ is defined, then the graph of f has a point of inflection at $x = c$. Because $f'(c) = \frac{1}{2}$, the slope of the inflectional tangent is $\frac{1}{2}$, and there is some interval that contains c with f increasing on that interval. However, we do not know whether f is increasing on $(-\infty, +\infty)$. Because $f''(x) > 0$ if $x < c$, the graph of f is concave upward at every point for which x is in $(-\infty, c)$. Because $f''(x) < 0$ if $x > c$, the graph of f is concave downward at every point for which x is in $(c, +\infty)$. A sketch of a possible graph is shown in Fig. 5.4.20. We show the case where f is increasing in $(c, +\infty)$, but not in $(-\infty, c)$. The sketch shows the inflectional tangent at the point $(c, f(c))$.

24. $\lim\limits_{x \to c^-} f'(x) = +\infty$; $\lim\limits_{x \to c^+} f'(x) = -\infty$; $f''(x) > 0$ if $x < c$; $f''(x) > 0$ if $x > c$.

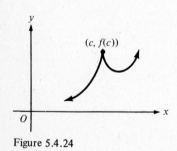

Figure 5.4.24

SOLUTION: Because $f'_-(c) = +\infty$, $f'_+(c) = -\infty$, and f is continuous at c, the tangent line to the graph of f at the point where $x = c$ exists and is a vertical line. Because $f''(x)$ does not change sign at the point where $x = c$, the graph of f does not have a point of inflection at this point. The graph is concave upward at every point for which $x < c$ and concave upward at every point for which $x > c$. Because $f'_-(c) = +\infty$, there is some interval whose right-hand endpoint is c on which f is increasing. However, we do not know whether f is increasing on $(-\infty, c)$. Because $f'_+(c) = -\infty$, there is some interval whose left-hand endpoint is c on which f is decreasing. We do not know if f is decreasing on $(c, +\infty)$. A sketch of a possible graph is shown in Fig. 5.4.24.

26. If $f(x) = 3x^2 + x|x|$, prove that $f''(0)$ does not exist, but the graph of f is concave upward everywhere.

SOLUTION: Because $|x| = x$ if $x > 0$ and $|x| = -x$ if $x < 0$, then

$$f(x) = \begin{cases} 2x^2 & \text{if } x < 0 \\ 4x^2 & \text{if } x > 0 \end{cases}$$

$$f'(x) = \begin{cases} 4x & \text{if } x < 0 \\ 8x & \text{if } x > 0 \end{cases}$$

$$f''(x) = \begin{cases} 4 & \text{if } x < 0 \\ 8 & \text{if } x > 0 \end{cases}$$

Because $f''_+(0) = 8$ and $f''_-(0) = 4$, then $f''(0)$ does not exist. Because $f'_+(0) = 0$ and $f'_-(0) = 0$, then $f'(0) = 0$. Therefore, the tangent line to the graph of f at the point where $x = 0$ has slope zero. Because $f(0) = 0$, this horizontal tangent is the line $y = 0$. Because $f(x) > 0$ if $x \neq 0$, the graph of f is above the x-axis, that is, above the line $y = 0$, for all $x \neq 0$. Therefore, by Definition 5.4.1 the graph of f is concave upward at the origin. Furthermore, because $f''(x) > 0$ if $x \neq 0$, the graph is concave upward at every point for which $x \neq 0$. Hence, the graph of f is concave upward everywhere.

5.5 APPLICATIONS TO DRAWING A SKETCH OF THE GRAPH OF A FUNCTION

The following steps may be followed when using the first and second derivatives to draw a sketch of the graph of a function f.

1. Find each number at which $f'(x)$ is either zero or undefined and find each number at which $f''(x)$ is either zero or undefined.
2. Use the numbers found in step 1, arranged in increasing order, to partition the number line into open intervals, and for each such interval determine whether $f'(x)$ and $f''(x)$ are positive or negative (or zero).
3. Use the information from step 2 to determine for each interval whether f is increasing or decreasing and whether the graph of f is concave upward or concave downward.
4. Identify each relative extremum and each point of inflection, and find the function value and the slope of the tangent line at each of these points.
5. Plot the points found in step 4 and use the information from steps 3 and 4 to draw a sketch of the graph of f.

Exercises 5.5

For each of the following functions find: the relative extrema of f, the points of inflection of the graph of f, the intervals on which f is increasing, the intervals on which f is decreasing, where the graph is concave upward, where the graph is concave downward, and the slope of any inflectional tangent. Draw a sketch of the graph.

4. $f(x) = 3x^4 + 2x^3$

SOLUTION:

$$f'(x) = 12x^3 + 6x^2 = 6x^2(2x + 1)$$
$$f''(x) = 36x^2 + 12x = 12x(3x + 1)$$

Because $f'(x) = 0$ when $x = 0$ or when $x = -\frac{1}{2}$ and $f''(x) = 0$ when $x = 0$ or $x = -\frac{1}{3}$, we use the numbers $-\frac{1}{2}$, $-\frac{1}{3}$, and 0 to partition the x-axis into intervals. Table 4a indicates whether $f'(x)$ and $f''(x)$ are positive or negative in each of the intervals. Note that $f'(x)$ does not change sign at $x = 0$ because the factor x^2, which appears in $f'(x)$, is always nonnegative. However, $f'(x)$ changes sign at $x = -\frac{1}{2}$, because the factor $2x + 1$ changes sign at $x = -\frac{1}{2}$. And $f''(x)$ changes sign at the numbers $x = -\frac{1}{3}$ and $x = 0$ because the factors $3x + 1$ and $12x$, which appear in $f''(x)$, change sign at these numbers. Table 4a also indicates for each interval whether f is increasing or decreasing and whether the graph of f is concave upward or concave downward.

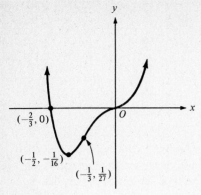

Figure 5.5.4

Because f is a polynomial function, we conclude that f has a relative extremum at each point where $f'(x)$ changes sign, and the graph of f has a point of inflection at each point where $f''(x)$ changes sign. We use the information in Tables 4a and 4b to draw a sketch of the graph of f, as illustrated in Fig. 5.5.4.

Table 4a

	$-\infty$	$-\frac{1}{2}$	$-\frac{1}{3}$	0	$+\infty$
$f'(x)$		$-$	$+$	$+$	$+$
f is		decreasing	increasing	increasing	increasing
$f''(x)$		$+$	$+$	$-$	$+$
Graph is		concave upward	concave upward	concave downward	concave upward

Table 4b

x	$f(x)$	$f'(x)$	Description of the Graph
$-\frac{1}{2}$	$-\frac{1}{16}$	0	f has a relative minimum value; graph has a horizontal tangent line.
$-\frac{1}{3}$	$-\frac{1}{27}$	$\frac{2}{9}$	Graph has a point of inflection; slope of tangent line is $\frac{2}{9}$.
0	0	0	Graph has a point of inflection; graph has a horizontal tangent line.

10. $f(x) = 3x^4 + 4x^3 + 6x^2 - 4$

SOLUTION:

$$f'(x) = 12x^3 + 12x^2 + 12x = 12x(x^2 + x + 1)$$
$$f''(x) = 36x^2 + 24x + 12 = 12(3x^2 + 2x + 1)$$

Because the equation $x^2 + x + 1 = 0$ has no real solution, $f'(x) = 0$ only when $x = 0$. Because the equation $3x^2 + 2x + 1 = 0$ has no real solution, there are no numbers for which $f''(x) = 0$. We use the number 0 to partition the real number line. Tables 10a and 10b summarize the conditions for each interval and at the critical number. A sketch of the graph of f is shown in Fig. 5.5.10. The sketch is drawn from the information in Tables 10a and 10b and by plotting a few points.

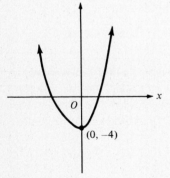

Figure 5.5.10

Table 10a

	$-\infty$	0	$+\infty$
$f'(x)$		$-$	$+$
f is		decreasing	increasing
$f''(x)$		$+$	$+$
Graph is		concave upward	concave upward

Table 10b

x	$f(x)$	$f'(x)$	Description of the Graph
0	-4	0	f has a relative minimum value; graph has a horizontal tangent line.

14. $f(x) = \begin{cases} 2(x-1)^3 & \text{if } x < 1 \\ (x-1)^4 & \text{if } x \geq 1 \end{cases}$

SOLUTION:

$$f'(x) = \begin{cases} 6(x-1)^2 & \text{if } x < 1 \\ 4(x-1)^3 & \text{if } x > 1 \end{cases}$$

$$f''(x) = \begin{cases} 12(x-1) & \text{if } x < 1 \\ 12(x-1)^2 & \text{if } x > 1 \end{cases}$$

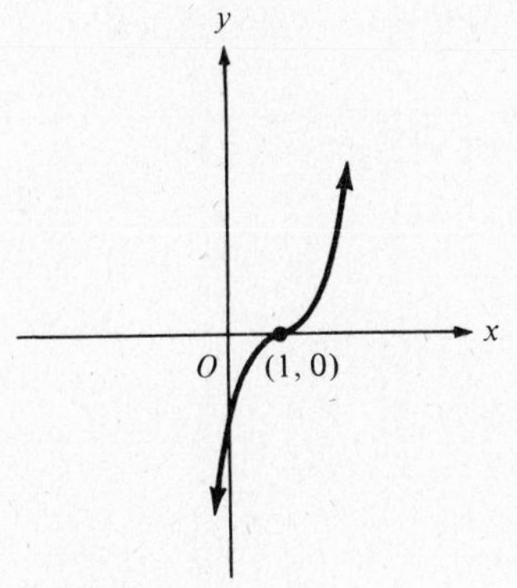

Because $f'_-(1) = f'_+(1) = 0$, then $f'(1) = 0$. Because $f''_-(1) = f''_+(1) = 0$, then $f''(1) = 0$. The only critical number is 1. A sketch of the graph, drawn from the information in Tables 14a and 14b, is shown in Fig. 5.5.14.

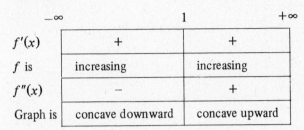

Figure 5.5.14

Table 14a

	$-\infty$	1	$+\infty$
$f'(x)$	+	+	
f is	increasing	increasing	
$f''(x)$	−	+	
Graph is	concave downward	concave upward	

Table 14b

x	$f(x)$	$f'(x)$	Description of the Graph
1	0	0	Graph has a point of inflection; graph has a horizontal tangent line.

20. $f(x) = 3x^{1/3} - x$

SOLUTION:

$$f'(x) = x^{-2/3} - 1 = \frac{1 - x^{2/3}}{x^{2/3}} = \frac{(1 + x^{1/3})(1 - x^{1/3})}{x^{2/3}}$$

$$f''(x) = -\frac{2}{3}x^{-5/3} = \frac{-2}{3x^{5/3}}$$

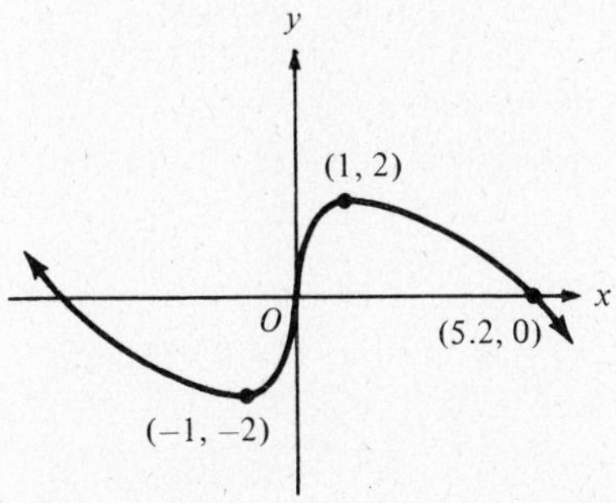

Because $f'(x) = 0$ when $x = \pm 1$ and $f'(x)$ is not defined when $x = 0$, and because $f''(x)$ is not defined when $x = 0$, we use the numbers $-1, 0$, and 1 to make Tables 20a and 20b. A sketch of the graph, drawn from the information in the tables, is shown in Fig. 5.5.20.

Figure 5.5.20

Table 20a

	$-\infty$	-1	0	1	$+\infty$
$f'(x)$	−	+	+	−	
f is	decreasing	increasing	increasing	decreasing	
$f''(x)$	+	+	−	−	
Graph is	concave upward	concave upward	concave downward	concave downward	

Table 20b

x	$f(x)$	$f'(x)$	Description of Graph
-1	-2	0	f has a relative minimum value; graph has a horizontal tangent line.
0	0	∞	Graph has a point of inflection; graph has a vertical tangent line.
1	2	0	f has a relative maximum value; graph has a horizontal tangent line.

26. $f(x) = 2 + (x - 3)^{2/3}$

SOLUTION:

$$f'(x) = \frac{2}{3}(x - 3)^{-1/3} = \frac{2}{3(x - 3)^{1/3}}$$

$$f''(x) = -\frac{2}{9}(x - 3)^{-4/3} = \frac{-2}{9(x - 3)^{4/3}}$$

The only number we need consider is $x = 3$. Tables 26a and 26b summarize the behavior of f and its graph. A sketch of the graph is drawn from the information in the tables and is shown in Fig. 5.5.26.

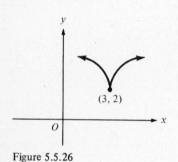

(3, 2)

Figure 5.5.26

Table 26a

	$-\infty$	3	$+\infty$
$f'(x)$	$-$	$+$	
f is	decreasing	increasing	
$f''(x)$	$-$	$-$	
Graph is	concave downward	concave downward	

Table 26b

x	$f(x)$	$f'(x)$	Description of Graph
3	2	∞	f has a relative minimum value; graph has a vertical tangent line.

32. $f(x) = \dfrac{9x}{x^2 + 9}$

SOLUTION: Because $\lim\limits_{x \to +\infty} f(x) = 0$ and $\lim\limits_{x \to -\infty} f(x) = 0$, the line $y = 0$ is a horizontal asymptote.

$$f'(x) = \frac{-9(x + 3)(x - 3)}{(x^2 + 9)^2}$$

$$f''(x) = \frac{18x(x + 3\sqrt{3})(x - 3\sqrt{3})}{(x^2 + 9)^3}$$

$f'(x) = 0$ if $x = \pm 3$; $f''(x) = 0$ if $x = 0$ or $x = \pm 3\sqrt{3}$; and both $f'(x)$ and $f''(x)$ are defined for all x. We construct Tables 32a, 32b, and 32c and use the information from the tables and the horizontal asymptotes to draw a sketch of the graph. See Fig. 5.5.32.

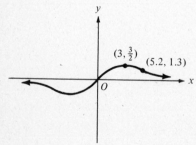

$(3, \frac{3}{2})$ (5.2, 1.3)

Figure 5.5.32

Table 32a

	$-\infty$	-3	3	$+\infty$
$f'(x)$	$-$	$+$	$-$	
f is	decreasing	increasing	decreasing	

Table 32b

	$-\infty$	$-3\sqrt{3}$	0	$3\sqrt{3}$	$+\infty$
$f''(x)$	$-$	$+$	$-$	$+$	
Graph is	concave downward	concave upward	concave downward	concave upward	

Table 32c

x	$f(x)$	$f'(x)$	Description of the Graph
$-3\sqrt{3}$	$-\dfrac{3}{4}\sqrt{3}$	$\dfrac{1}{8}$	Graph has a point of inflection; slope of the tangent line is $\frac{1}{8}$.
-3	$-\dfrac{3}{2}$	0	f has a relative minimum value; graph has a horizontal tangent line.
0	0	1	Graph has a point of inflection; slope of the tangent line is 1.
3	$\dfrac{3}{2}$	0	f has a relative maximum value; graph has a horizontal tangent line.
$3\sqrt{3}$	$\dfrac{3}{4}\sqrt{3}$	$-\dfrac{1}{8}$	Graph has a point of inflection; slope of the tangent line is $-\frac{1}{8}$.

5.6 AN APPLICATION OF THE DERIVATIVE TO ECONOMICS

The concepts in this section are not new. However, there is some special vocabulary, and there are some special formulas that are needed to solve the exercises.

The total cost equals the average cost per unit times the number of units. If $C(x)$ dollars is the total cost of producing x units, and $Q(x)$ dollars is the average cost per unit, then

$$Q(x) = \frac{C(x)}{x}$$

5.6.1 Definition If $C(x)$ is the number of dollars in the total cost of producing x units of a commodity, then the *marginal cost,* when $x = x_1$, is given by $C'(x_1)$ if it exists. The function C' is called the *marginal cost function.*

Thus, the marginal cost is the instantaneous rate of change of the total cost per one unit change in the number of units.

5.6.2 Definition If $Q(x)$ is the number of dollars in the average cost of producing one unit of x units of a commodity, then the *marginal average cost,* when $x = x_1$, is given by $Q'(x_1)$ if it exists, and Q' is called the *marginal average cost function.*

Thus the marginal average cost is the instantaneous rate of change of the average cost per unit per one unit change in the number of units.

An equation that expresses the relationship between the amount, given by x, of a commodity demanded and the price, given by p, is called a *demand equation.* If the equation is solved for p, we have the *price function P,* given by

$$p = P(x)$$

The total revenue equals the price per unit times the number of units demanded at that price. If x units are demanded when the price is $P(x)$ dollars per unit, then $R(x)$ dollars is the total revenue, and

$$R(x) = xP(x)$$

5.6.3 Definition   If $R(x)$ is the number of dollars in the total revenue obtained when x units of a commodity are demanded, then the *marginal revenue*, when $x = x_1$, is given by $R'(x_1)$, if it exists. The function R' is called the *marginal revenue function*.

Thus, the marginal revenue is the instantaneous rate of change of the revenue per one unit change in the number of units demanded.

The total profit equals the total revenue minus the total cost. If $S(x)$ dollars is the total profit on x units sold when $R(x)$ dollars is the total revenue on x units sold, and $C(x)$ dollars is the total cost of production for x units, then

$$S(x) = R(x) - C(x)$$

S is called the *profit function*.

If there is an additive tax $T(x)$ dollars on x units produced, then the profit is reduced by $T(x)$ dollars. Thus

$$S(x) = R(x) - C(x) - T(x)$$

Exercises 5.6

4. The number of dollars in the total cost of producing x units of a certain commodity is $C(x) = 40 + 3x + 9\sqrt{2x}$. Find: (a) the marginal cost when 50 units are produced and (b) the number of units produced when the marginal cost is $4.50.

SOLUTION: The marginal cost function is C', the rate of change of the total cost function.

$$C'(x) = 3 + \frac{9}{\sqrt{2x}}$$

(a) $C'(50) = 3 + \dfrac{9}{\sqrt{2 \cdot 50}} = 3.9$

The marginal cost when 50 units are produced is $3.90.

(b) Let $C'(x) = 4.5$.

$$3 + \frac{9}{\sqrt{2x}} = 4.5$$

$$\sqrt{2x} = 6$$
$$x = 18$$

If 18 units are produced, the marginal cost is $4.50.

8. If $C(x)$ dollars is the total cost of producing x units of a commodity and $C(x) = 2x^2 - 8x + 18$, find:

(a) the domain and range of C
(b) the average cost function
(c) the absolute minimum average unit cost
(d) the marginal cost function.
(e) Draw sketches of the total cost, average cost, and marginal cost curves on the same set of axes.

SOLUTION:

(a) Because C, the total cost function, must be increasing on its domain, we find where $C'(x) > 0$. Because $C'(x) = 4x - 8$, and $C'(x) > 0$ if $x > 2$, we conclude that the domain of C is $[2, +\infty)$. Because $C(2) = 10$ and $\lim\limits_{x \to +\infty} C(x) = +\infty$, the range of C is $[2, +\infty)$.

(b) The average cost function Q is defined by

$$Q(x) = \frac{C(x)}{x} = \frac{2x^2 - 8x + 18}{x} = 2x - 8 + \frac{18}{x}$$

(c) $Q'(x) = 2 - \dfrac{18}{x^2}$

If $Q'(x) = 0$, then $x = 3$. Because

$$Q''(x) = \frac{36}{x^3}$$

Then $Q''(3) = \frac{4}{3} > 0$. We conclude that Q has an absolute minimum value at $x = 3$. Since $Q(3) = 2 \cdot 3 - 8 + \frac{18}{3} = 4$, we conclude that \$4 is the absolute minimum average unit cost.

(d) The marginal cost function is C'. We found $C'(x)$ in part (a). It is $C'(x) = 4x - 8$.

(e) Fig. 5.6.8 illustrates the total cost curve (labeled TC), the average cost curve AC, and the marginal cost curve MC.

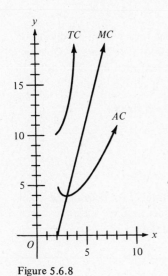

Figure 5.6.8

12. The demand equation for a particular commodity is $px^2 + 9p - 18 = 0$, where p dollars is the price per unit when $100x$ units are demanded. Find:

(a) the price function
(b) the total revenue function
(c) the marginal revenue function
(d) the absolute maximum total revenue.

SOLUTION:

(a) Solving the demand equation for p, we get

$$p = \frac{18}{x^2 + 9}$$

The price function is P where $P(x) = p$. Hence,

$$P(x) = \frac{18}{x^2 + 9} \tag{1}$$

(b) Because $100x$ units are demanded when the price is p dollars, the total revenue R is given by

$$R(x) = 100x \cdot P(x)$$

From Eq. (1), we get

$$R(x) = 100x \cdot \frac{18}{x^2 + 9} = \frac{1800x}{x^2 + 9} \tag{2}$$

(c) From Definition 5.6.3 the marginal revenue function is R'. From (2) we have

$$R'(x) = 1800 \frac{(x^2 + 9) \cdot 1 - x \cdot 2x}{(x^2 + 9)^2}$$

$$= \frac{-1800(x^2 - 9)}{(x^2 + 9)^2} \tag{3}$$

(d) We want to find the absolute maximum value of R. Because $x \geqslant 0$, from Eq. (3) we see that $R'(x) = 0$ only when $x = 3$. Furthermore, from Eq. (3) we have $R'(x) > 0$ if $0 < x < 3$, and $R'(x) < 0$ if $x > 3$. Therefore, R has a relative maximum value at $x = 3$, and because 3 is the only critical number of R, then R has an absolute maximum value at 3. From Eq. (2) we have $R(3) = 300$. Therefore, the absolute maximum total revenue is $300.

18. A monopolist determines that if $C(x)$ cents is the total cost of producing x units of a certain commodity, then $C(x) = 25x + 20,000$. The demand equation is $x + 50p = 5000$, where x units are demanded each week when the unit price is p cents. If the weekly profit is to be maximized, find:

(a) the number of units that should be produced each week
(b) the price of each unit
(c) the weekly profit.

SOLUTION:

(a) The profit equals the total revenue minus the total cost. Because the total revenue is the number of units times the price per unit, we solve the demand equation for p to find the price function. We obtain

$$P(x) = 100 - \frac{x}{50} \tag{1}$$

Therefore, if the total revenue is $R(x)$ dollars, then

$$R(x) = x \cdot P(x) = 100x - \frac{x^2}{50} \tag{2}$$

Hence, if the profit is $S(x)$ dollars, then

$$S(x) = R(x) - C(x) \tag{3}$$

Substituting from Eq. (2) and the given value for $C(x)$ into (3) we have

$$S(x) = 100x - \frac{x^2}{50} - (25x + 20,000)$$

$$= 75x - \frac{x^2}{50} - 20,000 \tag{4}$$

From Eq. (4) we obtain

$$S'(x) = 75 - \frac{x}{25}$$

If $S'(x) = 0$, then $x = 1875$. Because $S''(1875) = -\frac{1}{25} < 0$, we conclude that S has an absolute maximum value at $x = 1875$. Thus, 1875 units should be produced each week to maximize the weekly profit.
(b) From Eq. (1) we have $P(1875) = 62.5$. Therefore the price per unit is 62.5 cents.
(c) From Eq. (4) we get $S(1875) = 50312.5$. Therefore the weekly profit is 50312.5 cents or $503.125.

22. Find the maximum tax revenue that can be received by the government if an additive tax for each unit produced is levied on a monopolist for which the demand equation is $x + 3p = 75$, where x units are demanded when p dollars is the price of one unit, and $C(x) = 3x + 100$, where $C(x)$ dollars is the total cost of producing x units.

SOLUTION: Let

t = the number of dollars in the additive tax on each unit
T = the number of dollars in the total tax received from x units

Then T is a function of x,

$$T(x) = xt \tag{1}$$

and we want to find the absolute maximum value of T. Now the monopolist will set the price so that his profit is a maximum. From the demand equation we have

$$P(x) = 25 - \frac{x}{3}$$

Thus the revenue is $R(x)$ dollars, then

$$R(x) = x\,P(x) = 25x - \frac{x^2}{3} \tag{2}$$

Because the profit is found by subtracting the cost and the tax from the revenue, we have

$$S(x) = R(x) - C(x) - T(x)$$

From Eq. (2), the given value for C, and Eq. (1), we obtain

$$S(x) = \left(25x - \frac{x^2}{3}\right) - (3x + 100) - xt \tag{3}$$

For any constant value of t we have

$$S'(x) = 25 - \frac{2x}{3} - 3 - t$$

$$= 22 - \frac{2x}{3} - t$$

If $S'(x) = 0$, then

$$0 = 22 - \frac{2x}{3} - t$$

$$x = 33 - \frac{3}{2}t \tag{4}$$

Substituting the value of x from Eq. (4) into Eq. (1), we have

$$T = \left(33 - \frac{3}{2}t\right)t = 33t - \frac{3}{2}t^2$$

$$D_t T = 33 - 3t$$

If $D_t T = 0$, then $t = 11$. When $t = 11$, from Eq. (4) we get $x = \frac{33}{2}$. Substituting these values into Eq. (1), we obtain $T = 181.5$. Therefore, the maximum tax revenue is \$181.50.

Review Exercises

4. The demand in a certain market for a particular kind of breakfast cereal is given by the demand equation $px + 25p - 2000 = 0$, where p cents is the price of one box and x thousands of boxes is the quantity demanded per week. If the current price of the cereal is 40 cents per box and the price per box is increasing at the rate of 0.2 cents each week, find the rate of change in the demand.

SOLUTION: We are interested in a rate of change with respect to time. Thus, let

t = the number of weeks that have elapsed since some fixed moment of time
z = the number of boxes of cereal demanded per week after t weeks
p = the number of cents in the price per box after t weeks

We want to find $D_t z$ when $p = 40$. Because the price per box is increasing at the rate of 0.2 cents each week, we are given that $D_t p = 0.2$. Furthermore, we are given that

$$z = 1000x \tag{1}$$

where x satisfies the demand equation

$$px + 25p - 2000 = 0 \tag{2}$$

Solving Eq. (2) for x and substituting this value for x into Eq. (1), we obtain

$$z = 1000\left(\frac{2000}{p} - 25\right)$$

$$= 25000(80p^{-1} - 1) \tag{3}$$

Differentiating on both sides of Eq. (3) with respect to t gives

$$D_t z = 25000(-80p^{-2})D_t p \tag{4}$$

Substituting 40 for p and 0.2 for $D_t p$ in Eq. (4), we obtain

$$D_t z = 25000 \cdot \frac{-80}{40^2}(0.2)$$

$$= -250$$

We conclude that the demand is decreasing at the rate of 250 boxes per week.

Find the relative extrema of f, the points of inflection of the graph of f, the intervals on which f is increasing, the intervals on which f is decreasing, where the graph is concave upward, where the graph is concave downward, and the slope of any inflectional tangent. Draw a sketch of the graph.

8. $f(x) = x\sqrt{25 - x^2}$

SOLUTION: Because $25 - x^2 \geqslant 0$ if $-5 \leqslant x \leqslant 5$, the domain of f is $[-5, 5]$.

$$f'(x) = \frac{-(\sqrt{2}x + 5)(\sqrt{2}x - 5)}{(25 - x^2)^{1/2}}$$

$$f''(x) = \frac{x(2x^2 - 75)}{(25 - x^2)^{3/2}}$$

Because $f'(x) = 0$ if $x = \pm\frac{5}{2}\sqrt{2}$ and $f'(x)$ is not defined if $x = \pm 5$, there are four critical numbers of f. Now $f''(x) = 0$ if $x = 0$ or $x = \pm\frac{5}{2}\sqrt{6}$. However, neither of these latter numbers is in the domain of f. Therefore, we consider only 0. Table 8a indicates for each of the intervals determined by the numbers we are considering whether f is increasing or decreasing and whether the graph is concave upward or concave downward. Table 8b gives the function value, the value of $f'(x)$, and a description of the graph at each critical number and at 0. A sketch of the graph of f is shown in Fig. 5.8R.

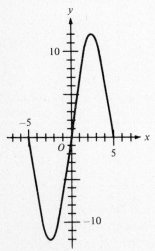

Figure 5.8.R

Table 8a

	-5	$-\frac{5}{2}\sqrt{2}$	0	$\frac{5}{2}\sqrt{2}$	5
$f'(x)$		$-$	$+$	$+$	$-$
f is		decreasing	increasing	increasing	decreasing
$f''(x)$		$+$	$+$	$-$	$-$
Graph is		concave upward	concave upward	concave downward	concave downward

Table 8b

x	$f(x)$	$f'(x)$	Description of the Graph
-5	0	∞	Endpoint; graph has a vertical tangent line.
$-\frac{5}{2}\sqrt{2}$	$-\frac{25}{2}$	0	f has a relative minimum value; graph has a horizontal tangent line.
0	0	5	Graph has a point of inflection; slope of the tangent line is 5.
$\frac{5}{2}\sqrt{2}$	$\frac{25}{2}$	0	f has a relative maximum value; graph has a horizontal tangent line.
5	0	∞	Endpoint; graph has a vertical tangent line.

12. Let $f(x) = x^n$ where n is a positive integer. Prove that the graph of f has a point of inflection at the origin if and only if n is an odd integer and $n > 1$. In addition show that if n is even, f has a relative minimum value at 0.

SOLUTION:

$$f'(x) = nx^{n-1}$$
$$f''(x) = n(n-1)x^{n-2}$$

Suppose that n is an odd integer greater than 1. Then $n - 1 > 0$, and thus $f'(0)$ exists. Hence, the graph of f has a tangent line at $x = 0$. Furthermore, because $n \geqslant 3$, then $n(n-1) > 0$. Thus,

$$f''(x) > 0 \quad \text{whenever} \quad x^{n-2} > 0 \tag{1}$$

and

$$f''(x) < 0 \quad \text{whenever} \quad x^{n-2} < 0 \tag{2}$$

Because n is an odd integer, $n - 2$ is also an odd integer. Thus,

$$x^{n-2} > 0 \quad \text{if} \quad x > 0 \tag{3}$$

and

$$x^{n-2} < 0 \quad \text{if} \quad x < 0 \tag{4}$$

From statements (1) and (3) we have

$$f''(x) > 0 \quad \text{if} \quad x > 0 \tag{5}$$

and from statements (2) and (4),

$$f''(x) < 0 \quad \text{if} \quad x < 0 \tag{6}$$

By statements (5) and (6) it follows from Definition 5.4.4 that the graph of f has a point of inflection at $x = 0$.

Now, suppose that n is an even integer. Because n is positive, then $n \geq 2$, and thus $n - 1 > 0$. Therefore $f'(0)$ exists, and hence f is continuous at $x = 0$. Because n is even, then $n - 1$ is odd. Therefore, $f'(x) > 0$ if $x > 0$, and $f'(x) < 0$ if $x < 0$. By the first derivative test we conclude that f has a relative minimum value at $x = 0$.

If $n = 1$, then $f(x) = x$ and the graph of f is a straight line. Therefore, the graph of f has a point of inflection at the origin if and only if n is odd and $n > 1$, and f has a relative minimum value at 0 if n is even.

14. Suppose the graph of a function f has a point of inflection at $x = c$. What can you conclude, if anything, about:

(a) the continuity of f at c
(b) the continuity of f' at c
(c) the continuity of f'' at c?

SOLUTION:

(a) If the graph of f has a point of inflection at c, then the graph of f has a tangent line at $x = c$; therefore, f is continuous at c.
(b) If the tangent line to the graph of f at c is a vertical line, then f' does not exist. Thus, f' may not be continuous at c.
(c) If f' does not exist at c, then f'' does not exist at c. Thus, f'' may not be continuous at c.

18. A tent is to be in the shape of a cone. Find the ratio of the measure of the radius to the measure of the altitude for a tent of given volume to require the least material.

SOLUTION: We want to minimize the lateral surface area of the cone. Let

r = the number of units in the radius of the cone
h = the number of units in the altitude of the cone
A = the number of square units in the lateral surface area of the cone

We have the formula

$$A = \pi r \sqrt{r^2 + h^2} \tag{1}$$

We want to find r/h so that A has an absolute minimum value. If V is the number of cubic units in the volume of the cone, then

$$V = \frac{1}{3} \pi r^2 h$$

$$D_r V = \frac{1}{3} \pi (r^2 D_r h + 2rh) \tag{2}$$

Because V is constant, $D_r V = 0$, and thus Eq. (2) gives

$$0 = \frac{1}{3} \pi (r^2 D_r h + 2rh)$$

$$D_r h = \frac{-2h}{r} \tag{3}$$

From Eq. (1) we obtain

$$D_r A = \pi \left[r \left(\frac{1}{2} \right) (r^2 + h^2)^{-1/2} (2r + 2h D_r h) + (r^2 + h^2)^{1/2} (1) \right]$$

$$= \pi (r^2 + h^2)^{-1/2} [r^2 + rh\, D_r h + (r^2 + h^2)] \tag{4}$$

Substituting the value for $D_r h$ given in Eq. (3) into Eq. (4), we have

$$D_r A = \pi (r^2 + h^2)^{-1/2} \left[r^2 + rh \left(-\frac{2h}{r} \right) + r^2 + h^2 \right]$$

$$= \pi (r^2 + h^2)^{-1/2} (2r^2 - h^2) \tag{5}$$

Setting $D_r A = 0$ in Eq. (5), we obtain

$$2r^2 - h^2 = 0 \tag{6}$$

We show that when Eq. (6) is satisfied, then A has a relative minimum value.
From Eq. (5) we get

$$D_r{}^2 A = \pi [(r^2 + h^2)^{-1/2} (4r - 2h \, D_r h) + (2r^2 - h^2) D_r (r^2 + h^2)^{-1/2}] \tag{7}$$

Substituting from Eq. (3) and Eq. (6) into Eq. (7) we obtain

$$D_r{}^2 A = \pi \left[(r^2 + h^2)^{-1/2} \left(4r - 2h \left(\frac{-2h}{r} \right) \right) + (0) D_r (r^2 + h^2)^{-1/2} \right]$$

$$= 4\pi \left(r + \frac{h^2}{r} \right) (r^2 + h^2)^{-1/2}$$

Because $r > 0$ and $h > 0$, then $D_r{}^2 A > 0$ whenever Eq. (6) holds. Thus A has a relative minimum value whenever Eq. (6) holds. Furthermore Eq. (6) is satisfied only for

$$\frac{r}{h} = \frac{1}{\sqrt{2}}$$

and A is a continuous function. Thus, we conclude that the relative minimum value is also an absolute minimum value. Therefore, the tent will require the least material if the ratio of radius to altitude is $\frac{1}{2}\sqrt{2}$.

20. The demand equation for a certain commodity is $p + 2\sqrt{x-1} = 6$ and the total cost function is given by $C(x) = 2x - 1$.

(a) Determine the permissible values of x.
(b) Find the marginal revenue and marginal cost functions.
(c) Find the value of x that yields the maximum profit.

SOLUTION:

(a) From the demand equation we have

$$p = 6 - 2\sqrt{x-1} \tag{1}$$

Hence, $x \geqslant 1$. Furthermore, because $p \geqslant 0$, then

$$6 - 2\sqrt{x-1} \geqslant 0$$
$$\sqrt{x-1} \leqslant 3$$
$$x - 1 \leqslant 9$$
$$x \leqslant 10$$

Therefore, the permissible values of x are $1 \leqslant x \leqslant 10$.

(b) The revenue function R is given by $R(x) = xP(x)$. From Eq. (1) we have $P(x) = 6 - 2\sqrt{x-1}$. Therefore,

$$R(x) = 6x - 2x\sqrt{x-1} \tag{2}$$

The marginal revenue function is R' and

$$R'(x) = 6 - 2x \left(\frac{1}{2} \right) (x-1)^{-1/2} + (x-1)^{1/2}(-2)$$

$$R'(x) = 6 - \frac{x}{\sqrt{x-1}} - 2\sqrt{x-1} \tag{3}$$

Because $C(x) = 2x - 1$, the marginal cost function is given by

$$C'(x) = 2 \tag{4}$$

(c) If S is the profit function then

$$S(x) = R(x) - C(x) \tag{5}$$

Therefore,

$$S'(x) = R'(x) - C'(x)$$

and $S'(x) = 0$ if

$$R'(x) = C'(x) \tag{6}$$

From Eq. (3), (4), and (6) we have

$$6 - \frac{x}{\sqrt{x-1}} - 2\sqrt{x-1} = 2$$

$$6\sqrt{x-1} - x - 2(x-1) = 2\sqrt{x-1}$$
$$4\sqrt{x-1} = 3x - 2$$
$$16(x-1) = (3x-2)^2$$

$$x = 2 \quad \text{or} \quad x = \frac{10}{9}$$

We must find the function value $S(x)$ for each critical number and for each endpoint of $[1, 10]$. By Eqs. (5) and (2) and the given value for $C(x)$, we have

$$S(x) = 6x - 2x\sqrt{x-1} - 2x + 1$$

Then $S(1) = 5$, $S(\frac{10}{9}) = \frac{127}{27} \approx 4.7$, $S(2) = 9$, and $S(10) = -19$. We conclude that the profit is a maximum when $x = 2$.

26. If $f(x) = |x|^a \cdot |x - 1|^b$, where a and b are positive rational numbers, prove that f has a relative maximum value of $a^a b^b / (a + b)^{a+b}$

SOLUTION: First, we note that

$$D_x|x| = D_x\sqrt{x^2} = \frac{1}{2}(x^2)^{-1/2}2x = \frac{x}{|x|}$$

Therefore,

$$D_x|x|^a = a|x|^{a-1}D_x|x|$$

$$= a|x|^{a-1}\frac{x}{|x|}$$

$$= ax|x|^{a-2} \tag{1}$$

Similarly,

$$D_x|x - 1|^b = b|x - 1|^{b-1}\frac{x-1}{|x-1|}$$

$$= b(x-1)|x - 1|^{b-2} \tag{2}$$

Now by the derivative of a product formula and (1) and (2) we have

$$f'(x) = |x|^a D_x|x - 1|^b + |x - 1|^b D_x|x|^a$$

$$= |x|^a \cdot b(x-1)|x - 1|^{b-2} + |x - 1|^b \cdot ax|x|^{a-2}$$

$$= |x|^{a-2}|x - 1|^{b-2}[|x|^2 b(x-1) + |x - 1|^2 ax]$$

$$= |x|^{a-2}|x - 1|^{b-2}[bx^2(x-1) + ax(x-1)^2]$$

$$= |x|^{a-2}|x-1|^{b-2}x(x-1)[bx+a(x-1)]$$

$$= |x|^{a-2}|x-1|^{b-2}x(x-1)[(a+b)x-a]$$

$$= |x|^{a-2}|x-1|^{b-2}x(x-1)(a+b)\left[x-\frac{a}{a+b}\right] \tag{3}$$

From Eq. (3) the critical numbers of f are 0, 1, and $a/(a+b)$. Table 26 shows the sign of $f'(x)$ for each of the intervals determined by these critical numbers. Note that because $a>0$ and $b>0$, then

$$0<\frac{a}{a+b}<1$$

The results of Table 26 and the first derivative test show that f has a relative maximum value at $x = a/(a+b)$. Moreover, because $a>0$ and $b>0$, then the maximum value is given by

$$f\left(\frac{a}{a+b}\right) = \left|\frac{a}{a+b}\right|^a \cdot \left|\frac{a}{a+b}-1\right|^b$$

$$= \left|\frac{a}{a+b}\right|^a \cdot \left|\frac{-b}{a+b}\right|^b$$

$$= \left|\frac{a}{a+b}\right|^a \cdot \left|\frac{b}{a+b}\right|^b$$

$$= \left(\frac{a}{a+b}\right)^a \cdot \left(\frac{b}{a+b}\right)^b$$

$$= \frac{a^a b^b}{(a+b)^{a+b}}$$

Table 26

	$-\infty$		0		$\dfrac{a}{a+b}$		1		$+\infty$
$f'(x)$		$-$		$+$		$-$		$+$	

6

The differential and antidifferentiation

6.1 THE DIFFERENTIAL

6.1.1 Definition If the function f is defined by $y = f(x)$, then the *differential of* y, denoted by dy, is given by

$$dy = f'(x)\, \Delta x$$

where x is in the domain of f' and Δx is an arbitrary increment of x.

   We note that dy is a function of two variables. We need a replacement for both x and Δx in order to calculate the function value dy.

6.1.2 Definition If the function f is defined by $y = f(x)$, then the *differential of* x, denoted by dx, is given by

$$dx = \Delta x$$

where Δx is an arbitrary increment of x, and x is any number in the domain of f'.

   When the absolute value of the ratio dx/x is "small," the value of dy is approximately the same as the value of Δy, where $\Delta y = f(x + \Delta x) - f(x)$. That is

$$dy \approx f(x + \Delta x) - f(x)$$

or, equivalently,

$$f'(x)dx \approx f(x + \Delta x) - f(x)$$

Exercises 6.1

4. Find
    (a) Δy (b) dy (c) $\Delta y - dy$

$$y = \frac{1}{x^2 + 1}$$

SOLUTION: Let $y = f(x) = \frac{1}{x^2 + 1}$. Then

(a) $\Delta y = f(x + \Delta x) - f(x)$

$$= \frac{1}{(x + \Delta x)^2 + 1} - \frac{1}{x^2 + 1}$$

$$= \frac{x^2 + 1 - x^2 - 2x\Delta x - (\Delta x)^2 - 1}{[(x + \Delta x)^2 + 1][x^2 + 1]}$$

$$= \frac{-2x\Delta x - (\Delta x)^2}{[(x + \Delta x)^2 + 1][x^2 + 1]}$$

$$= \frac{-\Delta x(2x + \Delta x)}{[(x + \Delta x)^2 + 1][x^2 + 1]}$$

(b) $dy = f'(x)\Delta x$

$$= \frac{-2x\Delta x}{(x^2 + 1)^2}$$

(c) $\Delta y - dy = \frac{-\Delta x(2x + \Delta x)}{[(x + \Delta x)^2 + 1][x^2 + 1]} + \frac{2x\Delta x}{(x^2 + 1)^2}$

$$= \frac{-\Delta x(2x + \Delta x)(x^2 + 1) + 2x\Delta x[(x + \Delta x)^2 + 1]}{[(x + \Delta x)^2 + 1][x^2 + 1]^2}$$

$$= \frac{(\Delta x)^2(3x^2 + 2x\Delta x - 1)}{[(x + \Delta x)^2 + 1][x^2 + 1]^2}$$

10. For the given values find:
   (a) Δy (b) dy (c) $\Delta y - dy$

$$y = x^3 + 1; \quad x = -1; \quad \Delta x = 0.1$$

SOLUTION: Let $y = f(x) = x^3 + 1$

(a) $\Delta y = f(x + \Delta x) - f(x)$
$$= [(x + \Delta x)^3 + 1] - [x^3 + 1]$$
$$= x^3 + 3x^2\Delta x + 3x(\Delta x)^2 + (\Delta x)^3 + 1 - x^3 - 1$$
$$= \Delta x[3x^2 + 3x\Delta x + (\Delta x)^2]$$

If $x = -1$ and $\Delta x = 0.1$, then

$$\Delta y = 0.1[3(-1)^2 + 3(-1)(0.1) + (0.1)^2]$$
$$= 0.271$$

(b) $dy = f'(x) \cdot \Delta x$
$$= 3x^2\Delta x$$

If $x = -1$ and $\Delta x = 0.1$, then
$$dy = 0.3$$

(c) If $x = -1$ and $\Delta x = 0.1$, then

$$\Delta y - dy = 0.271 - 0.3 = -0.029$$

In Exercises 13-20, use differentials to find an approximate value for the given quantity. Express your answer to three significant digits.

14. $\sqrt[3]{7.5}$

SOLUTION: Let $y = f(x) = x^{1/3}$. Because $\Delta y = f(x + \Delta x) - f(x)$, and $\Delta y \approx dy$, we have

$$f(x + \Delta x) \approx f(x) + dy \qquad (1)$$

we choose x and Δx so that $x + \Delta x = 7.5$, $f(x)$ is a rational number, and $|\Delta x|$ is as small as possible. Because 8 is the nearest perfect cube to 7.5, we let $x = 8$. Then we must have $\Delta x = -0.5$, so that $\Delta x + x = 7.5$. With these replacements, we have

$$f(x + \Delta x) = f(7.5) = \sqrt[3]{7.5} \qquad (2)$$
$$f(x) = f(8) = \sqrt[3]{8} = 2 \qquad (3)$$
$$dy = f'(x)\Delta x$$

$$= \frac{1}{3}x^{-2/3}\Delta x$$

$$= \frac{1}{12}\left(-\frac{1}{2}\right) = -\frac{1}{24} \qquad (4)$$

Substituting from (2), (3), and (4) into (1), we obtain

$$\sqrt[3]{7.5} \approx 2 + \left(-\frac{1}{24}\right) = \frac{47}{24} = 1.96$$

18. $\sqrt{0.042}$

SOLUTION: Let $y = f(x) = x^{1/2}$. Then, as in Exercise 14, we have

$$f(x + \Delta x) \approx f(x) + dy$$

$$(x + \Delta x)^{1/2} \approx x^{1/2} + \frac{1}{2}x^{-1/2}\Delta x \qquad (1)$$

We choose x and Δx so that $x + \Delta x = 0.042$, $f(x)$ is rational, and $|\Delta x|$ is as small as possible. Let $x = 0.04$ and $\Delta x = 0.002$. Then by Eq. (1) we have

$$(0.042)^{1/2} \approx (0.04)^{1/2} + \frac{1}{2}(0.04)^{-1/2}(0.002)$$

$$\sqrt{0.042} \approx .2 + 0.005 = 0.205$$

22. The altitude of a right circular cone is twice the radius of the base. The altitude is measured as 12 in, with a possible error of 0.005 in. Find the approximate error in the calculated volume of the cone.

SOLUTION: If r is the number of inches in the radius of the base, h is the number of inches in the altitude, and V is the number of cubic inches in the volume of a right circular cone, then $V = \frac{1}{3}\pi r^2 h$. We want to find $|\Delta V|$ when $h = 12$ and $|\Delta h| = 0.005$. Because $h = 2r$, or $r = \frac{1}{2}h$, we may express V as a function of h. We have

$$V = \frac{1}{12}\pi h^3$$

$$dV = \frac{1}{4}\pi h^2 \; \Delta h$$

$$|dV| = \frac{1}{4}\pi h^2 \; |\Delta h|$$

If $h = 12$ and $|\Delta h| = 0.005$, then

$$|dV| = \frac{1}{4}\pi(12)^2(0.005) = 0.18\pi = 0.565$$

Because $\Delta V \approx dV$, we conclude that 0.565 cu in is the approximate error in the calculated volume.

26. A contractor agrees to paint on both sides of 1000 circular signs each of radius 3 ft. Upon receiving the signs, he discovers that the radius is $\frac{1}{2}$ in too large. Use differentials to find the approximate percent increase of paint that will be needed.

SOLUTION: Let

r = the number of feet in the radius of each sign
A = the number of square feet in the total area that must be painted

Because there are 1000 signs, each to be painted on both sides, by the formula for the area of a circle we have

$$A = 2000\pi r^2 \tag{1}$$

Because ΔA is the increase in the paint required to do the job, then

$$\frac{100\Delta A}{A}\%$$

is the *percent* increase in paint needed. We want to find the percent increase when $r = 3$ and $\Delta r = \frac{1}{24}$, since $\frac{1}{2}$ in is $\frac{1}{24}$ ft.
Now by Eq. (1) we have

$$dA = 4000\pi r\,\Delta r$$

Thus

$$\frac{dA}{A} = \frac{4000\pi r\,\Delta r}{2000\pi r^2} = \frac{2\Delta r}{r}$$

When $r = 3$ and $\Delta r = \frac{1}{24}$, we get

$$\frac{dA}{A} = \frac{2\left(\frac{1}{24}\right)}{3} = \frac{1}{36} = 2.78\%$$

Because $\Delta A \approx dA$, we conclude that 2.78% is the approximate percent increase of paint needed.

28. If t sec is the time for one complete swing of a simple pendulum of length ℓ ft, then $4\pi^2\ell = gt^2$, where $g = 32.2$. A clock having a pendulum of length 1 ft gains 5 min each day. Find the approximate amount by which the pendulum should be lengthened in order to correct the inaccuracy.

SOLUTION: Let n be the number of complete swings that the pendulum makes in one day. Now 5 min is 5(60) sec, and one day is 24(60)^2 sec. Because the clock gains 5 min in one day, we are given that

$$n\,\Delta t = 5(60)$$
$$n t = 24(60)^2$$

Dividing the corresponding members of the above, we obtain

$$\frac{n\,\Delta t}{n t} = \frac{5(60)}{24(60)^2}$$

$$\frac{\Delta t}{t} = \frac{1}{288} \tag{1}$$

We want to find $\Delta \ell$ when $\ell = 1$. We have

$$4\pi^2 \ell = gt^2 \tag{2}$$

Taking the differential on both sides of (2) we get

$$4\pi^2 \, d\ell = 2gt \, \Delta t \tag{3}$$

Dividing the members of Eq. (3) by the corresponding members of (2) we obtain

$$\frac{4\pi^2 \, d\ell}{4\pi^2 \ell} = \frac{2gt \, \Delta t}{gt^2}$$

or, equivalently,

$$\frac{d\ell}{\ell} = \frac{2\Delta t}{t} \tag{4}$$

Substituting from Eq. (1) into Eq. (4) and replacing ℓ by 1, we obtain

$$d\ell = 2 \, \frac{1}{288} = \frac{1}{144}$$

Because $\Delta \ell \approx d\ell$, we conclude that we should lengthen the pendulum by approximately $\frac{1}{144}$ ft, which is $\frac{1}{12}$ in.

6.2 DIFFERENTIAL FORMULAS

6.2.1 Theorem   If $y = f(x)$, then when $f'(x)$ exists,

$$dy = f'(x)dx$$

whether or not x is an independent variable. Therefore, the ratio dy/dx is the derivative of y with respect to x. That is, if $y = f(x)$, then

$$f'(x) = D_x y = \frac{dy}{dx} \quad \text{if} \quad dx \neq 0$$

Using differentials, we express the chain rule for differentiation as follows.

$$\frac{dy}{dx} = \frac{dy}{du} \cdot \frac{du}{dx} \quad \text{if} \quad du \neq 0 \quad \text{and} \quad dx \neq 0$$

Although we use the symbol d^2y/dx^2 to represent $D_x^2 y$, we do not regard this fraction as the quotient of two differentials. Neither the numerator nor the denominator of the fraction d^2y/dx^2 has any meaning when considered alone. In general, we represent $D_x^n y$ by $d^n y/dx^n$, but the fraction does *not* represent the quotient of two differentials.

The following formulas are used to find differentials. In the formulas, c is any constant, n is any rational number, and u and v are any functions that are differentiable.

1. $d(c) = 0$
2. $d(x^n) = nx^{n-1}dx$
3. $d(cu) = c \, du$
4. $d(u + v) = du + dv$
5. $d(uv) = u \, dv + v \, du$
6. $d\left(\dfrac{u}{v}\right) = \dfrac{v \, du - u \, dv}{v^2}$
7. $d(u^n) = nu^{n-1}du$

You should realize that the formulas for finding differentials given here follow from the corresponding formulas for finding derivatives given in Chapter 3, because of Theorem 6.2.1. The differential formulas are particularly useful when doing implicit differentiation because we do not need to distinguish the independent variable from the dependent variable. This use of the differential formulas is illustrated in Exercises 12 and 16.

Exercises 6.2

In Exercises 1-8, find dy.

4. $y = \dfrac{3x}{x^2 + 2}$

SOLUTION:

$$dy = \frac{(x^2 + 2)d(3x) - 3x\, d(x^2 + 2)}{(x^2 + 2)^2}$$

$$= \frac{(x^2 + 2)(3dx) - 3x(2xdx)}{(x^2 + 2)^2}$$

$$= -\frac{3(x^2 - 2)dx}{(x^2 + 2)^2}$$

8. $y = \sqrt{3x + 4}\ \sqrt[3]{x^2 - 1}$

SOLUTION: $y = (3x + 4)^{1/2}(x^2 - 1)^{1/3}$

$$dy = (3x + 4)^{1/2}d[(x^2 - 1)^{1/3}] + (x^2 - 1)^{1/3}d[(3x + 4)^{1/2}]$$

$$= (3x + 4)^{1/2}\frac{1}{3}(x^2 - 1)^{-2/3}(2xdx) + (x^2 - 1)^{1/3}\frac{1}{2}(3x + 4)^{-1/2}(3dx)$$

$$= \frac{1}{6}(3x + 4)^{-1/2}(x^2 - 1)^{-2/3}[4x(3x + 4) + 9(x^2 - 1)]dx$$

$$= \frac{1}{6}(3x + 4)^{-1/2}(x^2 - 1)^{-2/3}(21x^2 + 16x - 9)dx$$

In Exercises 9-16, x and y are functions of a third variable t. Find dy/dx by finding the differential term by term.

12. $2x^2y - 3xy^3 + 6y^2 = 1$

SOLUTION: We take the differential on both sides of the given equation.

$$d(2x^2y) + d(-3xy^3) + d(6y^2) = d(1)$$
$$2x^2dy + yd(2x^2) + (-3x)d(y^3) + y^3d(-3x) + 6d(y^2) = 0$$
$$2x^2dy + y(4xdx) - 3x(3y^2dy) + y^3(-3dx) + 6(2ydy) = 0$$

We separate the terms that contain dx from those that contain dy and solve for dy/dx.

$$(2x^2 - 9xy^2 + 12y)dy = (3y^3 - 4xy)dx$$

$$\frac{dy}{dx} = \frac{3y^3 - 4xy}{2x^2 - 9xy^2 + 12y}$$

16. $x^2 + y^2 = \sqrt[3]{x + y}$

SOLUTION:

$$2xdx + 2ydy = \frac{1}{3}(x + y)^{-2/3}(dx + dy)$$

$$6x(x+y)^{2/3}dx + 6y(x+y)^{2/3}dy = dx + dy$$
$$[6y(x+y)^{2/3} - 1]dy = [1 - 6x(x+y)^{2/3}]dx$$
$$\frac{dy}{dx} = \frac{1 - 6x(x+y)^{2/3}}{6y(x+y)^{2/3} - 1}$$

In Exercises 17-24, find dy/dt.

20. $y = \sqrt[3]{5x - 1}$; $x = \sqrt{2t + 3}$

SOLUTION: Taking the differential on both sides of each equation, we have

$$dy = \frac{1}{3}(5x - 1)^{-2/3}(5dx) \tag{1}$$

$$dx = \frac{1}{2}(2t + 3)^{-1/2}(2dt) \tag{2}$$

Substituting the value for dx from Eq. (2) into Eq. (1), we have

$$dy = \frac{5}{3}(5x - 1)^{-2/3}(2t + 3)^{-1/2}dt$$

$$\frac{dy}{dt} = \frac{5}{3(5x - 1)^{2/3}(2t + 3)^{1/2}} \tag{3}$$

Substituting the given value for x into Eq. (3), we obtain

$$\frac{dy}{dt} = \frac{5}{3(5\sqrt{2t + 3} - 1)^{2/3}(2t + 3)^{1/2}}$$

24. $3x^2y - 4xy^2 + 7y^3 = 0$; $2x^3 - 3xt^2 + t^3 = 1$

SOLUTION: Because

$$3x^2y - 4xy^2 + 7y^3 = 0$$

then by taking the differential on each side, we have

$$3x^2dy + y(6xdx) - 4x(2ydy) + y^2(-4dx) + 21y^2dy = 0$$

$$\frac{dy}{dx} = \frac{4y^2 - 6xy}{3x^2 - 8xy + 21y^2} \tag{1}$$

Because

$$2x^3 - 3xt^2 + t^3 = 1$$

then by taking the differential on each side, we obtain

$$6x^2dx - 3x(2tdt) + t^2(-3dx) + 3t^2dt = 0$$

$$\frac{dx}{dt} = \frac{6xt - 3t^2}{6x^2 - 3t^2} = \frac{2xt - t^2}{2x^2 - t^2} \tag{2}$$

Because

$$\frac{dy}{dt} = \frac{dy}{dx} \cdot \frac{dx}{dt}$$

From Eq. (1) and Eq. (2) we get

$$\frac{dy}{dt} = \frac{4y^2 - 6xy}{3x^2 - 8xy + 21y^2} \cdot \frac{2xt - t^2}{2x^2 - t^2}$$

6.3 THE INVERSE OF DIFFERENTIATION

6.3.1 Definition A function F is called an *antiderivative* of a function f on an interval I if $F'(x) = f(x)$ for every value of x on I.

6.3.4 Theorem If F is any particular antiderivative of f on an interval I, then the most general antiderivative of f on I is given by $F(x) + C$, where C is an arbitrary constant, and all antiderivatives of f on I can be obtained from $F(x) + C$ by assigning particular values to C.

The symbol $\int$ denotes the operation of antidifferentiation. That is,

$$\int d(F(x)) = F(x) + C \tag{1}$$

Eq. (1) states that when we antidifferentiate the differential of a function, we obtain that function plus an arbitrary constant. We have the following formulas for antidifferentiation.

6.3.5 Formula 1 $\displaystyle\int dx = x + C$

6.3.6 Formula 2 $\displaystyle\int af(x)\, dx = a \int f(x)\, dx$, where a is a constant

6.3.7 Formula 3 $\displaystyle\int [f_1(x) + f_2(x)]\, dx = \int f_1(x)\, dx + \int f_2(x)\, dx$

6.3.8 Formula 4 $\displaystyle\int [c_1 f_1(x) + c_2 f_2(x) + \cdots + c_n f_n(x)]\, dx = c_1 \int f_1(x)\, dx + c_2 \int f_2(x)\, dx$

$$+ \cdots + c_n \int f_n(x)\, dx$$

6.3.9 Formula 5 $\displaystyle\int x^n\, dx = \frac{x^{n+1}}{n+1} + C \quad \text{if } n \neq -1$

6.3.11 Formula 6 If g is a differentiable function, then if $u = g(x)$,

$$\int [g(x)]^n g'(x)\, dx = \int u^n\, du = \frac{u^{n+1}}{n+1} + C = \frac{[g(x)]^{n+1}}{n+1} + C$$

We note that

$$\int f(x) \cdot g(x)\, dx \neq \int f(x)\, dx \cdot \int g(x)\, dx$$

$$\int \frac{f(x)}{g(x)}\, dx \neq \frac{\displaystyle\int f(x)\, dx}{\displaystyle\int g(x)\, dx}$$

To find the antiderivative of a product or quotient we may sometimes replace the given function by an equivalent sum and then use Formula 3 or Formula 4, as illustrated in Exercise 8. Sometimes we may be able to apply Formula 6, as illustrated in Exercises 20 and 26.

Exercises 6.3

In Exercises 1-26, find the most general antiderivative. In Exercises 1-10, check by finding the derivative of your answer.

4. $\int (ax^2 + bx + c)\, dx$

SOLUTION: Because a, b, and c are constants, then

$$\int (ax^2 + bx + c)\, dx = a \int x^2 dx + b \int x dx + c \int dx$$

$$= \frac{1}{3} ax^3 + \frac{1}{2} bx^2 + cx + C$$

CHECK: Let $F(x) = \frac{1}{3} ax^3 + \frac{1}{2} bx^2 + cx + C$, then

$$F'(x) = \frac{a}{3}(3x^2) + \frac{b}{2}(2x) + c$$

$$= ax^2 + bx + c$$

Because $F'(x)$ is the given function, the solution checks.

8. $\int \dfrac{27t^3 - 1}{\sqrt[3]{t}}\, dt$

SOLUTION: We replace the quotient by an equivalent sum.

$$\int \frac{27t^3 - 1}{\sqrt[3]{t}}\, dt = \int (27t^3 - 1) t^{-1/3} dt$$

$$= \int (27t^{8/3} - t^{-1/3}) dt$$

$$= 27 \int t^{8/3} dt - \int t^{-1/3} dt$$

$$= \frac{27 t^{8/3 + 1}}{\frac{8}{3} + 1} - \frac{t^{-1/3 + 1}}{-\frac{1}{3} + 1} + C$$

$$= \frac{81}{11} t^{11/3} - \frac{3}{2} t^{2/3} + C$$

CHECK: Let

$$f(t) = \frac{81}{11} t^{11/3} - \frac{3}{2} t^{2/3} + C$$

Then

$$f'(t) = \frac{81}{11} \cdot \frac{11}{3} t^{8/3} - \frac{3}{2} \cdot \frac{2}{3} t^{-1/3}$$

$$= 27 t^{8/3} - t^{-1/3}$$

$$= t^{-1/3}(27t^3 - 1)$$

$$= \frac{27t^3 - 1}{\sqrt[3]{t}}$$

12. $\int \sqrt{5r + 1}\, dr$

SOLUTION: We use Formula 6. Let $u = 5r + 1$. Then $du = 5 \cdot dr$, so $dr = \frac{1}{5} du$.

$$\int \sqrt{5r+1}\, dr = \int \sqrt{u}\,\frac{1}{5}\, du$$

$$= \frac{1}{5}\int u^{1/2}\, du$$

$$= \frac{1}{5}\,\frac{u^{1/2+1}}{\frac{1}{2}+1} + C$$

$$= \frac{1}{5}\,\frac{2}{3}\, u^{3/2} + C$$

$$= \frac{2}{15}(5r+1)^{3/2} + C$$

Note that we use the substitution $u = 5r + 1$ in the last step.

16. $\displaystyle\int (x^2 - 4x + 4)^{4/3}\, dx$

SOLUTION: Because $x^2 - 4x + 4 = (x-2)^2$, then

$$\int (x^2 - 4x + 4)^{4/3}\, dx = \int [(x-2)^2]^{4/3}\, dx$$

$$= \int (x-2)^{8/3}\, dx \qquad (1)$$

Let $u = x - 2$. Then $du = dx$, and by substituting in the right member of (1), we obtain

$$\int (x^2 - 4x + 4)^{4/3}\, dx = \int u^{8/3}\, du$$

$$= \frac{u^{8/3+1}}{\frac{8}{3}+1} + C$$

$$= \frac{3}{11}\, u^{11/3} + C$$

$$= \frac{3}{11}(x-2)^{11/3} + C$$

20. $\displaystyle\int (x^2 + 3)^{1/4}\, x^5 dx$

SOLUTION: Let

$$u = x^2 + 3 \qquad (1)$$

Then $du = 2x\, dx$, or

$$\frac{1}{2}\, du = x\, dx \qquad (2)$$

Because $(x^2 + 3)^{1/4} x^5 dx = (x^2 + 3)^{1/4} \cdot x^4 \cdot x\, dx$, we express x^4 as a function of u. From (1) we have

$$x^2 = u - 3$$
$$x^4 = u^2 - 6u + 9 \qquad (3)$$

Therefore, by (1), (2), and (3) we obtain

$$\int (x^2 + 3)^{1/4} x^5 \, dx = \int (x^2 + 3)^{1/4} \cdot x^4 (x \, dx)$$

$$= \int u^{1/4} (u^2 - 6u + 9)\left(\frac{1}{2} \, du\right)$$

$$= \frac{1}{2} \int (u^{9/4} - 6u^{5/4} + 9u^{1/4}) \, du$$

$$= \frac{1}{2}\left[\frac{4}{13} u^{13/4} - 6 \cdot \frac{4}{9} u^{9/4} + 9 \cdot \frac{4}{5} \cdot u^{5/4}\right] + C$$

$$= \frac{2}{13}(x^2 + 3)^{13/4} - \frac{4}{3}(x^2 + 3)^{9/4} + \frac{18}{5}(x^2 + 3)^{5/4} + C$$

26. $\int \left(t + \frac{1}{t}\right)^{3/2} \frac{t^2 - 1}{t^2} \, dt$

SOLUTION: Let

$$u = t + \frac{1}{t} \tag{1}$$

$$du = \left(1 - \frac{1}{t^2}\right) dt$$

$$= \frac{t^2 - 1}{t^2} \, dt \tag{2}$$

Therefore by (1) and (2) we have

$$\int \left(t + \frac{1}{t}\right)^{3/2} \frac{t^2 - 1}{t^2} \, dt = \int u^{3/2} \, du$$

$$= \frac{2}{5} u^{5/2} + C$$

$$= \frac{2}{5} \left(t + \frac{1}{t}\right)^{5/2} + C$$

28. Evaluate $\int \sqrt{x - 1} \, x^2 \, dx$ by two methods.

(a) Make the substitution $u = x - 1$
(b) Make the substitution $v = \sqrt{x - 1}$

SOLUTION:

(a) If $u = x - 1$, then $du = dx$, and $x^2 = (u + 1)^2$. Thus,

$$\int \sqrt{x - 1} \, x^2 \, dx = \int u^{1/2} (u + 1)^2 \, du$$

$$= \int u^{1/2} (u^2 + 2u + 1) \, du$$

$$= \int (u^{5/2} + 2u^{3/2} + u^{1/2}) \, du$$

$$= \frac{2}{7} u^{7/2} + 2 \cdot \frac{2}{5} u^{5/2} + \frac{2}{3} u^{3/2} + C$$

$$= \frac{2}{7}(x - 1)^{7/2} + \frac{4}{5}(x - 1)^{5/2} + \frac{2}{3}(x - 1)^{3/2} + C$$

(b) If $v = \sqrt{x-1}$, then $v^2 = x - 1$ or $x = v^2 + 1$. Thus, $dx = 2vdv$ and $x^2 = (v^2 + 1)^2$. Making these substitutions, we have

$$\int \sqrt{x-1}\, x^2 dx = \int v(v^2 + 1)^2\, 2vdv$$

$$= 2 \int v^2(v^4 + 2v^2 + 1)\, dv$$

$$= 2 \int (v^6 + 2v^4 + v^2)\, dv$$

$$= 2 \left[\frac{1}{7}v^7 + \frac{2}{5}v^5 + \frac{1}{3}v^3 \right] + C$$

$$= \frac{2}{7}(\sqrt{x-1})^7 + \frac{4}{5}(\sqrt{x-1})^5 + \frac{2}{3}(\sqrt{x-1})^3 + C$$

$$= \frac{2}{7}(x-1)^{7/2} + \frac{4}{5}(x-1)^{5/2} + \frac{2}{3}(x-1)^{3/2} + C$$

32. Show that the unit step function U (Exercise 21 in Exercises 1.8) does not have an antiderivative on $(-\infty, +\infty)$.

SOLUTION: Suppose that F is an antiderivative of U on $(-\infty, +\infty)$. Then

$$F'(x) = U(x) = \begin{cases} 0 & \text{if } x < 0 \\ 1 & \text{if } x \geqslant 0 \end{cases} \tag{1}$$

Because F' exists on $(-\infty, +\infty)$, then for any $x > 0$, the hypothesis of the mean-value theorem is satisfied for the function F and the interval $[0, x]$. Thus, there is some c in $(0, x)$ such that

$$\frac{F(x) - F(0)}{x - 0} = F'(c) \tag{2}$$

Because $c > 0$ and $F'(c) = U(c)$, by (1) we have $F'(c) = 1$. Hence, from (2) we get

$$F(x) = x + F(0) \quad \text{if } x > 0 \tag{3}$$

On the other hand, if $x < 0$, then the mean-value theorem is satisfied for the function F and the interval $[x, 0]$. Thus, there is some c in $(x, 0)$ such that Eq. (2) holds. Because $c < 0$ and $F'(c) = U(c)$, by (1) we have $F'(c) = 0$. Thus, from (2) we get

$$F(x) = F(0) \quad \text{if } x < 0 \tag{4}$$

Now $F(0)$ is a constant; so differentiating both sides of Eq. (3) we have

$$F'(x) = 1 \quad \text{if } x > 0 \tag{5}$$

and differentiating both sides of Eq. (4), we get

$$F'(x) = 0 \quad \text{if } x < 0 \tag{6}$$

By (5) and (6) we have $F'_+(0) \neq F'_-(0)$. Therefore $F'(0)$ does not exist. But this contradicts (1); hence there is no such function F which is an antiderivative for U on $(-\infty, +\infty)$.

6.4 DIFFERENTIAL EQUATIONS WITH VARIABLES SEPARABLE

A first-order differential equation in the two variables x and y is an equation involving x, y, dx, and dy. A solution of this differential equation is an equation involving x and y such that if we take the differential of each side, the result is the original differential equation. The following steps are used to find a solution of a first-order differential equation if we can separate the variables.

1. Separate the variables. That is, write the differential equation in the form

$$g(y)dy = f(x)dx$$

2. Antidifferentiate on both sides of the equation in step 1. This results in an equation of the form

$$G(y) = F(x) + C$$

where $G'(y) = g(y)$, $F'(x) = f(x)$, and C is an arbitrary constant. This equation is called the *complete solution* of the differential equation.

3. If boundary conditions are given (that is, replacements for x and y are given), substitute the given values into the complete solution, solve for C, and replace C in the complete solution by the value of C just found. This results in the *particular solution* of the differential equation.

The steps for finding the complete solution are illustrated in Exercise 4, and the steps for finding the particular solution are given in Exercise 10.

A second-order differential equation in x and y is an equation involving x, y, and d^2y/dx^2. The only type we consider is of the form

$$\frac{d^2y}{dx^2} = f(x)$$

To find a solution, first let $y' = dy/dx$. Then $d^2y/dx^2 = dy'/dx$, and thus the given differential equation is equivalent to

$$dy' = f(x)\,dx$$

This is a first-order differential equation in x and y' whose solution may be found by steps 2 and 3 above. The solution is an equation containing x and y'. Now replace y' by dy/dx and solve the resulting first-order differential equation in x and y. There will be two arbitrary constants in the complete solution of a second-order differential equation. The steps for finding the complete solution are illustrated in Exercise 8, and the steps for finding the particular solution are illustrated in Exercise 14.

Exercises 6.4

In Exercises 1-8, find the complete solution of the given differential equation

4. $\dfrac{ds}{dt} = 5\sqrt{s}$

SOLUTION: First, we separate the variables

$$\frac{ds}{\sqrt{s}} = 5dt$$

Next, we antidifferentiate on both sides.

$$\int \frac{ds}{\sqrt{s}} = \int 5dt$$

$$\int s^{-1/2}\,ds = 5\int dt$$

$$2s^{1/2} = 5t + C$$

We have found the complete solution.

8. $\dfrac{d^2y}{dx^2} = \sqrt{2x - 3}$

SOLUTION: Let $y' = dy/dx$. Then $d^2y/dx^2 = dy'/dx$, and the given differential equation is equivalent to

$$\frac{dy'}{dx} = \sqrt{2x - 3}$$

$$dy' = \sqrt{2x - 3}\, dx$$

$$\int dy' = \int \sqrt{2x - 3}\, dx$$

Let $u = 2x - 3$. Then $du = 2dx$, and thus

$$\int dy' = \int u^{1/2} \frac{1}{2}\, du$$

$$y' = \frac{1}{2} \cdot \frac{2}{3} u^{3/2} + C_1$$

$$y' = \frac{1}{3}(2x - 3)^{3/2} + C_1$$

Because $y' = dy/dx$, the above equation is equivalent to

$$\frac{dy}{dx} = \frac{1}{3}(2x - 3)^{3/2} + C_1$$

$$dy = \left[\frac{1}{3}(2x - 3)^{3/2} + C_1\right] dx$$

$$\int dy = \frac{1}{3}\int (2x - 3)^{3/2}\, dx + C_1 \int dx \qquad (1)$$

As before, we let $u = 2x - 3$ and $du = 2dx$. Thus (1) becomes

$$\int dy = \frac{1}{3}\int u^{3/2} \cdot \frac{1}{2}\, du + C_1 \int dx$$

$$y = \frac{1}{6} \cdot \frac{2}{5} u^{5/2} + C_1 x + C_2$$

$$y = \frac{1}{15}(2x - 3)^{5/2} + C_1 x + C_2$$

In Exercises 9-14, for each of the differential equations find the particular solution determined by the given boundary conditions.

10. $\dfrac{dy}{dx} = (x + 1)(x + 2)$; $y = -\frac{3}{2}$ when $x = -3$

SOLUTION:

$$dy = (x + 1)(x + 2)\, dx$$

$$\int dy = (x^2 + 3x + 2)\, dx$$

$$y = \frac{1}{3}x^3 + \frac{3}{2}x^2 + 2x + C \qquad (1)$$

We substitute the given replacements for x and y into Eq. (1)

$$-\frac{3}{2} = \frac{1}{3}(-3)^3 + \frac{3}{2}(-3)^2 + 2(-3) + C$$

$$C = 0$$

Replacing C by 0 in Eq. (1), we obtain

$$y = \frac{1}{3}x^3 + \frac{3}{2}x^2 + 2x$$

We have found the particular solution.

14. $\frac{d^2y}{dx^2} = \sqrt[3]{3x - 1}$; $y = 2$ and $y' = 5$ when $x = 3$.

SOLUTION: If $y' = dy/dx$, then $dy'/dx = d^2y/dx^2$. Therefore, the given differential equation is equivalent to

$$\frac{dy'}{dx} = \sqrt[3]{3x - 1}$$

$$dy' = \sqrt[3]{3x - 1}\,dx$$

$$\int dy' = \int \sqrt[3]{3x - 1}\,dx$$

Let $u = 3x - 1$ and $du = 3dx$.

$$\int dy' = \int u^{1/3} \cdot \frac{1}{3}\,du$$

$$y' = \frac{1}{3} \cdot \frac{3}{4}u^{4/3} + C_1$$

$$y' = \frac{1}{4}(3x - 1)^{4/3} + C_1 \qquad (1)$$

Because $y' = 5$ when $x = 3$, from Eq. (1), we have

$$5 = \frac{1}{4}(3 \cdot 3 - 1)^{4/3} + C_1$$

$$C_1 = 1$$

Thus, Eq. (1) with C_1 replaced by 1 becomes

$$y' = \frac{1}{4}(3x - 1)^{4/3} + 1$$

Because $y' = dy/dx$, then we have

$$\frac{dy}{dx} = \frac{1}{4}(3x - 1)^{4/3} + 1$$

$$dy = \left[\frac{1}{4}(3x - 1)^{4/3} + 1\right]dx$$

$$\int dy = \int \left[\frac{1}{4}(3x - 1)^{4/3}\,dx + 1\right]dx$$

$$= \frac{1}{4}\int (3x - 1)^{4/3}\,dx + \int dx$$

If $u = 3x - 1$, then $du = 3dx$, and

$$\int dy = \frac{1}{4}\int u^{4/3} \cdot \frac{1}{3}\,du + \int dx$$

$$y = \frac{1}{12} \cdot \frac{3}{7} u^{7/3} + x + C_2$$

$$y = \frac{1}{28} (3x - 1)^{7/3} + x + C_2 \qquad (2)$$

Because $y = 2$ when $x = 3$, from Eq. (2) we have

$$2 = \frac{1}{28}(3 \cdot 3 - 1)^{7/3} + 3 + C_2$$

$$C_2 = -\frac{39}{7}$$

With this replacement for C_2, Eq. (2) becomes

$$y = \frac{1}{28}(3x - 1)^{7/3} + x - \frac{39}{7}$$

16. The slope of the tangent line at any point (x, y) on a curve is $3\sqrt{x}$. If the point $(9, 4)$ is on the curve, find an equation of the curve.

SOLUTION: Because the slope of the tangent line is given by $D_x y$, which is equivalent to dy/dx, we are given that

$$\frac{dy}{dx} = 3\sqrt{x}$$

$$dy = 3\sqrt{x}\, dx$$

$$\int dy = \int 3\sqrt{x}\, dx$$

$$\int dy = 3\int x^{1/2}\, dx$$

$$y = 2x^{3/2} + C \qquad (1)$$

Because the point $(9, 4)$ is on the curve, then $y = 4$ when $x = 9$. Thus Eq. (1) yields

$$4 = 2(9)^{3/2} + C$$
$$C = -50$$

From Eq. (1) with $C = -50$ we find an equation of the curve. It is

$$y = 2x^{3/2} - 50$$

20. At any point (x, y) on a curve, $D_x^3 y = 2$, and $(1, 3)$ is a point of inflection at which the slope of the inflectional tangent is -2. Find an equation of the curve.

SOLUTION: If $y' = dy/dx$, then $dy'/dx = d^2y/dx^2$. Let $y'' = dy'/dx$, and then $dy''/dx = d^3y/dx^3 = D_x^3 y$. Because $D_x^3 y = 2$, we have

$$\frac{dy''}{dx} = 2$$

$$dy'' = 2dx$$

$$\int dy'' = \int 2dx$$

$$y'' = 2x + C_1 \qquad (1)$$

$$\frac{dy'}{dx} = 2x + C_1$$

$$dy' = (2x + C_1)dx$$

$$\int dy' = \int (2x + C_1)dx$$

$$y' = x^2 + C_1 x + C_2 \tag{2}$$

$$\frac{dy}{dx} = x^2 + C_1 x + C_2$$

$$dy = (x^2 + C_1 x + C_2)dx$$

$$\int dy = \int (x^2 + C_1 x + C_2)dx$$

$$y = \frac{1}{3}x^3 + \frac{1}{2}C_1 x^2 + C_2 x + C_3 \tag{3}$$

Because Eq. (3) defines a polynomial function, at a point of inflection of the graph of this function we must have $D_x{}^2 y = 0$. Because the curve has a point of inflection at the point $(1, 3)$, then $y'' = 0$ when $x = 1$. Substituting these values into Eq. (1), we have

$$0 = 2 + C_1$$
$$C_1 = -2$$

with $C_1 = -2$, Eq. (2) becomes

$$y' = x^2 - 2x + C_2 \tag{4}$$

Because the slope of the inflectional tangent is -2, then $y' = -2$ when $x = 1$. Thus, from (4) we have

$$-2 = 1^2 - 2 \cdot 1 + C_2$$
$$C_2 = -1$$

and from Eq. (3) with $C_1 = -2$ and $C_2 = -1$, we get

$$y = \frac{1}{3}x^3 - x^2 - x + C_3 \tag{5}$$

Because $(1, 3)$ is a point on the curve, then $y = 3$ when $x = 1$. Thus, from Eq. (5) we have

$$3 = \frac{1}{3} \cdot 1^3 - 1^2 - 1 + C_3$$

$$C_3 = \frac{14}{3}$$

Substituting this value for C_3 into Eq. (5), we get

$$y = \frac{1}{3}x^3 - x^2 - x + \frac{14}{3}$$

6.5 ANTIDIFFERENTIATION AND RECTILINEAR MOTION

If a particle moves in a straight line and if its directed distance from the origin is s units of distance at t units of time, then the units of velocity v and the units of acceleration a are given by

$$v = \frac{ds}{dt} \tag{1}$$

$$a = \frac{d^2 s}{dt^2} = \frac{dv}{dt} \qquad (2)$$

or

$$a = v \cdot \frac{dv}{ds} \qquad (3)$$

We use Eq. (1) and Eq. (2) when we are given the acceleration and wish to find the time, distance, and velocity, as illustrated in Exercise 8. We use Eq. (3) when we are interested only in acceleration, distance, and velocity, as illustrated in Exercises 4, 12, and 14.

If the only force acting on a particle is the force of gravity, then the acceleration is constant, and is approximately 32 ft/sec².

Exercises 6.5

4. A particle is moving in a straight line; s ft is the directed distance of the particle from the origin at t sec of time; v ft/sec is the velocity of the particle at t sec; and a ft/sec² is the acceleration of the particle at t sec. If $a = 2s + 1$ and $v = 2$ when $s = 1$, find an equation involving v and s.

SOLUTION: Because $a = dv/dt$ and $v = ds/dt$, then

$$a = \frac{dv}{dt} = \frac{dv}{ds} \cdot \frac{ds}{dt} = \frac{dv}{ds} \cdot v \qquad (1)$$

We are given that $a = 2s + 1$. From Eq. (1) we have

$$\frac{dv}{ds} \cdot v = 2s + 1$$

$$v\,dv = (2s + 1)ds$$

$$\int v\,dv = \int (2s + 1)ds$$

$$\frac{1}{2}v^2 = s^2 + s + C$$

Because $v = 2$ when $s = 1$, we get

$$\frac{1}{2} \cdot 2^2 = 1^2 + 1 + C$$

$$C = 0$$

Hence

$$\frac{1}{2}v^2 = s^2 + s$$

$$v^2 = 2s^2 + 2s$$

8. A stone is thrown vertically upward from the top of a house 60 ft above the ground with an initial velocity of 40 ft/sec. At what time will the stone reach its greatest height, and what is its greatest height? How long will it take the stone to pass the top of the house on its way down, and what is its velocity at that instant? How long will it take the stone to strike the ground, and with what velocity does it strike the ground?

SOLUTION: Let

t = the number of sec in the time that has elapsed since the stone was thrown

s = the number of ft in the distance of the stone above the ground at t sec

v = the number of ft/sec in the velocity of the stone at t sec

Because the positive direction is chosen as upward, there is constant negative accelera-tion due to the force of gravity, with $a = -32$. We are given that $s = 60$ and $v = 40$ when $t = 0$, because the top of the house is 60 ft above the ground, and the stone is is given an initial velocity of 40 ft/sec. Because $a = dv/dt$, we have

$$\frac{dv}{dt} = -32$$

$$dv = -32dt$$

$$\int dv = \int -32dt$$

$$v = -32t + C_1$$

Because $v = 40$ when $t = 0$, then $C_1 = 40$ and

$$v = -32t + 40 \tag{1}$$

Because $v = ds/dt$, from Eq. (1) we have

$$\frac{ds}{dt} = -32t + 40$$

$$ds = (-32t + 40)dt$$

$$\int ds = \int (-32t + 40)dt$$

$$s = -16t^2 + 40t + C_2$$

Because $s = 60$ when $t = 0$, then $C_2 = 60$, and

$$s = -16t^2 + 40t + 60 \tag{2}$$

The stone reaches its greatest height when $v = 0$. From Eq. (1), with $v = 0$, we get

$$0 = -32t + 40$$

$$t = \frac{5}{4}$$

Thus, it takes the stone $\frac{5}{4}$ sec to reach its greatest height. Replacing t by $\frac{5}{4}$ in Eq. (2), we have

$$s = -16\left(\frac{5}{4}\right)^2 + 40\left(\frac{5}{4}\right) + 60 = 85$$

Thus, the greatest height reached by the stone is 85 feet. When the stone passes the top of the house on its way down, then $s = 60$. With this replacement in Eq. (2) we have

$$60 = -16t^2 + 40t + 60$$
$$t(2t - 5) = 0$$

$$t = 0 \quad \text{or} \quad t = \frac{5}{2}$$

Because $t = 0$ when the stone is first thrown upward, we conclude that it takes the stone $\frac{5}{2}$ sec to pass the top of the house on its way down. Replacing t by $\frac{5}{2}$ in Eq. (1), we have

$$v = -32\left(\frac{5}{2}\right) + 40 = -40$$

Thus, the velocity of the stone is -40 ft/sec at the moment it passes the top of the house on its way down. When the stone strikes the ground, $s = 0$. With this replacement, Eq. (2) yields

$$0 = -16t^2 + 40t + 60$$
$$4t^2 - 10t - 15 = 0$$

$$t = \frac{10 \pm \sqrt{340}}{8} = \frac{5 \pm \sqrt{85}}{4}$$

Because $t \geqslant 0$ we disregard the solution that contains $-\sqrt{85}$. Thus,

$$t = \frac{5 + \sqrt{85}}{4} = 3.55$$

and we conclude that the stone takes approximately 3.55 sec to strike the ground. With this replacement for t in Eq. (1) we get

$$v = -32\left(\frac{5 + \sqrt{85}}{4}\right) + 40 = -8\sqrt{85} = -73.8$$

The velocity of the stone is approximately -73.8 ft/sec when it strikes the ground.

12. What constant acceleration (negative) will enable a driver to decrease his speed from 60 to 20 mph while traveling a distance of 300 ft?

SOLUTION: Let the origin be at the point where the speed is 60 mph with the positive direction toward the point where the speed is 20 mph. Let s ft be the directed distance of the driver from the origin at t sec of time, and let v ft/sec be his velocity at this moment. Because 60 mph $= 88$ ft/sec, we are given that $v = 88$ when $s = 0$. Because 20 mph $= \frac{88}{3}$ ft/sec, we are given that $v = \frac{88}{3}$ when $s = 300$. We want to find a, where a ft/sec^2 is the acceleration of the driver.

As in Exercise 4, we have

$$a = v\frac{dv}{ds}$$

Thus,

$$a\,ds = v\,dv$$

$$\int a\,ds = \int v\,dv$$

Because a is constant, the above is equivalent to

$$a\int ds = \int v\,dv$$

$$a\,s = \frac{1}{2}v^2 + C_1 \tag{1}$$

Because $v = 88$ when $s = 0$, we have

$$0 = \frac{1}{2}(88)^2 + C_1$$

$$C_1 = -3872$$

Substituting the value for C_1 into Eq. (1), we get

$$a\,s = \frac{1}{2}v^2 - 3872$$

Because $v = \frac{88}{3}$ when $s = 300$, the above equation yields

$$300a = \frac{1}{2}\left(\frac{88}{3}\right)^2 - 3872$$

$$a = -\frac{7744}{675} = -11.5$$

The constant acceleration required is approximately -11.5 ft/sec².

14. A ball started upward from the bottom of an inclined plane with an initial velocity of 6 ft/sec. If there is a downward acceleration of 4 ft/sec², how far up the plane will the ball go before rolling down?

SOLUTION: Let the origin be at the bottom of the inclined plane with positive direction up the surface of the plane. Let s ft be the directed distance of the ball from the origin at t sec of time, and let v ft/sec be the velocity of the ball at t sec of time. Because the ball has an initial velocity of 6 ft/sec, we are given that $v = 6$ when $s = 0$. Because the velocity is zero at the point where the ball reverses direction, we want to find s when $v = 0$. Let a ft/sec² be the acceleration of the ball.

As in Exercise 4, we have

$$a = v\frac{dv}{ds} \tag{1}$$

Because there is a downward acceleration of 4 ft/sec², we are given that $a = -4$. Thus, by Eq. (1) we have

$$-4 = v\frac{dv}{ds}$$

$$-4ds = v\,dv$$

$$\int -4ds = \int v\,dv$$

$$-4s = \frac{1}{2}v^2 + C \tag{2}$$

Because $v = 6$ when $s = 0$, we have

$$0 = \frac{1}{2}(6)^2 + C$$

$$C = -18$$

Thus, by Eq. (2) with $C = -18$, we have

$$-4s = \frac{1}{2}v^2 - 18$$

Substituting $v = 0$ in the above, we get

$$-4s = -18$$

$$s = \frac{9}{2}$$

Thus, the ball rolls $\frac{9}{2}$ ft up the plane.

6.6 APPLICATIONS OF ANTIDIFFERENTIATION IN ECONOMICS

If the marginal cost function C' is given, you may find the total cost function C by solving the indicated differential equation. The result is

$$C(x) = \int C'(x)\,dx$$

To find the arbitrary constant in the complete solution, use the fact that $C(0)$ is the overhead cost, if the overhead cost is known. Otherwise, you must be given the total cost of producing some particular number of units.

If the marginal revenue function R' is given, the total revenue function R is found by solving the indicated differential equation, resulting in

$$R(x) = \int R'(x)\, dx$$

To find the arbitrary constant, use the fact that the total revenue is zero when zero units are demanded; that is, $R(0) = 0$.

The price function P and the demand equation $p = P(x)$ are found from the total revenue function R by the formula

$$R(x) = xP(x)$$

To find the permissible values of x in $P(x)$, use the following two facts:

(i) The price is nonnegative; that is, $P(x) \geqslant 0$.
(ii) The price function is not increasing; thus $P'(x) \leqslant 0$.

Exercises 6.6

4. The marginal cost function is given by $3/\sqrt{2x+4}$. If the fixed cost is zero, find the total cost function.

SOLUTION: Let C be the total cost function. Because C' is the marginal cost function, we are given that

$$C'(x) = \frac{3}{\sqrt{2x+4}}$$

Hence

$$C(x) = \int \frac{3\,dx}{\sqrt{2x+4}} \tag{1}$$

Let $u = 2x + 4$, and $du = 2dx$. Thus,

$$\int \frac{3\,dx}{\sqrt{2x+4}} = \frac{3}{2}\int u^{-1/2}\, du$$

$$= 3u^{1/2} + k$$
$$= 3\sqrt{2x+4} + k \tag{2}$$

From Eqs. (1) and (2) we get

$$C(x) = 3\sqrt{2x+4} + k \tag{3}$$

Because the fixed cost is zero, then $C(0) = 0$. Hence, from Eq. (3) with x replaced by 0, we have

$$0 = 3\sqrt{2 \cdot 0 + 4} + k$$
$$k = -6$$

Therefore, by Eq. (3) the total cost function is defined by

$$C(x) = 3\sqrt{2x+4} - 6$$

6. Find the total revenue function, the demand equation, and the permissible values of x if the marginal revenue function is given by $\frac{3}{4}x^2 - 10x + 12$. Also draw sketches of the demand curve, the total revenue curve, and the marginal revenue curve on the same set of axes.

SOLUTION: Let R be the total revenue function. Because R' is the marginal revenue function, we are given that

$$R'(x) = \frac{3}{4}x^2 - 10x + 12 \tag{1}$$

Hence,

$$R(x) = \int \left(\frac{3}{4}x^2 - 10x + 12 \right) dx$$

$$= \frac{1}{4}x^3 - 5x^2 + 12x + C$$

Because $R(0) = 0$, then $C = 0$, and the total revenue function R is defined by

$$R(x) = \frac{1}{4}x^3 - 5x^2 + 12x \tag{2}$$

If P is the price function, then $R(x) = xP(x)$. Thus, from Eq. (2) we have

$$x P(x) = \frac{1}{4}x^3 - 5x^2 + 12x$$

$$P(x) = \frac{1}{4}x^2 - 5x + 12 \tag{3}$$

Because the demand equation is $p = P(x)$, from (3) the demand equation is

$$p = \frac{1}{4}x^2 - 5x + 12 \tag{4}$$

Because the price must decrease if the number of units demanded increases, then P is a decreasing function, and hence $P'(x) < 0$. From Eq. (3) we have

$$P'(x) = \frac{1}{2}x - 5$$

If $P'(x) < 0$, then $x < 10$. Because the price must be nonnegative, then $P(x) \geqslant 0$, and from Eq. (3) we get

$$\frac{1}{4}x^2 - 5x + 12 \geqslant 0$$

$$x^2 - 20x + 48 \geqslant 0$$
$$(x - 10)^2 \geqslant 52$$
$$|x - 10| \geqslant \sqrt{52}$$

Either $x \geqslant 10 + \sqrt{52}$ or $x \leqslant 10 - \sqrt{52}$. Because $x < 10$, we disregard $x \geqslant 10 + \sqrt{52}$. Therefore, we conclude that the permissible values of x are all x in $[0, 10 - \sqrt{52}]$ or, approximately, all x in $[0, 2.8]$.

Sketches of the demand curve, the total revenue curve TR, and the marginal revenue curve MR, are made from Eqs. (4), (2), and (1), respectively, and are shown in Fig. 6.6.6. Note that the total revenue has a maximum value when the marginal revenue is zero.

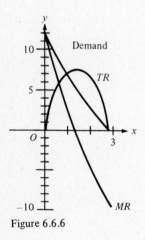

Figure 6.6.6

10. Suppose that a particular company estimates its growth in income from sales by the formula $D_t S = 2(t - 1)^{2/3}$, where S millions of dollars is the gross income from sales t years hence. If the gross income from the current year's sales is $8 million, what should be the expected gross income from sales two years from now?

SOLUTION: Because $D_t S = dS/dt$, we are given that

$$\frac{dS}{dt} = 2(t - 1)^{2/3}$$

Hence,

$$dS = 2(t-1)^{2/3} dt$$

$$\int dS = \int 2(t-1)^{2/3} dt$$

$$S = \frac{6}{5}(t-1)^{5/3} + C \tag{1}$$

Because the current gross income is \$8 million, $S = 8$ when $t = 0$. Thus, Eq. (1) yields

$$8 = \frac{6}{5}(-1)^{5/3} + C$$

$$C = \frac{46}{5}$$

Therefore, by Eq. (1), we have

$$S = \frac{6}{5}(t-1)^{5/3} + \frac{46}{5}$$

when $t = 2$, we get

$$S = \frac{6}{5}(2-1)^{5/3} + \frac{46}{5} = \frac{52}{5} = 10.4$$

We conclude that the gross income from sales will be \$10.4 million, or \$10,400,000, two years from now.

Review Exercises

4. Find the most general antiderivative.

$$\int (x^3 + x)\sqrt{x^2 + 3}\, dx$$

SOLUTION: Let

$$u = x^2 + 3 \tag{1}$$

Then $du = 2x\,dx$, or

$$\frac{1}{2} du = x\, dx \tag{2}$$

Hence, by making substitutions from Eqs. (1) and (2) we have

$$\int (x^3 + x)\sqrt{x^3 + 3}\, dx = \int (x^2 + 1)\sqrt{x^2 + 3}\,(x\,dx)$$

$$= \int (u - 2)\sqrt{u}\,\frac{1}{2}\,du$$

$$= \frac{1}{2}\int (u^{3/2} - 2u^{1/2})\,du$$

$$= \frac{1}{2}\left[\frac{2}{5}u^{5/2} - 2 \cdot \frac{2}{3}u^{3/2}\right] + C$$

$$= \frac{1}{5}(x^2 + 3)^{5/2} - \frac{2}{3}(x^2 + 3)^{3/2} + C$$

8. Find the complete solution of the given differential equation

$$\frac{dy}{dx} = \frac{x+1}{\sqrt{x}}$$

SOLUTION:

$$dy = \frac{x+1}{\sqrt{x}} \, dx$$

$$\int dy = \int (x^{1/2} + x^{-1/2}) \, dx$$

$$y = \frac{2}{3} x^{3/2} + 2x^{1/2} + C$$

12. Use differentials to find an approximate value of $\sqrt[3]{126}$.

SOLUTION: Let $y = f(x) = x^{1/3}$. Then

$$dy = \frac{1}{3} x^{-2/3} \Delta x$$

Because

$$f(x + \Delta x) = f(x) + \Delta y$$

then

$$f(x + \Delta x) \approx f(x) + dy$$

or, equivalently,

$$(x + \Delta x)^{1/3} \approx x^{1/3} + \frac{1}{3} x^{-2/3} \Delta x \qquad (1)$$

Let $x = 125$, $\Delta x = 1$, and $x + \Delta x = 126$. Then

$$(126)^{1/3} \approx (125)^{1/3} + \frac{1}{3}(125)^{-2/3} \cdot 1$$

$$\sqrt[3]{126} \approx 5 + \frac{1}{75} = 5.013$$

16. Find dy/dt if $y = x^3 - 2x + 1$ and $x^3 + t^3 - 3t = 0$

SOLUTION: Because $y = x^3 - 2x + 1$, then

$$\frac{dy}{dx} = 3x^2 - 2 \qquad (1)$$

Because $x^3 + t^3 - 3t = 0$, then

$$3x^2 \, dx + 3t^2 \, dt - 3 \, dt = 0$$

$$\frac{dx}{dt} = \frac{-t^2 + 1}{x^2} \qquad (2)$$

Because

$$\frac{dy}{dt} = \frac{dy}{dx} \cdot \frac{dx}{dt}$$

by Eq. (1) and Eq. (2), we have

$$\frac{dy}{dt} = \frac{(3x^2 - 2)(-t^2 + 1)}{x^2}$$

20. Find the particular solution of the differential equation $x^2 dy = y^3 dx$ for which $y = 1$ when $x = 4$.

SOLUTION:

$$x^2 dy = y^3 dx$$

$$\frac{dy}{y^3} = \frac{dx}{x^2}$$

$$\int y^{-3} dy = \int x^{-2} dx$$

$$\frac{y^{-3+1}}{-3+1} = \frac{x^{-2+1}}{-2+1} + C$$

$$-\frac{1}{2} y^{-2} = -x^{-1} + C \qquad (1)$$

Because $y = 1$ when $x = 4$, we have

$$-\frac{1}{2} = -4^{-1} + C$$

$$C = -\frac{1}{4}$$

From Eq. (1) we get

$$-\frac{1}{2} y^{-2} = -x^{-1} - \frac{1}{4}$$

$$4y^2 + xy^2 - 2x = 0$$

24. A ball is dropped from the top of a house 64 ft above the ground. Determine how long it takes for the ball to strike the ground and the velocity at impact.

SOLUTION: Let the origin be at ground level with positive direction upward. Let s be the number of ft in the directed distance of the ball from the origin t sec after it is dropped; let v ft/sec be the velocity of the ball t sec after it is dropped; and let a ft/sec^2 be the acceleration of the ball. Because the acceleration is caused by the force of gravity, which acts in the negative direction, we have $a = -32$. Because $a = dv/dt$, we have

$$\frac{dv}{dt} = -32$$

$$\int dv = \int -32 dt$$

$$v = -32t + C_1$$

Because the ball is dropped from rest, $v = 0$ when $t = 0$. Thus, $C_1 = 0$, and

$$v = -32t \qquad (1)$$

$$\frac{ds}{dt} = -32t$$

$$\int ds = \int -32t dt$$

$$s = -16t^2 + C_2$$

Because the top of the house is 64 ft above the ground, $s = 64$ when $t = 0$. Thus, $C_2 = 64$, and

$$s = -16t^2 + 64$$

When the ball strikes the ground, $s = 0$. Hence,

$$0 = -16t^2 + 64$$
$$t = 2$$

When $t = 2$, from Eq. (1) we get $v = -64$. Thus, the ball strikes the ground after 2 sec with a velocity of -64 ft/sec.

28. The measure of the radius of a right circular cone is $\frac{4}{3}$ times the measure of the altitude. How accurately must the altitude be measured if the error in the computed volume is not to exceed 3%?

SOLUTION: Let

$r = $ the number of units in the radius
$h = $ the number of units in the altitude
$V = $ the number of cubic units in the volume

We have the formula $V = \frac{1}{3}\pi r^2 h$. Because $r = \frac{4}{3}h$, we obtain

$$V = \frac{16}{27}\pi h^3 \tag{1}$$

$$dV = \frac{16}{9}\pi h^2 \, dh \tag{2}$$

Dividing the members of Eq. (2) by the members of Eq. (1), we obtain

$$\frac{dV}{V} = 3\frac{dh}{h}$$

or, equivalently,

$$\frac{|dh|}{h} = \frac{1}{3}\frac{|dV|}{V} \tag{3}$$

Because the error in the computed volume is not to exceed 3%, we have

$$\frac{|dV|}{V} \leq 0.03 \tag{4}$$

From (3) and (4) we have

$$\frac{|dh|}{h} \leq \frac{1}{3}(0.03) = 0.01$$

We conclude that the error in the measured altitude must not exceed 1%.

32. Let f and g be two functions such that for all x in $(-\infty, +\infty), f'(x) = g(x)$ and $g'(x) = -f(x)$. Furthermore, suppose that $f(0) = 0$ and $g(0) = 1$. Prove that $[f(x)]^2 + [g(x)]^2 = 1$.

SOLUTION: Let F and G be the functions defined by

$$F(x) = [f(x)]^2 \tag{1}$$
$$G(x) = -[g(x)]^2 \tag{2}$$

Then from (1) and the hypothesis $f'(x) = g(x)$, we obtain

$$F'(x) = 2f(x)\cdot f'(x) = 2f(x)\cdot g(x) \tag{3}$$

Similarly, from (2) and the hypothesis $g'(x) = -f(x)$, we obtain

$$G'(x) = -2g(x)\cdot g'(x) = 2f(x)\cdot g(x) \tag{4}$$

From Eqs. (3) and (4) we have

$$F'(x) = G'(x)$$

By Theorem 6.3.3, there is some k such that for all x in $(-\infty, +\infty)$, then

$$F(x) = G(x) + k \qquad\qquad (5)$$

And, in particular, when $x = 0$, we get

$$F(0) = G(0) + k \qquad\qquad (6)$$

Now by (1) and the hypothesis $f(0) = 0$, we have

$$F(0) = [f(0)]^2 = 0$$

And by (2) and the hypothesis $g(0) = 1$, we have

$$G(0) = -[g(0)]^2 = -1$$

Substituting these values for $F(0)$ and $G(0)$ into Eq. (6), we get $k = 1$. If $k = 1$, then Eq. (5) becomes

$$F(x) = G(x) + 1$$

Substituting for $F(x)$ and $G(x)$ from Eqs. (1) and (2) into the above, we have

$$[f(x)]^2 = -[g(x)]^2 + 1$$
$$[f(x)]^2 + [g(x)]^2 = 1$$

7
The definite integral

7.1 THE SIGMA NOTATION   If m and n are any integers with $m \leqslant n$ and F is a function defined for all integers in $[m, n]$, then

$$\sum_{i=m}^{n} F(i) = F(m) + F(m+1) + F(m+2) + \cdots + F(n)$$

We have the following general properties which you should memorize.

7.1.1 Property 1 $\displaystyle\sum_{i=1}^{n} c = cn$ where c is any constant

7.1.2 Property 2 $\displaystyle\sum_{i=1}^{n} cF(i) = c\sum_{i=1}^{n} F(i)$ where c is any constant

7.1.3 Property 3 $\displaystyle\sum_{i=1}^{n} [F(i) + G(i)] = \sum_{i=1}^{n} F(i) + \sum_{i=1}^{n} G(i)$

7.1.4 Property 4 $\displaystyle\sum_{i=1}^{n} [F(i) - F(i-1)] = F(n) - F(0)$

We have the following special formulas which you need not memorize.

7.1.5 Formula 1 $\displaystyle\sum_{i=1}^{n} i = \frac{n(n+1)}{2}$

7.1.6 Formula 2 $\displaystyle\sum_{i=1}^{n} i^2 = \frac{n(n+1)(2n+1)}{6}$

7.1.7 Formula 3 $\displaystyle\sum_{i=1}^{n} i^3 = \frac{n^2(n+1)^2}{4}$

7.1.8 Formula 4 $\displaystyle\sum_{i=1}^{n} i^4 = \frac{n(n+1)(6n^3+9n^2+n-1)}{30}$

Exercises 7.1

In Exercises 1-8, find the given sum.

4. $\displaystyle\sum_{j=3}^{6} \frac{2}{j(j-2)}$

SOLUTION: by definition of $\sum$ we have

$$\sum_{j=3}^{6} \frac{2}{j(j-2)} = \frac{2}{3(3-2)} + \frac{2}{4(4-2)} + \frac{2}{5(5-2)} + \frac{2}{6(6-2)}$$

$$= \frac{2}{3} + \frac{2}{8} + \frac{2}{15} + \frac{2}{24}$$

$$= \frac{17}{15}$$

8. $\displaystyle\sum_{k=-2}^{3} \frac{k}{k+3}$

SOLUTION:

$$\sum_{k=-2}^{3} \frac{k}{k+3} = \frac{-2}{-2+3} + \frac{-1}{-1+3} + \frac{0}{0+3} + \frac{1}{1+3} + \frac{2}{2+3} + \frac{3}{3+3}$$

$$= -2 - \frac{1}{2} + 0 + \frac{1}{4} + \frac{2}{5} + \frac{1}{2}$$

$$= \frac{-27}{20}$$

12. Prove Formula 2 (7.1.6) without using mathematical induction.

SOLUTION: Because

$$(i - 1)^3 = i^3 - 3i^2 + 3i - 1$$

then

$$i^3 - (i - 1)^3 = 3i^2 - 3i + 1$$

and thus

$$\sum_{i=1}^{n} [i^3 - (i-1)^3] = \sum_{i=1}^{n} (3i^2 - 3i + 1) \tag{1}$$

By Property 4 with $F(i) = i^3$, the left side of Eq. (1) has the value $F(n) - F(0)$, which is n^3. Thus, from Eq. (1) we have

$$n^3 = \sum_{i=1}^{n} (3i^2 - 3i + 1)$$

Using Properties 1, 2, and 3, on the right side of the above we have

$$n^3 = 3 \sum_{i=1}^{n} i^2 - 3 \sum_{i=1}^{n} i + n$$

By Formula 1, the above is equivalent to

$$n^3 = 3 \sum_{i=1}^{n} i^2 - 3 \cdot \frac{n(n+1)}{2} + n$$

$$2n^3 = 6 \sum_{i=1}^{n} i^2 - 3n(n+1) + 2n$$

$$6 \sum_{i=1}^{n} i^2 = 2n^3 + 3n(n+1) - 2n$$

$$= n(2n^2 + 3n + 3 - 2)$$
$$= n(n+1)(2n+1)$$

Dividing on both sides by 6, we get

$$\sum_{i=1}^{n} i^2 = \frac{n(n+1)(2n+1)}{6}$$

We have proved Formula 2.

In Exercises 17-25, evaluate the indicated sum by using Properties 1-4 and Formulas 1-4.

18. $\displaystyle\sum_{i=1}^{20} 3i(i^2 + 2)$

SOLUTION:

$$\sum_{i=1}^{20} 3i(i^2 + 2) = 3 \sum_{i=1}^{20} i^3 + 6 \sum_{i=1}^{20} i$$

$$= 3 \cdot \frac{(20)^2(21)^2}{4} + 6 \cdot \frac{(20)(21)}{2}$$

$$= 133{,}560$$

20. $\displaystyle\sum_{k=1}^{n} (2^{k-1} - 2^k)$

SOLUTION: We use Property 4 with $F(i) = 2^i$. Then

$$\sum_{k=1}^{n} (2^{k-1} - 2^k) = \sum_{i=1}^{n} (2^{i-1} - 2^i)$$

$$= \sum_{i=1}^{n} [F(i-1) - F(i)]$$

$$= -\sum_{i=1}^{n} [F(i) - F(i-1)]$$

$$= -[F(n) - F(0)] \quad \text{(By Property 4)}$$
$$= -[2^n - 2^0]$$
$$= 1 - 2^n$$

22. $\displaystyle\sum_{i=1}^{40} [\sqrt{2i+1} - \sqrt{2i-1}]$

SOLUTION: We use Property 4 with $F(i) = \sqrt{2i+1}$. Thus, $F(i-1) = \sqrt{2(i-1)+1} = \sqrt{2i-1}$ and,

$$\sum_{i=1}^{40} [\sqrt{2i+1} - \sqrt{2i-1}] = \sum_{i=1}^{40} [F(i) - F(i-1)]$$

$$= F(40) - F(0)$$
$$= \sqrt{2 \cdot 40 + 1} - \sqrt{2 \cdot 0 + 1}$$
$$= 8$$

26. Prove $\displaystyle\sum_{i=-n}^{n} \left[1 - \left(\frac{i}{n}\right)^2\right]^{1/2} = 2\sum_{i=1}^{n} \left[1 - \left(\frac{i}{n}\right)^2\right]^{1/2} + 1$

SOLUTION: Let F be the function defined by

$$F(i) = \left[1 - \left(\frac{i}{n}\right)^2\right]^{1/2} \tag{1}$$

We want to prove that

$$\sum_{i=-n}^{n} F(i) = 2 \sum_{i=1}^{n} F(i) + 1 \tag{2}$$

We may write

$$\sum_{i=-n}^{n} F(i) = \sum_{i=-n}^{-1} F(i) + F(0) + \sum_{i=1}^{n} F(i) \tag{3}$$

Replacing i by $-i$ we have

$$\sum_{i=-n}^{-1} F(i) = \sum_{-i=-n}^{-1} F(-i) \tag{4}$$

Because $i^2 = (-i)^2$, by Eq. (1), $F(-i) = F(i)$. Moreover, when $-i = -n, i = n$; and when $-i = -1, i = 1$. Thus,

$$\sum_{-i=-n}^{-1} F(-i) = \sum_{i=1}^{n} F(i) \tag{5}$$

Substituting from Eq. (5) into Eq. (4), we have

$$\sum_{i=-n}^{-1} F(i) = \sum_{i=1}^{n} F(i) \tag{6}$$

Substituting from Eq. (6) into Eq. (3), we obtain

$$\sum_{i=-n}^{n} F(i) = \sum_{i=1}^{n} F(i) + F(0) + \sum_{i=1}^{n} F(i) \tag{7}$$

By Eq. (1), $F(0) = 1$, and thus Eq. (7) is equivalent to

$$\sum_{i=-n}^{n} F(i) = 2 \sum_{i=1}^{n} F(i) + 1$$

or, equivalently,

$$\sum_{i=-n}^{n} \left[1 - \left(\frac{i}{n}\right)^2 \right]^{1/2} = 2 \sum_{i=1}^{n} \left[1 - \left(\frac{i}{n}\right)^2 \right]^{1/2} + 1$$

7.2 AREA　　If the function f is continuous on the closed interval $[a, b]$ and $f(x) \geqslant 0$ for all x in $[a, b]$, and if R is the region bounded by the curve $y = f(x)$, the x-axis, and the vertical lines $x = a$ and $x = b$, then the measure of the area of R, denoted by A, can be found by either of the following two formulas.

Area Formula 1　　$A = \lim\limits_{n \to +\infty} \sum\limits_{i=1}^{n} f(x_i) \, \Delta x$

Area Formula 2　　$A = \lim\limits_{n \to +\infty} \sum\limits_{i=1}^{n} f(x_{i-1}) \, \Delta x$

where

$$\Delta x = \frac{b-a}{n}$$

$$x_i = a + i\Delta x$$

If f is *increasing* on $[a, b]$, then the use of Area Formula 1 is called the method of circumscribed rectangles, and the use of Area Formula 2 is called the method of inscribed rectangles.

If f is *decreasing* on $[a, b]$, then the use of Area Formula 1 is called the method of inscribed rectangles, and the use of Area Formula 2 is called the method of circumscribed rectangles.

We must be able to compute the limits in the area formulas in order to calculate A. If f is a polynomial function of degree 4 or less, we may always use the properties and formulas for sigma notation given in Section 7.1 to find the limit. If f is not such a function, we may not be able to find A at this time. Later, however, we have theorems that will greatly enlarge the class of functions for which we can calculate the area.

Exercises 7.2

In Exercises 1-14, use the method of this section to find the area of the given region; use inscribed or circumscribed rectangles as indicated. For each exercise, draw a figure showing the region and the ith rectangle.

2. The region bounded by $y = x^2$, the x-axis, and the line, $x = 2$; circumscribed rectangles.

SOLUTION: A sketch of the region R is shown in Fig. 7.2.2. Because f, defined by $f(x) = x^2$, is increasing on $[0, 2]$, Area Formula 1 is the method of circumscribed rectangles. We have

$$\Delta x = \frac{b-a}{n} = \frac{2-0}{n} = \frac{2}{n} \tag{1}$$

$$x_i = a + i\,\Delta x = 0 + i\frac{2}{n} = \frac{2i}{n}$$

$$f(x_i) = (x_i)^2 = \left(\frac{2i}{n}\right)^2 = \frac{4i^2}{n^2} \tag{2}$$

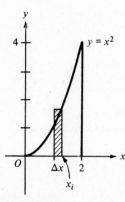

Figure 7.2.2

Thus, by Area Formula 1 and Eqs. (1) and (2), we have

$$A = \lim_{n \to +\infty} \sum_{i=1}^{n} f(x_i)\,\Delta x$$

$$= \lim_{n \to +\infty} \sum_{i=1}^{n} \frac{4i^2}{n^2} \cdot \frac{2}{n}$$

$$= \lim_{n \to +\infty} \frac{8}{n^3} \sum_{i=1}^{n} i^2$$

$$= \lim_{n \to +\infty} \frac{8}{n^3} \cdot \frac{n(n+1)(2n+1)}{6} \qquad \text{(By Sum Formula 2)}$$

$$= \lim_{n \to +\infty} \frac{8}{6}\left(1 + \frac{1}{n}\right)\left(2 + \frac{1}{n}\right)$$

$$= \frac{8}{3}$$

The area of region R is $\frac{8}{3}$ sq. units.

4. The region bounded by $y = 2x$, the x-axis, and the lines $x = 1$ and $x = 4$; inscribed rectangles.

SOLUTION: A sketch of the region is shown in Fig. 7.2.4. Because f, defined by $f(x) = 2x$, is increasing on $[1, 4]$, Area Formula 2 is the method of inscribed rectangles. We have

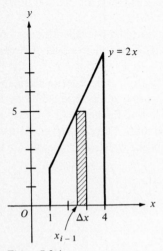

$$\Delta x = \frac{b - a}{n} = \frac{4 - 1}{n} = \frac{3}{n} \tag{1}$$

$$x_i = a + i\,\Delta x = 1 + \frac{3i}{n}$$

$$x_{i-1} = 1 + \frac{3(i-1)}{n}$$

$$f(x_{i-1}) = 2x_{i-1} = 2 + \frac{6i}{n} - \frac{6}{n} \tag{2}$$

Figure 7.2.4

Therefore, from Area Formula 2, Eq. (1), and Eq. (2), we have

$$A = \lim_{n \to +\infty} \sum_{i=1}^{n} f(x_{i-1})\,\Delta x$$

$$= \lim_{n \to +\infty} \sum_{i=1}^{n} \left(2 + \frac{6i}{n} - \frac{6}{n}\right)\frac{3}{n}$$

$$= \lim_{n \to +\infty} \left[\frac{6}{n}\sum_{i=1}^{n} 1 + \frac{18}{n^2}\sum_{i=1}^{n} i - \frac{18}{n^2}\sum_{i=1}^{n} 1\right]$$

$$= \lim_{n \to +\infty} \left[\frac{6}{n} \cdot n + \frac{18}{n^2} \cdot \frac{n(n+1)}{2} - \frac{18}{n^2} \cdot n\right] \quad \text{(By Sum Formulas 1 and 2)}$$

$$= \lim_{n \to +\infty} \left[6 + 9\left(1 + \frac{1}{n}\right) - \frac{18}{n}\right] = 15$$

The area of region P is 15 sq. units.

8. The region above the x-axis and to the left of the line $x = 1$ bounded by the curve $y = 4 - x^2$, the line $x = 1$, and the x-axis; inscribed rectangles.

SOLUTION: A sketch of the region R is shown in Fig. 7.2.8. The curve intersects the x-axis at $(-2, 0)$. Because f, defined by $f(x) = 4 - x^2$, is not monotonic on the interval $[-2, 1]$, we subdivide the region R into two parts. Let R_1 be the region to the left of the y-axis, bounded by the curve $y = 4 - x^2$, the x-axis, and the y-axis. Let R_2 be the region to the right of the y-axis bounded by the curve $y = 4 - x^2$, the x-axis, the y-axis, and the line $x = 1$.

To find A_1, the measure of the area of region R_1, we use Area Formula 2, with $[a, b] = [-2, 0]$.

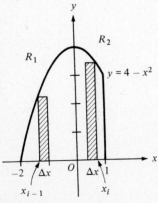

Figure 7.2.8

$$\Delta x = \frac{0 - (-2)}{n} = \frac{2}{n} \tag{1}$$

$$x_i = a + i\Delta x = -2 + \frac{2i}{n}$$

$$x_{i-1} = -2 + \frac{2(i-1)}{n}$$

$$f(x_{i-1}) = 4 - (x_{i-1})^2$$

$$= 4 - \left[-2 + \frac{2(i-1)}{n}\right]^2$$

$$= 4 - \left[4 - \frac{8(i-1)}{n} + \frac{4(i-1)^2}{n^2}\right]$$

$$= 4 - 4 + \frac{8(i-1)}{n} - \frac{4(i^2 - 2i + 1)}{n^2}$$

$$= \frac{8i}{n} - \frac{8}{n} - \frac{4i^2}{n^2} + \frac{8i}{n^2} - \frac{4}{n^2} \tag{2}$$

Hence, from Area Formula 2 and Eqs. (1) and (2) we obtain

$$A_1 = \lim_{n \to +\infty} \sum_{i=1}^{n} f(x_{i-1}) \, \Delta x$$

$$= \lim_{n \to +\infty} \sum_{i=1}^{n} \left[\frac{8i}{n} - \frac{8}{n} - \frac{4i^2}{n^2} + \frac{8i}{n^2} - \frac{4}{n^2}\right]\frac{2}{n}$$

$$= \lim_{n \to +\infty} \left[\frac{16}{n^2} \sum_{i=1}^{n} i - \frac{16}{n^2} \sum_{i=1}^{n} 1 - \frac{8}{n^3} \sum_{i=1}^{n} i^2 + \frac{16}{n^3} \sum_{i=1}^{n} i - \frac{8}{n^3} \sum_{i=1}^{n} 1\right]$$

$$= \lim_{n \to +\infty} \left[\frac{16}{n^2} \cdot \frac{n(n+1)}{2} - \frac{16}{n^2} \cdot n - \frac{8}{n^3} \cdot \frac{n(n+1)(2n+1)}{6} + \frac{16}{n^3} \cdot \frac{n(n+1)}{2}\right.$$

$$\left. - \frac{8}{n^3} \cdot n\right]$$

$$= \lim_{n \to +\infty} \left[8\left(1 + \frac{1}{n}\right) - \frac{16}{n} - \frac{4}{3}\left(1 + \frac{1}{n}\right)\left(2 + \frac{1}{n}\right) + 8\left(\frac{1}{n} + \frac{1}{n^2}\right) - \frac{8}{n^2}\right]$$

$$= 8 - \frac{4}{3} \cdot 2 = \frac{16}{3} \tag{3}$$

To find A_2, the measure of the area of region R_2, we use Area Formula 1, with $[a, b] = [0, 1]$. Thus,

$$\Delta x = \frac{1}{n}$$

$$x_i = i \cdot \Delta x = \frac{i}{n}$$

$$f(x_i) = 4 - (x_i)^2$$

$$= 4 - \frac{i^2}{n^2}$$

Hence,

$$A_2 = \lim_{n \to +\infty} \sum_{i=1}^{n} f(x_i)\,\Delta x$$

$$= \lim_{n \to +\infty} \sum_{i=1}^{n} \left[4 - \frac{i^2}{n^2}\right]\frac{1}{n}$$

$$= \lim_{n \to +\infty} \left[\frac{4}{n}\sum_{i=1}^{n} 1 - \frac{1}{n^3}\sum_{i=1}^{n} i^2\right]$$

$$= \lim_{n \to +\infty} \left[\frac{4}{n}\cdot n - \frac{1}{n^3}\cdot\frac{n(n+1)(2n+1)}{6}\right]$$

$$= \lim_{n \to +\infty} \left[4 - \frac{1}{6}\left(1 + \frac{1}{n}\right)\left(2 + \frac{1}{n}\right)\right]$$

$$= 4 - \frac{1}{6}\cdot 2 = \frac{11}{3} \tag{4}$$

If A is the measure of the area of region R, then substituting from (3) and (4) we have

$$A = A_1 + A_2 = \frac{16}{3} + \frac{11}{3} = 9$$

Thus, the area of region R is 9 sq units.

12. The region bounded by $y = x^3$, the x-axis, and the lines $x = -1$ and $x = 2$; circumscribed rectangles.

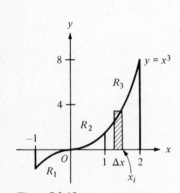

Figure 7.2.12

SOLUTION: A sketch of the region R is shown in Fig. 7.2.12. We divide the region into three parts: R_1 is the region to the left of the y-axis bounded by the curve $y = x^3$, the x-axis, and the line $x = -1$; R_2 is the region to the right of the y-axis bounded by the curve $y = x^3$, the x-axis, and the line $x = 1$; R_3 is the region to the right of the line $x = 1$ bounded by the curve $y = x^3$, the x-axis, and the lines $x = 1$ and $x = 2$. Because the curve $y = x^3$ lies below the x-axis in region R_1, we cannot use either area formula to calculate A_1, the measure of the area of region R_1. However, the curve $y = x^3$ is symmetric with respect to the origin. Therefore, $A_1 = A_2$ where A_2 is the measure of the area of region R_2. To find A_2, we use Area Formula 1 with $f(x) = x^3$ and $[a, b] = [0, 1]$. Thus, for A_2 we have

$$\Delta x = \frac{1}{n}$$

$$x_i = i\Delta x = \frac{i}{n}$$

$$f(x_i) = (x_i)^3 = \frac{i^3}{n^3}$$

Hence,

$$A_2 = \lim_{n \to +\infty} \sum_{i=1}^{n} f(x_i)\,\Delta x$$

$$= \lim_{n \to +\infty} \sum_{i=1}^{n} \frac{i^3}{n^3}\cdot\frac{1}{n}$$

$$= \lim_{n \to +\infty} \frac{1}{n^4} \sum_{i=1}^{n} i^3$$

$$= \lim_{n \to +\infty} \frac{1}{n^4} \frac{n^2(n+1)^2}{4} \qquad \text{(By Sum Formula 3)}$$

$$= \lim_{n \to +\infty} \frac{1}{4}\left(1 + \frac{1}{n}\right)^2$$

$$= \frac{1}{4}$$

Thus $A_1 = A_2 = \frac{1}{4}$.

To find A_3, the measure of the area of region R_3, we use Area Formula 1 with $f(x) = x^3$ and $[a, b] = [1, 2]$. Thus,

$$\Delta x = \frac{1}{n}$$

$$x_i = 1 + i\Delta x = 1 + \frac{i}{n}$$

$$f(x_i) = \left(1 + \frac{i}{n}\right)^3 = 1 + \frac{3i}{n} + \frac{3i^2}{n^2} + \frac{i^3}{n^3}$$

Hence,

$$A_3 = \lim_{n \to +\infty} \sum_{i=1}^{n} f(x_i)\, \Delta x$$

$$= \lim_{n \to +\infty} \sum_{i=1}^{n} \left[1 + \frac{3i}{n} + \frac{3i^2}{n^2} + \frac{i^3}{n^3}\right] \frac{1}{n}$$

$$= \lim_{n \to +\infty} \left[\frac{1}{n}\sum_{i=1}^{n} 1 + \frac{3}{n^2}\sum_{i=1}^{n} i + \frac{3}{n^3}\sum_{i=1}^{n} i^2 + \frac{1}{n^4}\sum_{i=1}^{n} i^3\right]$$

$$= \lim_{n \to +\infty} \left[\frac{1}{n}\cdot n + \frac{3}{n^2}\cdot\frac{n(n+1)}{2} + \frac{3}{n^3}\frac{n(n+1)(2n+1)}{6} + \frac{1}{n^4}\frac{n^2(n+1)^2}{4}\right]$$

$$= \lim_{n \to +\infty} \left[1 + \frac{3}{2}\left(1 + \frac{1}{n}\right) + \frac{1}{2}\left(1 + \frac{1}{n}\right)\left(2 + \frac{1}{n}\right) + \frac{1}{4}\left(1 + \frac{1}{n}\right)^2\right]$$

$$= 1 + \frac{3}{2} + \frac{1}{2}\cdot 2 + \frac{1}{4} = \frac{15}{4}$$

Because $A = A_1 + A_2 + A_3 = \frac{1}{4} + \frac{1}{4} + \frac{15}{4} = \frac{17}{4}$, the area of region R is $\frac{17}{4}$ sq. units.

18. R is the region bounded by $y = x^2$, the x-axis, and the line $x = 2$. Find the area of the region R by taking as the measure of the altitude of the ith rectangle $f(m_i)$, where m_i is the midpoint of the ith subinterval. (*Hint:* $m_i = \frac{1}{2}(x_i + x_{i-1})$.)

SOLUTION: A sketch of the graph of the region R is shown in Fig. 7.2.2. As in Exercise 2, we have

$$\Delta x = \frac{2}{n}$$

$$x_i = \frac{2i}{n}$$

Thus,

$$x_{i-1} = \frac{2(i-1)}{n}$$

$$m_i = \frac{1}{2}(x_i + x_{i-1}) = \frac{1}{2}\left[\frac{2i}{n} + \frac{2(i-1)}{n}\right] = \frac{2i}{n} - \frac{1}{n}$$

$$f(m_i) = (m_i)^2 = \left(\frac{2i}{n} - \frac{1}{n}\right)^2 = \frac{4i^2}{n^2} - \frac{4i}{n^2} + \frac{1}{n^2}$$

Hence,

$$A = \lim_{n \to +\infty} \sum_{i=1}^{n} f(m_i)\, \Delta x$$

$$= \lim_{n \to +\infty} \sum_{i=1}^{n} \left(\frac{4i^2}{n^2} - \frac{4i}{n^2} + \frac{1}{n^2}\right)\frac{2}{n}$$

$$= \lim_{n \to +\infty} \left[\frac{8}{n^3}\sum_{i=1}^{n} i^2 - \frac{8}{n^3}\sum_{i=1}^{n} i + \frac{2}{n^3}\sum_{i=1}^{n} 1\right]$$

$$= \lim_{n \to +\infty} \left[\frac{8}{n^3}\cdot\frac{n(n+1)(2n+1)}{6} - \frac{8}{n^3}\cdot\frac{n(n+1)}{2} + \frac{2}{n^3}\cdot n\right]$$

$$= \lim_{n \to +\infty} \left[\frac{4}{3}\left(1 + \frac{1}{n}\right)\left(2 + \frac{1}{n}\right) - 4\left(\frac{1}{n} + \frac{1}{n^2}\right) + \frac{2}{n^2}\right]$$

$$= \frac{4}{3}\cdot 2 = \frac{8}{3}$$

The area of region R is $\frac{8}{3}$ sq. units. We note that this method gives the same value for the area as does the method of circumscribed rectangles of Exercise 2.

7.3 THE DEFINITE INTEGRAL

A *Riemann sum* for the function f on the closed interval $[a, b]$ is any sum of the form

$$\sum_{i=1}^{n} f(\xi_i)\, \Delta_i x$$

where

(i) $a = x_1 < x_2 < \cdots < x_i < \cdots < x_n = b$
(ii) ξ_i is any number in the closed interval $[x_{i-1}, x_i]$
(iii) $\Delta_i x = x_i - x_{i-1}$

The set $\Delta = \{x_1, x_2, \cdots, x_n\}$ is called a *partition* of the interval $[a, b]$. The largest of the numbers $\Delta_1 x, \Delta_2 x, \cdots, \Delta_n x$ is called the *norm* of the partition and is represented by $\|\Delta\|$.

Each of the sums that appear in the area formulas of Section 7.2 is a Riemann sum.

7.3.1 Definition

Let f be a function whose domain includes the closed interval $[a, b]$. Then f is said to be *integrable* on $[a, b]$ if there is a number L satisfying the condition that, for every $\epsilon > 0$, there exists a $\delta > 0$ such that

$$\left|\sum_{i=1}^{n} f(\xi_i)\, \Delta_i x - L\right| < \epsilon$$

for every partition Δ for which $\|\Delta\| < \delta$, and for any ξ_i in the closed interval $[x_{i-1}, x_i]$, $i = 1, 2, \ldots, n$.

7.3.2 Definition   If f is a function defined on the closed interval $[a, b]$, then the *definite integral* of f from a to b, denoted by $\int_a^b f(x)\, dx$, is given by

$$\int_a^b f(x)\, dx = \lim_{\|\Delta\| \to 0} \sum_{i=1}^n f(\xi_i)\, \Delta_i x$$

if the limit exists.

In Definition 7.3.2, which is one of the most important definitions in calculus, the sum is a Riemann sum for the function f on $[a, b]$. Thus, ξ_i and $\Delta_i x$ are defined as indicated above for a Riemann sum. And $\|\Delta\|$ is defined to be the largest $\Delta_i x$ for $i = 1, 2, 3, \ldots, n$.

If the limit in Definition 7.3.2 exists, then f is said to be integrable on $[a, b]$. Furthermore, if $f(x) \geqslant 0$ for all x in $[a, b]$, then the limit in Definition 7.3.2, that is, $\int_a^b f(x)\, dx$, gives the measure of the area of the region bounded by the curve $y = f(x)$, the x-axis, and the lines $x = a$ and $x = b$. And if $f(x) \leqslant 0$ for all x in $[a, b]$, then $-\int_a^b f(x)\, dx$ gives the measure of the area of the region.

If f is a polynomial function of degree 4 or less, we may calculate $\int_a^b f(x)\, dx$ by using area Formula 1 given in Section 7.2. That is,

$$\int_a^b f(x)\, dx = \lim_{n \to +\infty} \sum_{i=1}^n f(x_i)\, \Delta x$$

where

$$\Delta x = \frac{b - a}{n} \quad \text{and} \quad x_i = a + i\, \Delta x$$

If n is a large positive integer, the Riemann sum given by

$$R = \sum_{i=1}^n f(x_i)\, \Delta_i x$$

where

$$\Delta_i x = \frac{b - a}{n} \quad \text{and} \quad x_i = a + i\Delta_i x$$

is an approximation for the definite integral $\int_a^b f(x)\, dx$. Table 7.3a shows a flowchart that can be used to program a computer to calculate the Riemann sum R for any function F, any closed interval $[A, B]$, and any positive integer N. In Table 7.3b we show an actual computer program written in BASIC that uses R with 100 terms to approximate the definite integral

$$\int_{-1}^2 (4x^3 - 3x^2)\, dx$$

of Exercise 14. We read in $A = -1$, $B = 2$, and $N = 100$.

By modifying instructions 20 and 70 in the program of Table 7.3b, we may approximate any definite integral. For example, to approximate

$$\int_0^4 (x^2 + x - 6)\, dx$$

with a Riemann sum with 500 terms, we make the following changes in the program.

20 DATA   0,4,500
70 LET R = R + x↑2 + x − 6

Table 7.3a
Riemann sum

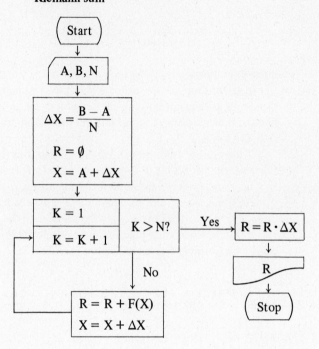

Table 7.3b

```
LIST
1Ø  READ A,B,N
2Ø  DATA -1,2,1ØØ
3Ø  LET D=(B-A)/N
4Ø  LET R=Ø
5Ø  LET X=A+D
6Ø  FOR K=1 TO N
7Ø  LET R=R+4*X↑3-3*X↑2
8Ø  LET X=X+D
9Ø  NEXT K
1ØØ  PRINT "R = ";R*D
999  END
```

```
RUN
R =   6.4Ø632
```

Exercises 7.3

4. Find the Riemann sum for the function on the interval, using the given partition Δ and the given values of ξ_i.

$$f(x) = x^3, -1 \leqslant x \leqslant 2; \text{ for } \Delta: x_0 = -1,$$

$$x_1 = -\frac{1}{3}, x_2 = \frac{1}{2}, x_3 = 1, x_4 = 1\frac{1}{4}, x_5 = 2;$$

$$\xi_1 = -\frac{1}{2}, \xi_2 = 0, \xi_3 = \frac{2}{3}, \xi_4 = 1, \xi_5 = 1\frac{1}{2}$$

SOLUTION: Table 4 summarizes the calculations necessary to find the Riemann sum. Note that $f(\xi_i) = \xi_i^3$, and

$$\Delta_1 x = x_1 - x_0 = -\frac{1}{3} - (-1) = \frac{2}{3}$$

$$\Delta_2 x = x_2 - x_1 = \frac{1}{2} - \left(-\frac{1}{3}\right) = \frac{5}{6}$$

and so on.

Table 4

i	x_i	ξ_i	$f(\xi_i)$	$\Delta_i x$	$f(\xi_i) \cdot \Delta_i x$
0	-1				
1	$-\dfrac{1}{3}$	$-\dfrac{1}{2}$	$-\dfrac{1}{8}$	$\dfrac{2}{3}$	$-\dfrac{1}{12}$
2	$\dfrac{1}{2}$	0	0	$\dfrac{5}{6}$	0
3	1	$\dfrac{2}{3}$	$\dfrac{8}{27}$	$\dfrac{1}{2}$	$\dfrac{4}{27}$
4	$\dfrac{5}{4}$	1	1	$\dfrac{1}{4}$	$\dfrac{1}{4}$
5	2	$\dfrac{3}{2}$	$\dfrac{27}{8}$	$\dfrac{3}{4}$	$\dfrac{81}{32}$

$$\sum_{i=1}^{5} f(\xi_i)\,\Delta_i x = \frac{2459}{864} = 2.85$$

In Exercises 7-14, find the exact value of the definite integral.

8. $\displaystyle\int_{2}^{4} x^2\,dx$

SOLUTION: We have $f(x) = x^2$ and $[a, b] = [2, 4]$. We use the formula

$$\int_{a}^{b} f(x)\,dx = \lim_{n \to +\infty} \sum_{i=1}^{n} f(\xi_i)\,\Delta x$$

where

$$\Delta x = \frac{b-a}{n} = \frac{2}{n}$$

$$\xi_i = a + i\Delta x = 2 + \frac{2i}{n}$$

$$f(\xi_i) = \xi_i^2 = 4 + \frac{8i}{n} + \frac{4i^2}{n^2}$$

Thus,

$$\int_2^4 x^2\,dx = \lim_{n \to +\infty} \sum_{i=1}^{n}\left(4 + \frac{8i}{n} + \frac{4i^2}{n^2}\right)\frac{2}{n}$$

$$= \lim_{n \to +\infty}\left[\frac{8}{n}\sum_{i=1}^{n}1 + \frac{16}{n^2}\sum_{i=1}^{n}i + \frac{8}{n^3}\sum_{i=1}^{n}i^2\right]$$

$$= \lim_{n \to +\infty}\left[\frac{8}{n}\cdot n + \frac{16}{n^2}\frac{n(n+1)}{2} + \frac{8}{n^3}\frac{n(n+1)(2n+1)}{6}\right]$$

$$= \lim_{n \to +\infty}\left[8 + 8\left(1 + \frac{1}{n}\right) + \frac{4}{3}\left(1 + \frac{1}{n}\right)\left(2 + \frac{1}{n}\right)\right]$$

$$= 8 + 8 + \frac{8}{3} = \frac{56}{3}$$

12. $\displaystyle\int_0^4 (x^2 + x - 6)\,dx$

SOLUTION: $f(x) = x^2 + x - 6$ and $[a, b] = [0, 4]$. As in Exercise 8, we have

$$\Delta x = \frac{b - a}{n} = \frac{4}{n}$$

$$\xi_i = a + i\Delta x = \frac{4i}{n}$$

$$f(\xi_i) = \xi_i^2 + \xi_i - 6 = \frac{16i^2}{n^2} + \frac{4i}{n} - 6$$

Thus,

$$\int_0^4 (x^2 + x - 6)\,dx = \lim_{n \to +\infty} \sum_{i=1}^{n} f(\xi_i)\,\Delta x$$

$$= \lim_{n \to +\infty} \sum_{i=1}^{n}\left(\frac{16i^2}{n^2} + \frac{4i}{n} - 6\right)\cdot\frac{4}{n}$$

$$= \lim_{n \to +\infty}\left[\frac{64}{n^3}\sum_{i=1}^{n}i^2 + \frac{16}{n^2}\sum_{i=1}^{n}i - \frac{24}{n}\sum_{i=1}^{n}1\right]$$

$$= \lim_{n \to +\infty}\left[\frac{64}{n^3}\frac{n(n+1)(2n+1)}{6} + \frac{16}{n^2}\cdot\frac{n(n+1)}{2} - \frac{24}{n}\cdot n\right]$$

$$= \lim_{n \to +\infty}\left[\frac{32}{3}\left(1 + \frac{1}{n}\right)\left(2 + \frac{1}{n}\right) + 8\left(1 + \frac{1}{n}\right) - 24\right]$$

$$= \frac{64}{3} + 8 - 24 = \frac{16}{3}$$

In Exercises 15-20, find the exact area of the region in the following way:

(a) Express the measure of the area as the limit of a Riemann sum with regular partitions.

(b) Express this limit as a definite integral.

(c) Evaluate the definite integral by the method of this section and a suitable choice of ξ_i. Draw a figure showing the region.

18. Bounded by the curve $y = (x + 3)^2$, the x-axis, and the lines $x = -3$ and $x = 0$.

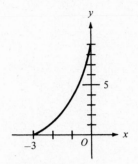

Figure 7.3.18

SOLUTION: A sketch of the region R is shown in Fig. 7.3.18. We have $f(x) = (x + 3)^2$, and $[a, b] = [-3, 0]$. Thus

(a) $A = \lim\limits_{n \to +\infty} \sum\limits_{i=1}^{n} f(\xi_i)\, \Delta_i x = \lim\limits_{n \to +\infty} \sum\limits_{i=1}^{n} (\xi_i + 3)^2\, \Delta_i x$

(b) $\lim\limits_{n \to +\infty} \sum\limits_{i=1}^{n} (\xi_i + 3)^2\, \Delta_i x = \int_{-3}^{0} (x + 3)^2\, dx$

(c) Let

$$\Delta_i x = \frac{b - a}{n} = \frac{3}{n}$$

$$\xi_i = a + i\Delta_i x = -3 + \frac{3i}{n}$$

$$(\xi_i + 3)^2 = \frac{9i^2}{n^2}$$

Hence,

$$A = \lim\limits_{n \to +\infty} \sum\limits_{i=1}^{n} (\xi_i + 3)^2 \cdot \Delta_i x$$

$$= \lim\limits_{n \to +\infty} \sum\limits_{i=1}^{n} \frac{9i^2}{n^2} \cdot \frac{3}{n}$$

$$= \lim\limits_{n \to +\infty} \left[\frac{27}{n^3} \sum\limits_{i=1}^{n} i^2 \right]$$

$$= \lim\limits_{n \to +\infty} \left[\frac{27}{n^3} \frac{n(n+1)(2n+1)}{6} \right]$$

$$= \lim\limits_{n \to +\infty} \left[\frac{9}{2} \left(1 + \frac{1}{n} \right) \left(2 + \frac{1}{n} \right) \right]$$

$$= 9$$

The area of the region is 9 sq. units.

20. Bounded by the curve $y = 6x + x^2 - x^3$, the x-axis, and the lines $x = -1$ and $x = 3$.

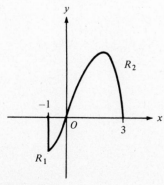

Figure 7.3.20

SOLUTION: A sketch of the graph of the region R is shown in Fig. 7.3.20. We have $f(x) = 6x + x^2 - x^3$. Because $f(x) \leq 0$ if $-1 \leq x \leq 0$ and $f(x) \geq 0$ if $0 \leq x \leq 3$, we divide the region R into two regions: R_1 is that part of R to the left of the y-axis, and R_2 is that part of R to the right of the y-axis. For region R_1 the measure of the area is given by

$$A_1 = - \lim_{n \to +\infty} \sum_{i=1}^{n} f(\xi_i) \, \Delta_i x$$

Thus,

(a)　$$A_1 = - \lim_{n \to +\infty} \sum_{i=1}^{n} (6\xi_i + \xi_i^2 - \xi_i^3) \, \Delta_i x$$

(b)　$$- \lim_{n \to +\infty} \sum_{i=1}^{n} (6\xi_i + \xi_i^2 - \xi_i^3) \, \Delta_i x = - \int_{-1}^{0} (6x + x^2 - x^3) \, dx$$

(c)　$$\Delta_i x = \frac{1}{n}$$

$$\xi_i = -1 + \frac{i}{n}$$

$$6\xi_i + \xi_i^2 - \xi_i^3 = 6\left(-1 + \frac{i}{n}\right) + \left(-1 + \frac{i}{n}\right)^2 - \left(-1 + \frac{i}{n}\right)^3$$

$$= -6 + \frac{6i}{n} + 1 - \frac{2i}{n} + \frac{i^2}{n^2} + 1 - \frac{3i}{n} + \frac{3i^2}{n^2} - \frac{i^3}{n^3}$$

$$= -4 + \frac{i}{n} + \frac{4i^2}{n^2} - \frac{i^3}{n^3}$$

Thus,

$$A_1 = - \lim_{n \to +\infty} \sum_{i=1}^{n} (6\xi_i + \xi_i^2 - \xi_i^3) \, \Delta_i x$$

$$= - \lim_{n \to +\infty} \sum_{i=1}^{n} \left(-4 + \frac{i}{n} + \frac{4i^2}{n^2} - \frac{i^3}{n^3}\right)\frac{1}{n}$$

$$= - \lim_{n \to +\infty} \left[-\frac{4}{n}\sum_{i=1}^{n} 1 + \frac{1}{n^2}\sum_{i=1}^{n} i + \frac{4}{n^3}\sum_{i=1}^{n} i^2 - \frac{1}{n^4}\sum_{i=1}^{n} i^3\right]$$

$$= - \lim_{n \to +\infty} \left[-\frac{4}{n}n + \frac{1}{n^2}\frac{n(n+1)}{2} + \frac{4}{n^3}\frac{n(n+1)(2n+1)}{6} - \frac{1}{n^4}\frac{n^2(n+1)^2}{4}\right]$$

$$= - \lim_{n \to +\infty} \left[-4 + \frac{1}{2}\left(1 + \frac{1}{n}\right) + \frac{2}{3}\left(1 + \frac{1}{n}\right)\left(2 + \frac{1}{n}\right) - \frac{1}{4}\left(1 + \frac{1}{n}\right)^2\right]$$

$$= - \left[-4 + \frac{1}{2} + \frac{4}{3} - \frac{1}{4}\right] = \frac{29}{12}$$

For R_2 the measure of the area is given by

$$A_2 = \lim_{n \to +\infty} \sum_{i=1}^{n} f(\xi_i) \, \Delta_i x$$

Thus,

(a)　$$A_2 = \lim_{n \to +\infty} \sum_{i=1}^{n} (6\xi_i + \xi_i^2 - \xi_i^3) \, \Delta_i x$$

(b) $\displaystyle\lim_{n \to +\infty} \sum_{i=1}^{n} (6\xi_i + \xi_i^2 - \xi_i^3)\,\Delta_i x = \int_0^3 (6x + x^2 - x^3)\,dx$

(c) $\displaystyle \Delta_i x = \frac{3}{n}$

$$\xi_i = \frac{3i}{n}$$

$$6\xi_i + \xi_i^2 - \xi_i^3 = \frac{18i}{n} + \frac{9i^2}{n^2} - \frac{27i^3}{n^3}$$

Thus,

$$A_2 = \lim_{n \to +\infty} \sum_{i=1}^{n} (6\xi_i + \xi_i^2 - \xi_i^3)\,\Delta_i x$$

$$= \lim_{n \to +\infty} \sum_{i=1}^{n} \left(\frac{18i}{n} + \frac{9i^2}{n^2} - \frac{27i^3}{n^3} \right) \frac{3}{n}$$

$$= \lim_{n \to +\infty} \left[\frac{54}{n^2} \sum_{i=1}^{n} i + \frac{27}{n^3} \sum_{i=1}^{n} i^2 - \frac{81}{n^4} \sum_{i=1}^{n} i^3 \right]$$

$$= \lim_{n \to +\infty} \left[\frac{54}{n^2} \cdot \frac{n(n+1)}{2} + \frac{27}{n^3} \cdot \frac{n(n+1)(2n+1)}{6} - \frac{81}{n^4} \cdot \frac{n^2(n+1)^2}{4} \right]$$

$$= \lim_{n \to +\infty} \left[27\left(1 + \frac{1}{n}\right) + \frac{9}{2}\left(1 + \frac{1}{n}\right)\left(2 + \frac{1}{n}\right) - \frac{81}{4}\left(1 + \frac{1}{n}\right)^2 \right]$$

$$= 27 + 9 - \frac{81}{4} = \frac{63}{4}$$

Because $A = A_1 = A_2 = \frac{29}{12} + \frac{63}{4} = \frac{218}{12} = \frac{109}{6}$, the area of the region R is $\frac{109}{6}$ sq. units.

24. Let $[a, b]$ be any interval such that $a < 0 < b$. Prove that even though the unit step function (Exercise 21 in Exercises 1.8) is discontinuous on $[a, b]$, it is integrable on $[a, b]$ and $\int_a^b U(x)\,dx = b$.

SOLUTION: Because

$$U(x) = \begin{cases} 0 & \text{if } x < 0 \\ 1 & \text{if } x \geqslant 0 \end{cases}$$

Then U is discontinuous at 0, and thus U is discontinuous on $[a, b]$. To prove that $\int_a^b U(x)\,dx = b$, we use Definitions 7.3.1 and 7.3.2. For any $\epsilon > 0$ we must find a $\delta > 0$, such that

$$\left| \sum_{i=1}^{n} U(\xi_i)\,\Delta_i x - b \right| < \epsilon \tag{1}$$

where Δ is a partition of $[a, b]$ with $\|\Delta\| < \delta$, and ξ_i is in $[x_{i-1}, x_i]$ for $i = 1, 2, \ldots, n$. Now for any partition of $[a, b]$, we have the following facts.

First, we note that because ξ_i is in $[x_{i-1}, x_i]$ and $x_i - x_{i-1} = \Delta_i x \leqslant \|\Delta\|$, then for $i = 1, 2, \ldots, n$,

$$x_i - \xi_i \leqslant \|\Delta\| \quad \text{and} \quad \xi_i - x_{i-1} \leqslant \|\Delta\| \tag{2}$$

Next, we note that

$$\sum_{i=j}^{n} \Delta_i x = \Delta_j x + \Delta_{j+1} x + \cdots + \Delta_n x$$

$$= (x_j - x_{j-1}) + (x_{j+1} - x_j) + \cdots + (x_n - x_{n-1})$$

$$= x_n - x_{j-1} \tag{3}$$

Then let Δ be any partition of $[a, b]$, and consider the following three cases.

Case I: $\xi_i < 0$ for $i = 1, 2, \ldots, n$

Because $U(\xi_i) = 0$ for $i = 1, 2, \ldots, n$, and $b > 0$, then

$$\left| \sum_{i=1}^{n} U(\xi_i) \Delta_i x - b \right| = |-b| = b \tag{4}$$

Because $b = x_n$, then by (2) we have

$$b - \xi_n \leqslant \|\Delta\|$$

Because $\xi_n < 0$, then by Theorem 1.1.23 (if $a < b$ and $c < d$, then $a + c < b + d$) we have

$$(b - \xi_n) + \xi_n < \|\Delta\| + 0$$
$$b < \|\Delta\| \tag{5}$$

Thus, from (4) and (5) we have

$$\left| \sum_{i=1}^{n} U(\xi_i) \Delta_i x - b \right| = b < \|\Delta\| \tag{6}$$

Case II: $\xi_i \geqslant 0$ for $i = 1, 2, \ldots, n$

Because $U(\xi_i) = 1$ for $i = 1, 2, \ldots, n$, and $a < 0$, then by Eq. (3) we have

$$\left| \sum_{i=1}^{n} U(\xi_i) \Delta_i x - b \right| = \left| \sum_{i=1}^{n} \Delta_i x - b \right|$$

$$= |(x_n - x_0) - b|$$
$$= |(b - a) - b|$$
$$= -a \tag{7}$$

Because $a = x_0$, then by (2) we have

$$\xi_1 - a \leqslant \|\Delta\|$$

Because $-\xi_1 \leqslant 0$, then by Theorem 1.1.23 we have

$$(\xi_1 - a) - \xi_1 \leqslant \|\Delta\| + 0$$
$$-a \leqslant \|\Delta\| \tag{8}$$

Thus, by (7) and (8) we have

$$\left| \sum_{i=1}^{n} U(\xi_i) \Delta_i x - b \right| = -a \leqslant \|\Delta\| \tag{9}$$

Case III: $\xi_i < 0$ for $i = 1, 2, \ldots, j$ and $\xi_i \geqslant 0$ for $i = j+1, j+2, \ldots, n$. Then $u(\xi_i) = 0$ for $i = 1, 2, \cdots, j$, and $u(\xi_i) = 1$ for $i = j+1, j+2, \cdots, n$. Thus,

$$\left| \sum_{i=1}^{n} U(\xi_i) \Delta_i x - b \right| = \left| \sum_{i=j+1}^{n} \Delta_i x - b \right|$$

$$= |(x_n - x_j) - b|$$
$$= |x_j| \tag{10}$$

Because $x_j - \xi_j \leqslant \|\Delta\|$ and $\xi_j < 0$, then

$$x_j \leqslant \|\Delta\| \tag{11}$$

Because $\xi_{j+1} - x_j \leqslant \|\Delta\|$ and $-\xi_{j+1} \leqslant 0$, then

$$-x_j \leqslant \|\Delta\| \tag{12}$$

Therefore, from (11) and (12) we have

$$|x_j| \leqslant \|\Delta\| \tag{13}$$

and by (10) and (13) we have

$$\left| \sum_{i=1}^{n} U(\xi_i) \Delta_i x - b \right| = |x_j| \leqslant \|\Delta\| \tag{14}$$

Therefore, we take $\delta = \epsilon$ whenever $\|\Delta\| < \delta$, then $\|\Delta\| < \epsilon$, and (6), (9), and (14) imply (1). Hence,

$$\int_a^b U(x)\, dx = b$$

7.4 PROPERTIES OF THE DEFINITE INTEGRAL

The following theorems are the ones in this section that we most frequently use for problem solving.

7.4.3 Theorem If the function f is integrable on the closed interval $[a, b]$ and if k is any constant, then

$$\int_a^b kf(x)\, dx = k \int_a^b f(x)\, dx$$

7.4.4 Theorem If the functions f and g are integrable on $[a, b]$, then $f + g$ is integrable on $[a, b]$ and

$$\int_a^b [f(x) + g(x)]\, dx = \int_a^b f(x)\, dx + \int_a^b g(x)\, dx$$

7.4.6 Theorem If f is integrable on a closed interval containing the three numbers, a, b, and c, then

$$\int_a^b f(x)\, dx = \int_a^c f(x)\, dx + \int_c^b f(x)\, dx$$

regardless of the order of $a, b,$ and c.

7.4.8 Theorem If the functions f and g are integrable on the closed interval $[a, b]$ and if $f(x) \geqslant g(x)$ for all x in $[a, b]$, then

$$\int_a^b f(x)\, dx \geqslant \int_a^b g(x)\, dx$$

7.4.9 Theorem Suppose that the function f is continuous on the closed interval $[a, b]$. If m and M are, respectively, the absolute minimum and absolute maximum function values of f on $[a, b]$ so that

$$m \leqslant f(x) \leqslant M \quad \text{for} \quad a \leqslant x \leqslant b$$

then

$$m(b - a) \leqslant \int_a^b f(x)\, dx \leqslant M(b - a)$$

Exercises 7.4

4. Prove that Theorem 7.4.6 is valid when $c < a < b$. Use the result of Theorem 7.4.5.

SOLUTION: We want to prove that if f is integrable on a closed interval containing the three numbers $a, b,$ and c, then

$$\int_a^b f(x) = \int_a^c f(x)\, dx + \int_c^b f(x)\, dx \tag{1}$$

Because, $c < a < b$, then by Theorem 7.4.5, we have

$$\int_c^b f(x)\, dx = \int_c^a f(x)\, dx + \int_a^b f(x)\, dx \tag{2}$$

By Definition 7.3.4,

$$\int_c^a f(x)\, dx = -\int_a^c f(x)\, dx \tag{3}$$

Substituting from Eq. (3) into Eq. (2), we get

$$\int_c^b f(x)\, dx = -\int_a^c f(x)\, dx + \int_a^b f(x)\, dx$$

or, equivalently,

$$\int_a^c f(x)\, dx + \int_c^b f(x)\, dx = \int_a^b f(x)\, dx \tag{4}$$

Because Eq. (4) is equivalent to Eq. (1), we have proved the theorem.

In Exercises 8-17, apply Theorem 7.4.9 to find a smallest and a largest possible value of the given integral.

10. $\displaystyle\int_{-2}^1 (x + 1)^{2/3}\, dx$

SOLUTION: Let f be the function defined by $f(x) = (x + 1)^{2/3}$. Because $f'(x) = \frac{2}{3}(x + 1)^{-1/3}$, the only critical number of the function f is -1. Because $f(-1) = 0$, $f(1) = 2^{2/3}$, and $f(-2) = 1$, we conclude that 0 is the absolute minimum value of f on the closed interval $[-2, 1]$ and $2^{2/3}$ is the absolute maximum value of f on $[-2, 1]$. We use Theorem 7.4.9 with $m = 0$, $M = 2^{2/3}$, and $[a, b] = [-2, 1]$.

Thus, $b - a = 3$, and

$$0 \leqslant \int_{-2}^{1} (x + 1)^{2/3}\, dx \leqslant 3 \cdot 2^{2/3}$$

14. $\displaystyle \int_{-5}^{2} \frac{x + 5}{x - 3}\, dx$

SOLUTION: Let f be the function defined by

$$f(x) = \frac{x + 5}{x - 3}$$

Because,

$$f'(x) = \frac{-8}{(x - 3)^2}$$

f has no critical numbers in the closed interval $[-5, 2]$. Because $f(2) = -7$ and $f(-5) = 0$, we may use Theorem 7.4.9 with $m = -7$ and $M = 0$. Since $b - a = 2 - (-5) = 7$, we have

$$-7(7) \leqslant \int_{-5}^{2} \frac{x + 5}{x - 3} dx \leqslant 0(7)$$

or, equivalently,

$$-49 \leqslant \int_{-5}^{2} \frac{x + 5}{x - 3} dx \leqslant 0$$

18.   Show that if f is continuous on $[-1, 2]$, then

$$\int_{-1}^{2} f(x)\, dx + \int_{2}^{0} f(x)\, dx + \int_{0}^{1} f(x)\, dx + \int_{1}^{-1} f(x)\, dx = 0$$

SOLUTION: Because f is continuous on $[-1, 2]$, then by Theorem 7.3.3 f is integrable on $[-1, 2]$. Therefore, by Theorem 7.4.6

$$\int_{-1}^{2} f(x)\, dx + \int_{2}^{0} f(x)\, dx = \int_{-1}^{0} f(x)\, dx \tag{1}$$

and

$$\int_{0}^{1} f(x)\, dx + \int_{1}^{-1} f(x)\, dx = \int_{0}^{-1} f(x)\, dx \tag{2}$$

By Definition 7.3.4,

$$\int_{-1}^{0} f(x)\, dx = -\int_{0}^{-1} f(x)\, dx \tag{3}$$

By substituting from Eqs. (1) and (2) into Eq. (3), we obtain

$$\int_{-1}^{2} f(x)\, dx + \int_{2}^{0} f(x)\, dx = -\left[\int_{0}^{1} f(x)\, dx + \int_{1}^{-1} f(x)\, dx \right]$$

or, equivalently,

$$\int_{-1}^{2} f(x)\,dx + \int_{2}^{0} f(x)\,dx + \int_{0}^{1} f(x)\,dx + \int_{1}^{-1} f(x)\,dx = 0$$

20. If f is continuous on $[a, b]$, prove that

$$\left| \int_{a}^{b} f(x)\,dx \right| \leqslant \int_{a}^{b} |f(x)|\,dx$$

SOLUTION: Because $|a| \leqslant b$ if and only if $-b \leqslant a \leqslant b$, we must show that

$$-\int_{a}^{b} |f(x)|\,dx \leqslant \int_{a}^{b} f(x)\,dx \leqslant \int_{a}^{b} |f(x)|\,dx$$

Because f is continuous on $[a, b]$, then $|f|$ is also continuous on $[a, b]$, and thus both f and $|f|$ are integrable on $[a, b]$. Because $-|f(x)| \leqslant f(x) \leqslant |f(x)|$ for all x in $[a, b]$, by Theorem 7.4.8 we conclude that

$$\int_{a}^{b} -|f(x)|\,dx \leqslant \int_{a}^{b} f(x)\,dx \leqslant \int_{a}^{b} |f(x)|\,dx$$

which, by Theorem 7.4.3, is equivalent to

$$-\int_{a}^{b} |f(x)|\,dx \leqslant \int_{a}^{b} f(x)\,dx \leqslant \int_{a}^{b} |f(x)|\,dx$$

Therefore,

$$\left| \int_{a}^{b} f(x)\,dx \right| \leqslant \int_{a}^{b} |f(x)|\,dx$$

7.5 THE MEAN-VALUE THEOREM FOR INTEGRALS

7.5.1 Theorem (*Intermediate-value theorem*) If the function f is continuous on the closed interval $[a, b]$ and if $f(a) \neq f(b)$, then for any number k between $f(a)$ and $f(b)$ there exists a number c between a and b such that $f(c) = k$.

7.5.2 Theorem (*Mean-value theorem for integrals*) If the function f is continuous on the closed interval $[a, b]$, then there exists a number X such that $a \leqslant X \leqslant b$, and

$$\int_{a}^{b} f(x)\,dx = f(X)(b - a)$$

The two theorems in this section are not often used for problem solving. However, they are used later to prove theorems that are quite useful for problem solving. The number $f(X)$ in Theorem 7.5.2 is called the "average value" of f on $[a, b]$.

7.5.3 Definition If the function f is integrable on the closed interval $[a, b]$, the *average value* of f on $[a, b]$ is

$$\frac{\displaystyle\int_{a}^{b} f(x)\,dx}{b - a}$$

Exercises 7.5

In Exercises 1-8, a function f and a closed interval $[a, b]$ are given. Determine if the intermediate-value theorem holds for the given value of k. If the theorem holds, find a number c such that $f(c) = k$. If the theorem does not hold, give the reason. Draw a sketch of the curve and the line $y = k$.

4. $f(x) = -\sqrt{100 - x^2}$; $[a, b] = [0, 8]$; $k = -8$.

SOLUTION: Because $f(a) = f(0) = -10$ and $f(b) = f(8) = -6$, then $f(a) \neq f(b)$, and k is between $f(a)$ and $f(b)$. Furthermore, f is continuous on $[0, 8]$. Thus, the hypothesis of the intermediate-value theorem (7.5.1) is satisfied, and we can find a number c such that $f(c) = -8$. Because $f(c) = -\sqrt{100 - c^2}$, we solve the equation

$$-\sqrt{100 - c^2} = -8$$
$$100 - c^2 = 64$$
$$c = \pm 6$$

Because -6 is not between 0 and 8, the only suitable choice of c is 6. A sketch of the graph of f and the line $y = -8$ is shown in Fig. 7.5.4.

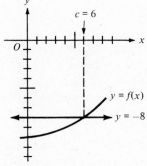

Figure 7.5.4

8. $f(x) = \dfrac{5}{2x - 1}$; $[a, b] = [0, 1]$; $k = 2$

SOLUTION: Because f is discontinuous at $\frac{1}{2}$, then f is discontinuous on the closed interval $[0,1]$. Thus the hypothesis of the intermediate-value theorem is not satisfied, and the theorem does not hold. A sketch of the graph of the function f and the line $y = 2$ is shown in Fig. 7.5.8. Because the line does not intersect the curve, there is no number c that satisfies the conclusion of the intermediate-value theorem.

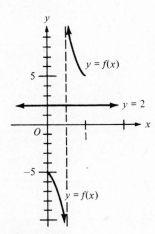

Figure 7.5.8

12. Find the value of χ satisfying the mean-value theorem for integrals. For the value of the definite integral, use the results of the corresponding Exercise in Exercises 7.3.

$$\int_0^4 (x^2 + x - 6)\, dx$$

SOLUTION: We have $f(x) = x^2 + x - 6$ and $[a, b] = [0, 4]$. Thus we wish to find a number χ with $0 \leqslant \chi \leqslant 4$, such that

$$\int_0^4 (x^2 + x - 6)\, dx = f(\chi) \cdot (4 - 0) \tag{1}$$

In Exercise 12 of Exercises 7.3, we found that the value of the definite integral in Eq. (1) is $\frac{16}{3}$. Therefore, Eq. (1) is equivalent to

$$\frac{16}{3} = (\chi^2 + \chi - 6) \cdot 4$$

$$0 = 3\chi^2 + 3\chi - 22$$

$$\chi = \frac{-3 \pm \sqrt{273}}{6}$$

We reject $(-3 - \sqrt{273})/6$ since it is not in the interval $[0, 4]$. Thus, the number χ which satisfies the mean-value theorem is $(-3 + \sqrt{273})/6 \approx 2.25$.

16. Suppose $C(x)$ hundreds of dollars is the total cost of producing $10x$ units of a certain commodity, and $C(x) = x^2$. Find the average total cost when the number of units produced takes on all values from 0 to 20. Use the result of Exercise 7 in Exercises 7.3 for the value of the definite integral.

SOLUTION: Because $10x$ is the number of units, we have $0 \leqslant 10x \leqslant 20$ or, equivalently, $0 \leqslant x \leqslant 2$. Let A.V. be the average value of C on $[0, 2]$, where $C(x) = x^2$. By Definition 7.5.3 we have

$$\text{A.V.} = \frac{1}{2-0} \int_0^2 x^2 \, dx \tag{1}$$

In Exercise 7 of Exercises 7.3, the value of the definite integral in (1) was found to be $\frac{8}{3}$. Thus the expression (1) has value $\frac{4}{3}$, and hence the average value of C on $[0, 2]$ is $\frac{4}{3}$. Because $100C$ is the number of dollars in the total cost, and when $C = \frac{4}{3}$, $100C = \frac{400}{3} = 133.33$, we conclude that $133.33 is the approximate average total cost when the number of units produced takes on all values from 0 to 20.

20. Suppose f is a function for which $0 \leqslant f(x) \leqslant 1$ if $0 \leqslant x \leqslant 1$. Prove that if F is continuous on $[0, 1]$ there is at least one number c in $[0, 1]$ such that $f(c) = c$.

SOLUTION: If $f(0) = 0$, then $c = 0$ satisfies the conclusion. If $f(1) = 1$, then $c = 1$ satisfies the conclusion. Suppose that $f(0) \neq 0$ and $f(1) \neq 1$. Then because $0 \leqslant f(x) \leqslant 1$ if $0 \leqslant x \leqslant 1$, we have $f(0) > 0$ and $f(1) < 1$. Let g be the function defined by

$$g(x) = f(x) - x \tag{1}$$

Then $g(0) = f(0) > 0$ and $g(1) = f(1) - 1 < 0$. Because g is continuous on $[0, 1]$, and $g(0) \neq g(1)$, and 0 is between $g(0)$ and $g(1)$, by the intermediate-value theorem there exists a number c between 0 and 1 such that

$$g(c) = 0 \tag{2}$$

By Eq. (1), we have

$$g(c) = f(c) - c \tag{3}$$

and from (2) and (3) we get

$$f(c) = c$$

7.6 THE FUNDAMENTAL THEOREM OF THE CALCULUS

7.6.1 Theorem Let the function f be continuous on the closed interval $[a, b]$ and let x be any number in $[a, b]$. If F is the function defined by

$$F(x) = \int_a^x f(t) \, dt$$

then

$$F'(x) = f(x) \tag{2}$$

(If $x = a$, the derivative in (2) may be a derivative from the right, and if $x = b$, the derivative in (2) may be a derivative from the left.)

7.6.2 Theorem (*Fundamental theorem of the calculus*) Let the function f be continuous on the closed interval $[a, b]$ and let g be a function such that

$$g'(x) = f(x) \tag{9}$$

for all x in $[a, b]$. Then

$$\int_a^b f(t)\, dt = g(b) - g(a)$$

(If $x = a$, the derivative in (9) may be a derivative from the right, and if $x = b$, the derivative in (9) may be a derivative from the left.)

There is no theorem in the book that we use more often than Theorem 7.6.2. Note that g may be any function that is an antiderivative of f. Usually, we choose the antiderivative whose constant term is zero. We denote

$$[g(b) - g(a)] \quad \text{by} \quad [g(x)]_a^b \quad \text{or} \quad g(x)]_a^b$$

Note the distinction between the *indefinite integral,* $\int f(x)\, dx$, which is defined to be a function g such that $g'(x) = f(x)$, and the *definite integral,* $\int_a^b f(x)\, dx$, which is a *number* defined to be the limit of a Riemann sum. The fundamental theorem of the calculus states that we may sometimes use the indefinite integral to calculate the number represented by the definite integral.

Exercises 7.6

In Exercises 1-18, evaluate the definite integral by using the fundamental theorem of the calculus.

4. $\displaystyle \int_{-1}^{3} \frac{dy}{(y+2)^3}$

SOLUTION:

$$\int_{-1}^{3} \frac{dy}{(y+2)^3} = \int_{-1}^{3} (y+2)^{-3}\, dy$$

$$= \left. \frac{(y+2)^{-2}}{-2} \right]_{-1}^{3}$$

$$= -\frac{1}{2}[5^{-2} - 1^{-2}]$$

$$= \frac{12}{25}$$

8. $\displaystyle \int_{4}^{5} x^2 \sqrt{x-4}\, dx$

SOLUTION: Let $u = \sqrt{x-4}$. Then $u^2 = x - 4$; $x = u^2 + 4$; $dx = 2u\, du$. Furthermore, when $x = 4$, $u = 0$, and when $x = 5$, $u = 1$. Thus,

$$\int_{4}^{5} x^2 \sqrt{x-4}\, dx = \int_{0}^{1} (u^2 + 4)^2\, u\, (2u\, du)$$

$$= 2\int_{0}^{1} (u^6 + 8u^4 + 16u^2)\, du$$

$$= 2\left[\frac{1}{7}u^7 + \frac{8}{5}u^5 + \frac{16}{3}u^3 \right]_{0}^{1}$$

$$= 2\left[\frac{1}{7} + \frac{8}{5} + \frac{16}{3} - 0\right]$$

$$= \frac{1486}{105}$$

12. $\displaystyle\int_{-3}^{3} \sqrt{3 + |x|}\, dx$

SOLUTION: Let $f(x) = \sqrt{3 + |x|}$. Then

$$f(x) = \begin{cases} \sqrt{3 + x} & \text{if } x \geqslant 0 \\ \sqrt{3 - x} & \text{if } x < 0 \end{cases}$$

Applying Theorem 7.4.6, we have

$$\int_{-3}^{3} \sqrt{3 + |x|}\, dx = \int_{-3}^{0} \sqrt{3 - x}\, dx + \int_{0}^{3} \sqrt{3 + x}\, dx \qquad (1)$$

We evaluate the integrals on the right of Eq. (1) separately. If $u = 3 - x$, then $dx = -du$; when $x = -3$, $u = 6$; when $x = 0$, $u = 3$. Therefore,

$$\int_{-3}^{0} \sqrt{3 - x}\, dx = -\int_{6}^{3} \sqrt{u}\, du$$

$$= \int_{3}^{6} u^{1/2}\, du$$

$$= \frac{2}{3} u^{3/2} \Big]_{3}^{6}$$

$$= \frac{2}{3}[6^{3/2} - 3^{3/2}]$$

$$= 2[2\sqrt{6} - \sqrt{3}] \qquad (2)$$

And,

$$\int_{0}^{3} \sqrt{3 + x}\, dx = \frac{2}{3}(3 + x)^{3/2} \Big]_{0}^{3}$$

$$= \frac{2}{3}[6^{3/2} - 3^{3/2}]$$

$$= 2[2\sqrt{6} - \sqrt{3}] \qquad (3)$$

Substituting from (2) and (3) into (1), we obtain

$$\int_{-3}^{3} \sqrt{3 + |x|}\, dx = 4[2\sqrt{6} - \sqrt{3}]$$

16. $\displaystyle\int_{-3}^{2} \frac{3x^3 - 24x^2 + 48x + 5}{x^2 - 8x + 16}\, dx$

SOLUTION: Dividing the numerator by the denominator, we obtain

$$\frac{3x^3 - 24x^2 + 48x + 5}{x^2 - 8x + 16} = 3x + \frac{5}{x^2 - 8x + 16}$$

$$= 3x + 5(x - 4)^{-2}$$

Therefore,

$$\int_{-3}^{2} \frac{3x^3 - 24x^2 + 48x + 5}{x^2 - 8x + 16} \, dx = \int_{-3}^{2} 3x \, dx + \int_{-3}^{2} 5(x-4)^{-2} \, dx$$

$$= \left[\frac{3}{2} x^2\right]_{-3}^{2} + \left[\frac{5(x-4)^{-1}}{-1}\right]_{-3}^{2}$$

$$= \frac{3}{2}[4 - 9] - 5[(-2)^{-1} - (-7)^{-1}]$$

$$= \frac{-15}{2} + \frac{25}{14}$$

$$= -\frac{40}{7}$$

20. Use Theorem 7.6.1 to find the indicated derivative.

$$D_x \int_{x}^{5} \sqrt{1 + t^4} \, dt$$

SOLUTION: Because Theorem 7.6.1 requires that the integral have a constant for its lower limit, we first let F be the function defined by

$$F(x) = \int_{5}^{x} \sqrt{1 + t^4} \, dt$$

We use Theorem 7.6.1 with $f(t) = \sqrt{1 + t^4}$ and $a = 5$ to obtain

$$F'(x) = f(x) = \sqrt{1 + x^4} \qquad\qquad (1)$$

Because

$$\int_{x}^{5} \sqrt{1 + t^4} \, dt = -\int_{5}^{x} \sqrt{1 + t^4} \, dt = -F(x) \qquad\qquad (2)$$

Differentiating on both sides of (2) and then substituting from (1) into the resulting equation, we obtain

$$D_x \int_{x}^{5} \sqrt{1 + t^4} \, dt = -F'(x) = -\sqrt{1 + x^4}$$

In Exercises 23-26 find the area of the region bounded by the given curve and lines. Draw a figure showing the region and a rectangular element of area. Express the measure of the area as the limit of a Riemann sum and then as a definite integral. Evaluate the definite integral by the fundamental theorem of the calculus.

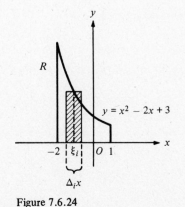

$y = x^2 - 2x + 3$

$-2 \quad \xi_i \quad O \quad 1$

$\Delta_i x$

Figure 7.6.24

24. $y = x^2 - 2x + 3$; x-axis; $x = -2$, $x = 1$

SOLUTION: Fig. 7.6.24 shows a sketch of the region R and a rectangular element of area. In Definition 7.3.6 let $f(x) = x^2 - 2x + 3$ and $[a, b] = [-2, 1]$. Then if A is the measure of the area of the region, we have

$$A = \lim_{\|\Delta\| \to 0} \sum_{i=1}^{n} (\xi_i^2 - 2\xi_i + 3) \, \Delta_i x$$

$$= \int_{-2}^{1} (x^2 - 2x + 3)\, dx$$

$$= \left[\frac{1}{3}x^3 - x^2 + 3x \right]_{-2}^{1}$$

$$= \left[\frac{1}{3} - 1 + 3 \right] - \left[-\frac{8}{3} - 4 - 6 \right]$$

$$= 15$$

The area of the region R is 15 sq. units.

28. $y = \dfrac{1}{x^2} - x$; x-axis; $x = 2$, $x = 3$

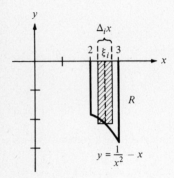

Figure 7.6.28

SOLUTION: Fig. 7.6.28 shows a sketch of the region R and a rectangular element of area. Let $f(x) = 1/x^2 - x$. Because $f(x) < 0$ for x in $[2, 3]$, the altitude of the ith rectangle is $-f(\xi_i)$ units, and the area of the ith rectangle is $-f(\xi_i)\,\Delta_i x$ sq. units. If A is the measure of the area of the region R, then

$$A = \lim_{\|\Delta\| \to 0} \sum_{i=1}^{n} -f(\xi_i)\Delta_i x$$

$$= \lim_{\|\Delta\| \to 0} \sum_{i=1}^{n} \left(-\frac{1}{\xi_i^2} + \xi_i \right) \Delta_i x$$

$$= \int_{2}^{3} \left(-\frac{1}{x^2} + x \right) dx$$

$$= x^{-1} + \frac{1}{2}x^2 \Big|_{2}^{3}$$

$$= \left[\frac{1}{3} + \frac{9}{2} \right] - \left[\frac{1}{2} + 2 \right]$$

$$= \frac{7}{3}$$

The area of the region R is $\frac{7}{3}$ sq. units.

32. Find the average value of the function f on the given interval $[a, b]$.

$$f(x) = 3x\sqrt{x^2 - 16}; [a, b] = [4, 5]$$

SOLUTION: Let A.V. = the average value of f on $[4, 5]$. From Definition 7.5.3 we have

$$\text{A.V.} = \frac{1}{5 - 4} \int_{4}^{5} 3x\sqrt{x^2 - 16}\, dx$$

$$= \frac{3}{2} \int_{4}^{5} (x^2 - 16)^{1/2} (2x\,dx)$$

$$= \frac{3}{2} \left[\frac{2}{3} (x^2 - 16)^{3/2} \right]_{4}^{5}$$

$$= 9^{3/2} - 0$$
$$= 27$$

36. Let f be a function whose derivative f' is continuous on $[a, b]$. Find the average value of the slope of the tangent line to the graph of f on $[a, b]$ and give a geometric interpretation of the result.

SOLUTION: If $m(x)$ is the slope of the tangent line to the graph of $y = f(x)$ at the point $(x, f(x))$, then

$$m(x) = f'(x) \tag{1}$$

Because f' is continuous on $[a, b]$, then f' is integrable on $[a, b]$ and hence m is integrable on $[a, b]$. Let A.V. be the average value of m on $[a, b]$. Then

$$\text{A.V.} = \frac{1}{b-a} \int_a^b m(x)\, dx \tag{2}$$

By Eqs. (1), (2), and the fundamental theorem of the calculus, we have

$$\text{A.V.} = \frac{1}{b-a} [f(x)]_a^b$$

$$= \frac{f(b) - f(a)}{b-a} \tag{3}$$

The fraction in Eq. (3) is also the slope of the line through the points $(a, f(a))$, and $(b, f(b))$. Thus, this slope is the average value of the slope of the tangent line to the graph of f on $[a, b]$.

Review Exercises

2. Find the exact value of the definite integral by making use of the definition of the definite integral; do not use the fundamental theorem of the calculus.

$$\int_{-1}^{3} (x^2 - 1)^2 \, dx$$

SOLUTION: We have $f(x) = (x^2 - 1)^2$ and $[a, b] = [-1, 3]$. Let

$$\Delta x = \frac{b-a}{n} = \frac{4}{n}$$

$$\xi_i = a + i\Delta x = -1 + \frac{4i}{n}$$

Thus,

$$f(\xi_i) = (\xi_i^2 - 1)^2$$

$$= \left[\left(-1 + \frac{4i}{n}\right)^2 - 1 \right]^2$$

$$= \left[\left(1 - \frac{8i}{n} + \frac{16i^2}{n^2}\right) - 1 \right]^2$$

$$= \left(\frac{-8}{n}\right)^2 \left(i - \frac{2i^2}{n}\right)^2$$

$$= \frac{64}{n^2} \left(i^2 - \frac{4i^3}{n} + \frac{4i^4}{n^2}\right)$$

Hence,

$$\int_{-1}^{3} (x^2 - 1)^2 \, dx = \lim_{n \to +\infty} \sum_{i=1}^{n} f(\xi_i) \cdot \Delta x$$

$$= \lim_{n \to +\infty} \sum_{i=1}^{n} \frac{64}{n^2} \left(i^2 - \frac{4i^3}{n} + \frac{4i^4}{n^2} \right) \frac{4}{n}$$

$$= \lim_{n \to +\infty} \left[\frac{256}{n^3} \sum_{i=1}^{n} i^2 - \frac{1024}{n^4} \sum_{i=1}^{n} i^3 + \frac{1024}{n^5} \sum_{i=1}^{n} i^4 \right]$$

$$= \lim_{n \to +\infty} \left[\frac{256}{n^3} \cdot \frac{n(n+1)(2n+1)}{6} - \frac{1024}{n^4} \frac{n^2(n+1)^2}{4} \right.$$

$$\left. + \frac{1024}{n^5} \frac{n(n+1)(6n^3 + 9n^2 + n - 1)}{30} \right]$$

$$= \lim_{n \to +\infty} \left[\frac{128}{3} \left(1 + \frac{1}{n} \right) \left(2 + \frac{1}{n} \right) - 256 \left(1 + \frac{1}{n} \right)^2 \right.$$

$$\left. + \frac{512}{15} \left(1 + \frac{1}{n} \right) \left(6 + \frac{9}{n} + \frac{1}{n^2} - \frac{1}{n^3} \right) \right]$$

$$= \frac{256}{3} - 256 + \frac{1024}{5}$$

$$= \frac{512}{15}$$

6. Express as a definite integral and evaluate the definite integral:

$$\lim_{n \to +\infty} \sum_{i=1}^{n} \frac{8\sqrt{i}}{n^{3/2}}$$

(*Hint:* Consider the function f for which $f(x) = \sqrt{x}$.)

SOLUTION: We must express the given sum as a Riemann sum for the function f on some closed interval $[a, b]$. For simplicity, let $a = 0$; let $\Delta_i x = (b - a)/n = b/n$; and let $\xi_i = a + i\Delta_i x = bi/n$. Then the Riemann sum is

$$\sum_{i=1}^{n} f(\xi_i) \Delta_i x = \sum_{i=1}^{n} \sqrt{\frac{bi}{n}} \cdot \frac{b}{n}$$

$$= \sum_{i=1}^{n} \frac{b^{3/2}\sqrt{i}}{n^{3/2}} \tag{1}$$

Comparing the sum on the right side of (1) with the given sum, we see that $b^{3/2} = 8$, or $b = 4$. Thus, the given sum is a Riemann sum for the function f on $[0, 4]$. Hence, by Definition 7.3.2

$$\lim_{n \to +\infty} \sum_{i=1}^{n} \frac{8\sqrt{i}}{n^{3/2}} = \int_{0}^{4} f(x) \, dx$$

$$= \int_0^4 \sqrt{x}\, dx$$

$$= \frac{2}{3}x^{3/2}\Big]_0^4$$

$$= \frac{16}{3}$$

In Exercises 7-16, evaluate the definite integral by using the fundamental theorem of the calculus.

10. $\displaystyle\int_{-5}^{5} 2x \sqrt[3]{x^2 + 2}\, dx$

SOLUTION: If $u = x^2 + 2$, then $du = 2x\,dx$. Hence,

$$\int_{-5}^{5} 2x \sqrt[3]{x^2 + 2}\, dx = \int_{-5}^{5} (x^2 + 2)^{1/3} d(x^2 + 2)$$

$$= \frac{3}{4}[(x^2 + 2)^{4/3}]_{-5}^{5}$$

$$= \frac{3}{4}[27^{4/3} - 27^{4/3}]$$

$$= 0$$

16. $\displaystyle\int_{1}^{2} \frac{x\,dx}{\sqrt{5 - x}}$

SOLUTION: Let $u = \sqrt{5 - x}$. Then $u^2 = 5 - x$, $x = 5 - u^2$, and $dx = -2u\,du$. When $x = 1$, $u = 2$; when $x = 2$, $u = \sqrt{3}$. Thus,

$$\int_{1}^{2} \frac{x\,dx}{\sqrt{5 - x}} = \int_{2}^{\sqrt{3}} \frac{(5 - u^2)(-2u\,du)}{u}$$

$$= 2 \int_{\sqrt{3}}^{2} (5 - u^2)\, du$$

$$= 2\left[5u - \frac{1}{3}u^3\right]_{\sqrt{3}}^{2}$$

$$= 2\left[\left(10 - \frac{8}{3}\right) - (5\sqrt{3} - \sqrt{3})\right]$$

$$= \frac{44}{3} - 8\sqrt{3}$$

16. $\displaystyle\int_{-2}^{2} x\,|x - 3|\,dx$

SOLUTION: Because $|x - 3| = -(x - 3)$ if $x < 3$, and $x < 3$ whenever x is in the closed interval $[-2, 2]$,

$$\int_{-2}^{2} x|x - 3|\,dx = \int_{-2}^{2} -x(x - 3)\,dx$$

$$= \int_{-2}^{2} (-x^2 + 3x)\,dx$$

$$= -\frac{1}{3}x^3 + \frac{3}{2}x^2\bigg]_{-2}^{2}$$

$$= \left[\left(-\frac{8}{3} + 6\right) - \left(\frac{8}{3} + 6\right)\right]$$

$$= -\frac{16}{3}$$

22. Find the area of the region bounded by the given curve and lines. Draw a figure showing the region and a rectangular element of area. Express the measure of the area as the limit of a Riemann sum and then as a definite integral. Evaluate the definite integral by the fundamental theorem of the calculus.

$$y = 16 - x^2; \ x\text{-axis}; \ x = -4, \ x = 4$$

SOLUTION: A sketch of the region R showing a rectangular element of area is shown in Fig. 7.22R. We have $f(x) = 16 - x^2$, and $[a, b] = [-4, 4]$. If A is the measure of the area of R, then

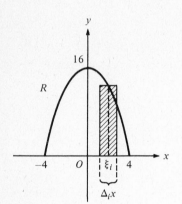

Figure 7.22R

$$A = \lim_{\|\Delta\| \to 0} \sum_{i=1}^{n} (16 - \xi_i^2)\,\Delta_i x$$

$$= \int_{-4}^{4} (16 - x^2)\,dx$$

$$= 16x - \frac{1}{3}x^3\bigg]_{-4}^{4}$$

$$= \left(64 - \frac{64}{3}\right) - \left(-64 + \frac{64}{3}\right)$$

$$= \frac{256}{3}$$

26. Interpret the mean-value theorem for integrals (7.5.2) in terms of an average function value.

SOLUTION: If f is continuous on the closed interval $[a, b]$, then by the mean-value theorem there exists a number χ such that $a \leqslant \chi \leqslant b$, and

$$\int_{a}^{b} f(x)\,dx = f(\chi)(b - a)$$

or, equivalently,

$$\frac{\int_{a}^{b} f(x)\,dx}{b - a} = f(\chi) \tag{1}$$

Because the fraction on the left side of Eq. (1) is by Definition 7.5.3 the average value of f on $[a, b]$, then $f(\chi)$ is the average value of f on $[a, b]$. That is, the number χ in the conclusion of the mean-value theorem for integrals is the number at which the function value of f is the average function value of f on the closed interval $[a, b]$.

28. Suppose a ball is dropped from rest and after t sec its directed distance from the starting point is s ft and its velocity is v ft/sec. Neglect air resistance. When $t = t_1, s = s_1$, and $v = v_1$.

(a) Express v as a function of t as $v = f(t)$ and find the average value of f on $[0, t_1]$.

(b) Express v as a function of s as $v = g(s)$ and find the average velocity of g on $[0, s_1]$.

(c) Write the results of parts (a) and (b) in terms of t_1 and determine which average velocity is larger.

SOLUTION: Let the origin be at the point where the ball is dropped and let the positive direction be downward. If a ft/sec^2 is the acceleration of the ball, then $a = 32$, because the acceleration is due to the force of gravity.

(a) Because $a = dv/dt$, we have

$$\frac{dv}{dt} = 32$$

$$dv = 32dt$$

$$\int dv = \int 32dt$$

$$v = 32t + C_1$$

Because the ball is dropped from rest, $v = 0$ when $t = 0$. Thus $C_1 = 0$, and hence

$$v = f(t) = 32t \tag{1}$$

Let $\bar{f}[0, t_1]$ be the average value of f on $[0, t_1]$. Then

$$\bar{f}[0, t_1] = \frac{1}{t_1} \int_0^{t_1} f(t)\, dt$$

$$= \frac{1}{t_1} \int_0^{t_1} (32t)\, dt$$

$$= \frac{1}{t_1} [16t^2]_0^{t_1}$$

$$= \frac{1}{t_1} [16t_1^2]$$

$$= 16t_1$$

(b) Because

$$a = \frac{dv}{ds} \cdot \frac{ds}{dt} = \frac{dv}{ds} \cdot v$$

and $a = 32$, we have

$$32ds = vdv$$

$$\int 32ds = \int vdv$$

$$32s = \frac{1}{2}v^2 + C_2$$

Because $v = 0$ when $s = 0$, we have $C_2 = 0$, and thus

$$32s = \frac{1}{2}v^2 \tag{2}$$

$$v = g(s) = 8\sqrt{s}$$

Let $\bar{g}[0, s_1]$ be the average value of g on $[0, s_1]$. Then

$$\bar{g}[0, s_1] = \frac{1}{s_1} \int_0^{s_1} g(s)\, ds$$

$$= \frac{1}{s_1} \int_0^{s_1} 8\sqrt{s}\, ds$$

$$= \frac{8}{s_1} \left[\frac{2}{3} s^{3/2} \right]_0^{s_1}$$

$$= \frac{16}{3}\sqrt{s_1} \tag{3}$$

(c) From (2) we have

$$s = \frac{v^2}{64} \tag{4}$$

Substituting from Eq. (1) into (4) we get

$$s = \frac{(32t)^2}{64}$$

Thus,

$$s_1 = 16t_1^2 \tag{5}$$

And by substituting from Eq. (5) into (3), we obtain

$$\bar{g}[0, s_1] = \frac{16}{3}\sqrt{16t_1^2}$$

$$= \frac{64}{3} t_1$$

Because $\frac{64}{3}t_1 > 16t$, we conclude that the average velocity on $[0, s_1]$ is larger than the average velocity on $[0, t_1]$.

32. Make up an example of a discontinuous function for which the mean-value theorem for integrals (a) does not hold, and (b) does hold.

SOLUTION:

(a) Let f be the unit step function on the closed interval $[-1, 1]$. That is,

$$f(x) = \begin{cases} 0 & \text{if } -1 \leqslant x < 0 \\ 1 & \text{if } 0 \leqslant x \leqslant 1 \end{cases} \tag{1}$$

In Exercise 24 of Exercises 7.3 let $[a, b] = [-1, 1]$. Then, by the result of the exercise, we have

$$\int_{-1}^{1} f(x)\, dx = 1 \tag{2}$$

If the mean-value theorem holds for f on $[-1, 1]$, then there must be some number χ such that

$$\int_{-1}^{1} f(x)\, dx = f(\chi) \cdot [1 - (-1)] \tag{3}$$

From (2) and (3), we have

$$f(\chi) = \frac{1}{2}$$

But this contradicts (1), and hence the mean-value theorem does not hold.

(b) Let g be the function defined on $[-1, 1]$ as follows:

$$g(x) = \begin{cases} 0 & \text{if } -1 \leqslant x < 0 \\ 1 & \text{if } x = 0 \\ 0 & \text{if } 0 < x \leqslant 1 \end{cases}$$

In Example 1 of Section 7.3 of the text we showed that g is discontinuous at 0 and that

$$\int_{-1}^{1} g(x)\, dx = 0 \tag{4}$$

Because $g(\chi) = 0$ for any χ in $[-1, 1]$ except 0, then let $\chi = 1$, and

$$g(\chi)[1 - (-1)] = 0 \tag{5}$$

By (4) and (5) the conclusion of the mean-value theorem holds for the function g on the closed interval $[-1, 1]$ if we let $\chi = 1$.

8
Applications of the definite integral

8.1 AREA OF A REGION IN A PLANE

Let R be a region in the xy plane. R is bounded on the left by the line $x = a$ and bounded on the right by the line $x = b$ if and only if every line of the form $x = x_1$ with $a \leqslant x_1 \leqslant b$ intersects the region R in one or more points and no line of the form $x = x_1$ with $x_1 > b$ or $x_1 < a$ intersects R. See Fig. 8.1(a). Furthermore, R is bounded above by the curve $y = f(x)$ and bounded below by the curve $y = g(x)$ if and only if every nonempty intersection of R and a line perpendicular to the x-axis is a line segment that has its upper endpoint on the curve $y = f(x)$ and its lower endpoint on the curve $y = g(x)$, and thus $f(x) \geqslant g(x)$ for all x in $[a, b]$.

If R is bounded on the left by the line $x = a$, bounded on the right by the line $x = b$, bounded above by the curve $y = f(x)$, and bounded below by the curve $y = g(x)$, then the area of R is A square units, where

$$A = \int_a^b [f(x) - g(x)]\, dx \tag{1}$$

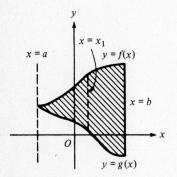

Figure 8.1(a)

Note that we do not need to consider whether R lies above or below the x-axis. If R is bounded below by the x-axis, then $g(x) = 0$ in Formula (1), and the integral reduces to the integral given in Definition 7.3.6. If R is bounded above by the x-axis, then $f(x) = 0$ in (1) and the integral reduces to that given in the formula in Section 7.6 for finding the area of such a region.

We may interchange the roles of x and y in the above discussion. Fig. 8.1(b) illustrates a region R that is bounded below by the line $y = a$ and bounded above by the line $y = b$. Every line of the form $y = y_1$ intersects R if $a \leqslant y_1 \leqslant b$ and no line of the form $y = y_1$ intersects R if $y_1 > b$ or $y_1 < a$. Furthermore, R is bounded on the right by the curve $x = f(y)$ and bounded on the left by the curve

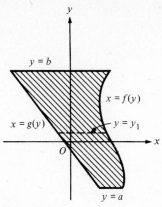

Figure 8.1(b)

$x = g(y)$. Every nonempty intersection of R and a line perpendicular to the y-axis is a line segment that has its right endpoint on the curve $x = f(y)$ and its left endpoint on the curve $x = g(y)$, and thus $f(y) \geqslant g(y)$ for all y in $[a, b]$. The area of any such region is A square units, where

$$A = \int_a^b [f(y) - g(y)]\, dy \tag{2}$$

The following steps are used to find the area of a region R in the xy plane if the equations for the boundaries are given.

1. Draw a sketch of the region R.
2. If the intersection of R and a vertical line is always a line segment that has its upper endpoint on the boundary curve $y = f(x)$ and its lower endpoint on the boundary curve $y = g(x)$, then find the numbers a and b such that R is bounded on the left by the line $x = a$ and bounded on the right by the line $x = b$, and use Formula (1) to find the area of R.
3. If the intersection of R and a horizontal line is always a line segment that has its right endpoint on the boundary curve $x = f(y)$ and its left endpoint on the boundary curve $x = g(y)$, then find the numbers a and b such that R is bounded below by the line $y = a$ and bounded above by the line $y = b$, and use Formula (2) to find the area of R.
4. If neither of the conditions described in steps (2) and (3) is satisfied by the region R, then divide R into two or more subregions that each satisfy either (2) or (3) and find the areas of these subregions separately. The area of R is the sum of the areas of the subregions.

Sometimes a given region R satisfies the conditions for both of the types discussed above. In that case we may use either Formula (1) or Formula (2) to find its area. We try to choose whichever formula gives the integral that is easier to evaluate.

Exercises 8.1

In Exercises 1-20, find the area of the region bounded by the given curves. In each problem do the following:

(a) Draw a figure showing the region and a rectangular element of area.
(b) Express the area of the region as the limit of a Riemann sum.
(c) Find the limit in part (b) by evaluating a definite integral by the fundamental theorem of the calculus.

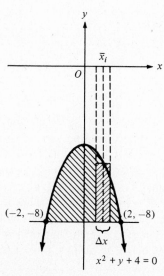

Figure 8.1.4

4. $x^2 + y + 4 = 0$; $y = -8$. Take the elements of area parallel to the y-axis.

SOLUTION: The region R is shown in Fig. 8.1.4. R is bounded above by the curve $x^2 + y + 4 = 0$. Solving for y, we obtain $y = f(x) = -x^2 - 4$. R is bounded below by the line $y = -8$. Thus $g(x) = -8$. The elements of area are rectangles with width Δx units and length $[f(\bar{x}_i) - g(\bar{x}_i)]$ units. Thus, the area of R is A sq. units where

$$A = \lim_{\Delta x \to 0} \sum_{i=1}^{n} [f(\bar{x}_i) - g(\bar{x}_i)]\, \Delta x$$

Solving the given equations simultaneously, we find that the points of intersection of the boundaries are $(-2, -8)$ and $(2, -8)$. Thus, R is bounded on the left by the line $x = -2$ and bounded on the right by the line $x = 2$. Therefore, the limit of the above Riemann sum can be found as follows.

$$A = \int_{-2}^{2} [f(x) - g(x)]\, dx$$

$$= \int_{-2}^{2} [(-x^2 - 4) - (-8)]\, dx$$

$$= \int_{-2}^{2} (-x^2 + 4)\, dx$$

$$= -\frac{1}{3}x^3 + 4x \Big]_{-2}^{2}$$

$$= \left(-\frac{8}{3} + 8\right) - \left(\frac{8}{3} - 8\right)$$

$$= \frac{32}{3}$$

The area of region R is $\frac{32}{3}$ sq. units.

6. $y^3 - 4x;\ x = 0;\ y = -2$

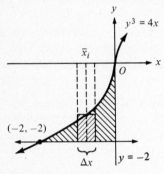

Figure 8.1.6

SOLUTION: The region R is shown in Fig. 8.1.6. The intersection of R with a vertical line is always a line segment that has its upper endpoint on the curve $y^3 = 4x$ and its lower endpoint on the line $y = -2$. Thus, we may take rectangular elements of area perpendicular to the x-axis with width Δx units. Solving the equation $y^3 = 4x$ for y, we obtain $y = f(x) = \sqrt[3]{4x}$. From the line $y = -2$ we have $g(x) = -2$. Thus, the length of each element of area is $[f(\bar{x}_i) - g(\bar{x}_i)]$ units and

$$A = \lim_{\Delta x \to 0} \sum_{i=1}^{n} [f(\bar{x}_i) - g(\bar{x}_i)]\, \Delta x$$

Because the curve $y^3 = 4x$ and the line $y = -2$ intersect at the point $(-2, -2)$, R is bounded on the left by the line $x = -2$. R is bounded on the right by the line $x = 0$. Thus,

$$A = \int_{-2}^{0} [\sqrt[3]{4x} - (-2)]\, dx$$

$$= 2^{2/3} \int_{-2}^{0} x^{1/3}\, dx + 2 \int_{-2}^{0} dx$$

$$= 2^{2/3} \cdot \frac{3}{4} x^{4/3} \Big]_{-2}^{0} + 2[x]_{-2}^{0}$$

$$= 2^{2/3} \cdot 3 \cdot 2^{-2} [0 - (-2)^{4/3}] + 2[0 - (-2)]$$
$$= 1$$

The area of region R is 1 sq. unit.

ALTERNATE SOLUTION: Because the intersection of R with any horizontal line is a line segment that has its right endpoint on the line $x = 0$ and its left endpoint on the curve $y^3 = 4x$, then R is bounded on the right by $x = f(y) = 0$ and bounded on the left by $x = g(y) = \frac{1}{4}y^3$. Furthermore, R is bounded below by the line $y = -2$ and bounded above by the line $y = 0$. Therefore,

$$A = \int_{-2}^{0} [f(y) - g(y)] \, dy$$

$$= \int_{-2}^{0} -\frac{1}{4} y^3 \, dy$$

$$= -\frac{1}{16} y^4 \Big]_{-2}^{0}$$

$$= 1$$

12. $x = 4 - y^2;\ x = 4 - 4y$

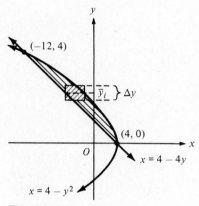

Figure 8.1.12

SOLUTION: The region R is shown in Fig. 8.1.12. We take elements of area perpendicular to the y-axis with width Δy units. Because R is bounded on the right by the curve $x = 4 - y^2$, we let $x = f(y) = 4 - y^2$. Because R is bounded on the left by the line $x = 4 - 4y$, we let $x = g(y) = 4 - 4y$. Then the length of each element of area is $[f(\bar{y}_i) - g(\bar{y}_i)]$ units, and

$$A = \lim_{\Delta y \to 0} \sum_{i=1}^{n} [f(\bar{y}_i) - g(\bar{y}_i)] \, \Delta y$$

Furthermore, because the given curve and the given line intersect at the points $(4, 0)$ and $(-12, 4)$, the region R is bounded below by the line $y = 0$ and bounded above by the line $y = 4$. Thus,

$$A = \int_{0}^{4} [f(y) - g(y)] \, dy$$

$$= \int_{0}^{4} [(4 - y^2) - (4 - 4y)] \, dy$$

$$= \int_{0}^{4} (-y^2 + 4y) \, dy$$

$$= -\frac{1}{3} y^3 + 2y^2 \Big]_{0}^{4}$$

$$= -\frac{1}{3} (4^3) + 2(4^2)$$

$$= \frac{32}{3}$$

The area of the region is $\frac{32}{3}$ sq. units.

14. $xy^2 = y^2 - 1;\ x = 1;\ y = 1;\ y = 4$

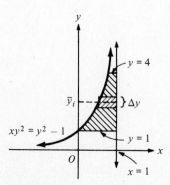

Figure 8.1.14

SOLUTION: The region R is shown in Fig. 8.1.14. We take elements of area perpendicular to the y-axis with width Δy units. Because the region R is bounded on the right by the line $x = 1$, let $x = f(y) = 1$. Because the region R is bounded on the left by the curve $xy^2 = y^2 - 1$, we solve for x and obtain $x = g(y) = 1 - y^{-2}$. Then we have

$$A = \lim_{\Delta y \to 0} \sum_{i=1}^{n} [f(\bar{y}_i) - g(\bar{y}_i)] \, \Delta y$$

Furthermore, R is bounded below by the line $y = 1$ and bounded above by the line $y = 4$. Therefore,

$$A = \int_1^4 [f(y) - g(y)] \, dy$$

$$= \int_1^4 [1 - (1 - y^{-2})] \, dy$$

$$= \int_1^4 y^{-2} \, dy$$

$$= -\frac{1}{y} \Big]_1^4$$

$$= -\frac{1}{4} - (-1)$$

$$= \frac{3}{4}$$

The area of region R is $\frac{3}{4}$ sq. units.

18. $3y = x^3 - 2x^2 - 15x$; $y = x^3 - 4x^2 - 11x + 30$

SOLUTION: A sketch of the region R is shown in Fig. 8.1.18. We take elements of area perpendicular to the x-axis with width Δx units. Solving the equation $3y = x^3 - 2x^2 - 15x$ for y, we obtain $y = f(x) = \frac{1}{3}x^3 - \frac{2}{3}x^2 - 5x$. Let $g(x) = x^3 - 4x^2 - 11x + 30$. Because the curves $y = f(x)$ and $y = g(x)$ intersect at the point where

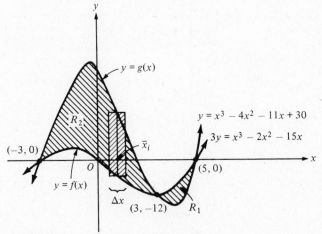

Figure 8.1.18

$x = 3$, we divide R into two subregions: R_1 is the part of R that is to the right of the line $x = 3$, and R_2 is the part of R that is to the left of the line $x = 3$. Region R_1 is bounded above by the curve $y = f(x)$ and bounded below by the curve $y = g(x)$. Thus,

$$A_1 = \lim_{\Delta x \to 0} \sum_{i=1}^{n} [f(\bar{x}_i) - g(\bar{x}_i)] \, \Delta x$$

Furthermore, because the curves intersect at the points $(3, -12)$ and $(5, 0)$ in region R_1, then R_1 is bounded on the left by the line $x = 3$ and bounded on the right by the line $x = 5$. Therefore,

$$A_1 = \int_{3}^{5} [f(x) - g(x)] \, dx$$

$$= \int_{3}^{5} \left[\left(\frac{1}{3} x^3 - \frac{2}{3} x^2 - 5x \right) - (x^3 - 4x^2 - 11x + 30) \right] dx$$

$$= \int_{3}^{5} \left(-\frac{2}{3} x^3 + \frac{10}{3} x^2 + 6x - 30 \right) dx$$

$$= -\frac{1}{6} [x^4]_3^5 + \frac{10}{9} [x^3]_3^5 + 3[x^2]_3^5 - 30[x]_3^5$$

$$= -\frac{1}{6} (5^4 - 3^4) + \frac{10}{9} (5^3 - 3^3) + 3(5^2 - 3^2) - 30(5 - 3)$$

$$= \frac{56}{9}$$

Region R_2 is bounded above by the curve $y = g(x)$ and bounded below by the curve $y = f(x)$. Thus

$$A_2 = \lim_{\Delta x \to 0} \sum_{i=1}^{n} [g(\bar{x}_i) - f(\bar{x}_i)] \, \Delta x$$

Furthermore, because the curves intersect at the points $(-3, 0)$ and $(3, -12)$ in R_2, then R_2 is bounded on the left by the line $x = -3$ and on the right by the line $x = 3$. Therefore,

$$A_2 = \int_{-3}^{3} [g(x) - f(x)] \, dx \tag{3}$$

$$= \int_{-3}^{3} \left[(x^3 - 4x^2 - 11x + 30) - \left(\frac{1}{3} x^3 - \frac{2}{3} x^2 - 5x \right) \right] dx$$

$$= \int_{-3}^{3} \left(\frac{2}{3} x^3 - \frac{10}{3} x^2 - 6x + 30 \right) dx$$

$$= \frac{1}{6} [x^4]_{-3}^{3} - \frac{10}{9} [x^3]_{-3}^{3} - 3[x^2]_{-3}^{3} + 30[x]_{-3}^{3}$$

$$= \frac{1}{6} [3^4 - (-3)^4] - \frac{10}{9} [3^3 - (-3)^3] - 3[3^2 - (-3)^2] + 30[3 - (-3)]$$

$$= \frac{1}{6} (0) - 60 - 3(0) + 30(6)$$

$$= 120$$

Then $A = A_1 + A_2 = \frac{56}{9} + 120 = \frac{1136}{9} = 126\frac{2}{9}$. The area of region R is $126\frac{2}{9}$ sq. units.

22. Find by integration the area of the triangle having vertices at $(3, 4), (2, 0)$, and $(0, 1)$.

SOLUTION: A sketch of the triangle is shown in Fig. 8.1.22. Because an equation of the line that contains $(0, 1)$ and $(3, 4)$ is $y = x + 1$, we let $f(x) = x + 1$. Because an equation of the line that contains $(3, 4)$ and $(2, 0)$ is $y = 4x - 8$, we let $g(x) = 4x - 8$. Because an equation of the line that contains $(0, 1)$ and $(2, 0)$ is $y = -\frac{1}{2}x + 1$, we let $h(x) = -\frac{1}{2}x + 1$. We divide the triangle into two parts: R_1 is that part of the triangle to the right of the vertex $(2, 0)$, and R_2 is that part of the triangle to the left of $(2, 0)$. Region R_1 is bounded above by the line $y = f(x)$, below by the line $y = g(x)$, on the left by the line $x = 2$, and on the right by the line $x = 3$. Therefore,

$$A_1 = \lim_{\Delta x \to 0} \sum_{i=1}^{n} [f(\bar{x}_i) - g(\bar{x}_i)] \, \Delta x$$

$$= \int_{2}^{3} [f(x) - g(x)] \, dx$$

$$= \int_{2}^{3} [(x + 1) - (4x - 8)] \, dx$$

$$= -3 \int_{2}^{3} (x - 3) \, dx$$

$$= -3 \left[\frac{1}{2}x^2 - 3x \right]_{2}^{3}$$

$$= -3 \left[\left(\frac{9}{2} - 9 \right) - (2 - 6) \right]$$

$$= \frac{3}{2}$$

Region R_2 is bounded above by the line $y = f(x)$, bounded below by the line $y = h(x)$, bounded on the left by the line $x = 0$, and bounded on the right by the line $x = 2$. Therefore,

$$A_2 = \lim_{\Delta x \to 0} \sum_{i=1}^{n} [f(\bar{x}_i) - h(\bar{x}_i)] \, x$$

$$= \int_{0}^{2} [f(x) - h(x)] \, dx$$

$$= \int_{0}^{2} \left[(x + 1) - \left(-\frac{1}{2}x + 1 \right) \right] dx$$

$$= \frac{3}{2} \int_{0}^{2} x \, dx$$

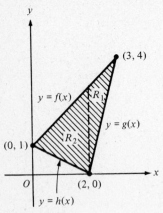

Figure 8.1.22

$$= \frac{3}{4} x^2 \Big]_0^2$$

$$= 3$$

Because $A = A_1 + A_2$, then $A = \frac{3}{2} + 3 = \frac{9}{2}$. Thus, the area of the triangle is $\frac{9}{2}$ sq. units.

24. Find the area of the region bounded by the three curves $y = x^2$, $x = y^3$, and $x + y = 2$.

SOLUTION: A sketch of the region R is shown in Fig. 8.1.24. Although there is more than one region that is bounded by two of the given curves, the only region bounded by all three curves is that indicated by the shading in the figure. We divide R into two parts: R_1 is the part to the right of the y-axis, and R_2 is the part to the left of the y-axis. Because R_1 is bounded above by the line $x + y = 2$, we let $y = f(x) = -x + 2$. Because R_1 is bounded below by the curve $x = y^3$, we let $y = g(x) = \sqrt[3]{x}$. Furthermore, R_1 is bounded on the left by the line $x = 0$ and bounded on the right by the line $x = 1$. Therefore,

$$A_1 = \lim_{\Delta x \to 0} \sum_{i=1}^{n} [f(\bar{x}_i) - g(\bar{x}_i)] \, dx$$

$$= \int_0^1 [f(x) - g(x)] \, dx$$

$$= \int_0^1 [(-x + 2) - \sqrt[3]{x}\,] \, dx$$

$$= -\frac{1}{2} x^2 + 2x - \frac{3}{4} x^{4/3} \Big]_0^1$$

$$= \left(-\frac{1}{2} + 2 - \frac{3}{4} \right) - 0$$

$$= \frac{3}{4}$$

Because R_2 is bounded above by the line $y = f(x)$, bounded below by the curve $y = h(x) = x^2$, bounded on the left by the line $x = -2$, and bounded on the right by the line $x = 0$, we have

$$A_2 = \lim_{\Delta x \to 0} \sum_{i=1}^{n} [f(\bar{x}_i) - h(\bar{x}_i)] \, \Delta x$$

$$= \int_{-2}^0 [f(x) - h(x)] \, dx$$

$$= \int_{-2}^0 [(-x + 2) - x^2] \, dx$$

$$= -\frac{1}{2} x^2 + 2x - \frac{1}{3} x^3 \Big]_{-2}^0$$

$$= \frac{10}{3}$$

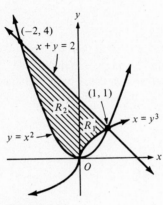

$(-2, 4)$

$x + y = 2$

$(1, 1)$

R_2

R_1

$x = y^3$

$y = x^2$

O

Figure 8.1.24

Because $A = A_1 + A_2$, then $A = \frac{3}{4} + \frac{10}{3} = \frac{49}{12}$. Thus, the area of region R is $\frac{49}{12}$ sq. units.

30. Determine m so that the region above the line $y = mx$ and below the parabola $y = 2x - x^2$ has an area of 36 square units.

SOLUTION: A sketch of the region R is shown in Fig. 8.1.30. To find the points of intersection for the line and the parabola we find the simultaneous solution of the given equations. Thus,

$$mx = 2x - x^2$$
$$x(x + m - 2) = 0$$
$$x = 0 \qquad x = 2 - m$$

Because the slope of the tangent line to the curve $y = 2x - x^2$ at the origin is 2, the slope of the line $y = mx$ must be less than 2, and thus $2 - m > 0$. We conclude that the region R is bounded on the left by the line $x = 0$ and on the right by the line $x = 2 - m$. Let $f(x) = 2x - x^2$ and $g(x) = mx$. Then the area of region R is given by

$$A = \lim_{\Delta x \to 0} \sum_{i=1}^{n} [f(\bar{x}_i) - g(\bar{x}_i)] \, \Delta x$$

$$= \int_0^{2-m} [f(x) - g(x)] \, dx$$

$$= \int_0^{2-m} [(2x - x^2) - mx] \, dx$$

$$= (2 - m) \int_0^{2-m} x \, dx - \int_0^{2-m} x^2 \, dx$$

$$= \frac{1}{2}(2 - m)x^2 \Big]_0^{2-m} - \frac{1}{3}x^3 \Big]_0^{2-m}$$

$$= \frac{1}{2}(2 - m)(2 - m)^2 - \frac{1}{3}(2 - m)^3$$

$$= \frac{1}{6}(2 - m)^3 \qquad \qquad \textbf{(1)}$$

Because we are given that $A = 36$, from Eq. (1) we have

$$36 = \frac{1}{6}(2 - m)^3$$
$$6^3 = (2 - m)^3$$
$$6 = 2 - m$$
$$m = -4$$

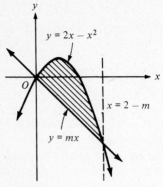

Figure 8.1.30

8.2 VOLUME OF A SOLID OF REVOLUTION: CIRCULAR-DISK AND CIRCULAR-RING METHODS

The formulas given in the text for finding the volume of a solid of revolution apply only to the case in which the x-axis is the axis of revolution. When these formulas do not apply, use the following more general methods which are based on the formula for the volume of a circular ring (or washer):

$$V = \pi(R^2 - r^2) \, \Delta h$$

where R units is the outer radius of the ring, r units is the inner radius, and Δh

units is the thickness of the ring (the distance between the circular faces of the ring). If the inner radius is zero $(r = 0)$, then the circular ring is a circular disk, and the formula reduces to

$$V = \pi R^2 \Delta h$$

Let G be a region in the xy plane that is bounded above by the curve $y = f(x)$, bounded below by the curve $y = g(x)$, bounded on the left by the line $x = a$, and bounded on the right by the line $x = b$. To find the volume of the solid of revolution that results if G is revolved about a line L parallel to the x-axis (where L does not intersect the interior of G), take an element of area that is perpendicular to the x-axis of width $\Delta_i x$ units and revolve the element of area about the line L. This results in an element of volume that is a circular ring whose thickness is given by $\Delta_i h = \Delta_i x$, whose outer radius is given by $R_i = R(\bar{x}_i)$, and whose inner radius is given by $r_i = r(\bar{x}_i)$.

The volume of the solid is given by

$$V = \lim_{\|\Delta\| \to 0} \sum_{i=1}^{n} \pi (R_i{}^2 - r_i{}^2) \, \Delta_i h$$

$$= \lim_{\|\Delta\| \to 0} \sum_{i=1}^{n} \pi ([R(\bar{x}_i)]^2 - [r(\bar{x}_i)]^2 \, \Delta_i x$$

$$= \pi \int_a^b ([R(x)]^2 - [r(x)]^2) \, dx$$

Let H be a region in the xy plane that is bounded on the right by the curve $x = f(y)$, bounded on the left by the curve $x = g(y)$, bounded below by the line $y = a$, and bounded above by the line $y = b$. To find the volume of the solid of revolution that results if H is revolved about a line L parallel to the y-axis (where L does not intersect the interior of H), take an element of area that is perpendicular to the y-axis of width $\Delta_i y$ units and revolve the element of area about the line L. This results in an element of volume that is a circular ring whose thickness is given by $\Delta_i h = \Delta_i y$, whose outer radius is given by $R_i = R(\bar{y}_i)$, and whose inner radius is given by $r_i = r(\bar{y}_i)$. The volume of the solid of revolution is given by

$$V = \lim_{\|\Delta\| \to 0} \sum_{i=1}^{n} \pi (R_i{}^2 - r_i{}^2) \Delta_i h$$

$$= \lim_{\|\Delta\| \to 0} \sum_{i=1}^{n} \pi ([R(\bar{y}_i)]^2 - [r(\bar{y}_i)]^2) \Delta_i y$$

$$= \pi \int_a^b ([R(y)]^2 - [r(y)]^2) \, dy$$

Exercises 8.2

In Exercises 1–8, find the volume of the solid of revolution when the given region of Fig. 8.2 is revolved about the indicated line.

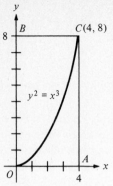

Figure 8.2

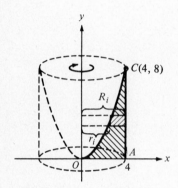

Figure 8.2.4

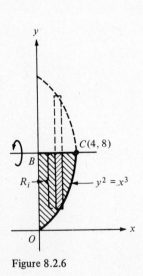

Figure 8.2.6

4. *OAC* about the *y*-axis

SOLUTION: Fig. 8.2.4 shows the region *OAC*, which is bounded by the curve $y^2 = x^3$, the *x*-axis, and the line $x = 4$, and a plane section of the solid of revolution. Because the *y*-axis is the axis of revolution, we take an element of area perpendicular to the *y*-axis of width $\Delta_i y$ units and revolve the element of area about the *y*-axis. This results in an element of volume that is a circular ring with units of thickness given by $\Delta_i h = \Delta_i y$. Because the outer radius of the ring is the horizontal distance between the *y*-axis and the line $x = 4$, then $R_i = 4$. Because the inner radius of the ring is the horizontal distance between the *y*-axis and a point on the curve $y^2 = x^3$, we solve the equation for *x* and obtain $x = f(y) = y^{2/3}$. Thus $r_i = f(\bar{y}_i)$. Furthermore, the region *OAC* is bounded above by the line $y = 8$. Therefore,

$$V = \lim_{\|\Delta\| \to 0} \sum_{i=1}^{n} \pi(R_i^2 - r_i^2) \Delta_i h$$

$$= \lim_{\|\Delta\| \to 0} \sum_{i=1}^{n} \pi(4^2 - [f(\bar{y}_i)]^2) \Delta_i y$$

$$= \pi \int_0^8 (4^2 - [f(y)]^2) dy$$

$$= \pi \int_0^8 (16 - y^{4/3}) \, dy$$

$$= \pi \left[16y - \frac{3}{7} y^{7/3} \right]_0^8$$

$$= \frac{512\pi}{7}$$

The volume of the solid of revolution is $\frac{512}{7}\pi$ cubic units.

6. *OBC* about the line *BC*

SOLUTION: Fig. 8.2.6 shows the region *OBC*, which is bounded by the curve $y^2 = x^3$, the *y*-axis, and the line $y = 8$, and a plane section of the solid of revolution. Because the line *BC* is the axis of revolution, we take an element of area perpendicular to the *x*-axis of width $\Delta_i x$ units and revolve the element about the line *BC*. This results in an element of volume that is a circular disk with units of thickness given by $\Delta_i h = \Delta_i x$. Because the radius of the disk is the vertical distance between the line $y = 8$ and a point on the curve $y^2 = x^3$, we solve the equation for *y*, obtaining $y = f(x) = x^{3/2}$. Thus $R_i = 8 - f(\bar{x}_i)$. Furthermore, the region *OBC* is bounded on the left by the line $x = 0$ and bounded on the right by the line $x = 4$. Therefore,

$$V = \lim_{\|\Delta\| \to 0} \sum_{i=1}^{n} \pi R_i^2 \Delta_i h$$

$$= \lim_{\|\Delta\| \to 0} \sum_{i=1}^{n} \pi[8 - f(\bar{x}_i)]^2 \Delta_i x$$

$$= \pi \int_0^4 [8 - f(x)]^2 \, dx$$

$$= \pi \int_0^4 (8 - x^{3/2})^2 \, dx$$

$$= \pi \int_0^4 (64 - 16x^{3/2} + x^3) \, dx$$

$$= \pi \left[64x - \frac{32}{5} x^{5/2} + \frac{1}{4} x^4 \right]_0^4$$

$$= \frac{576\pi}{5}$$

Thus the volume of the solid of revolution is $\frac{576}{5}\pi$ cubic units.

10. Find by integration the volume of a right circular cone of altitude h units and base radius a units.

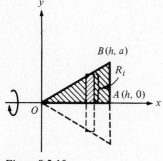

Figure 8.2.10

SOLUTION: The cone is generated by revolving the right triangle OAB about the x-axis where $0 = (0, 0)$, $A = (h, 0)$, $B = (h, a)$, as illustrated in Fig. 8.2.10. We take an element of area perpendicular to the x-axis of width $\Delta_i x$ units and revolve the element about the x-axis. This results in an element of volume that is a circular disk with units of thickness given by $\Delta_i h = \Delta_i x$. Because an equation of line OB is $y = ax/h$, we let $f(x) = ax/h$, and the radius of the disk is given by $R_i = f(\bar{x}_i)$. Furthermore, the triangle is bounded on the left by the line $x = 0$ and bounded on the right by the line $x = h$. Therefore,

$$V = \lim_{\|\Delta\| \to 0} \sum_{i=1}^n \pi R_i^2 \, \Delta_i h$$

$$= \lim_{\|\Delta\| \to 0} \sum_{i=1}^n \pi \, [f(\bar{x}_i)]^2 \, \Delta_i x$$

$$= \pi \int_0^h [f(x)]^2 \, dx$$

$$= \pi \int_0^h \frac{a^2 x^2}{h^2} \, dx$$

$$= \pi \left[\frac{a^2 x^3}{3h^2} \right]_0^h$$

$$= \pi \left[\frac{a^2 h^3}{3h^2} \right]$$

$$= \frac{1}{3} \pi a^2 h$$

Thus the volume of the cone is $\frac{1}{3}\pi a^2 h$ cubic units.

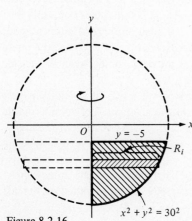

Figure 8.2.16

16. An oil drum in the shape of a sphere has a diameter of 60 ft. How much oil does the tank contain if the depth of the oil is 25 ft?

SOLUTION: Let the center of the tank be at the origin. The oil in the tank is in the shape of a solid of revolution if the region G, indicated by shading in Fig. 8.2.16, is revolved about the y-axis. Region G is bounded by the circle $x^2 + y^2 = 900$ because

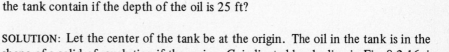

the radius of the sphere is 30 ft, bounded by the line $y = -5$ because the oil is 25 ft deep, and bounded by the y-axis. We take an element of area perpendicular to the y-axis of width $\Delta_i y$ feet and revolve the element about the y-axis. This results in an element of volume that is a circular disk with thickness given by $\Delta_i h = \Delta_i y$. Because the radius of the disk is the horizontal distance between the y-axis and a point on the circle, we solve the equation of the circle for x, obtaining $x = \pm\sqrt{900 - y^2}$. If we let $f(y) = \sqrt{900 - y^2}$, then $R_i = f(\bar{y}_i)$. Furthermore, the region G is bounded below by the line $y = -30$ and bounded above by the line $y = -5$. Therefore,

$$V = \lim_{\|\Delta\| \to 0} \sum_{i=1}^{n} \pi R_i^2 \, \Delta_i h$$

$$= \lim_{\|\Delta\| \to 0} \sum_{i=1}^{n} \pi [f(\bar{y}_i)]^2 \, \Delta_i y$$

$$= \pi \int_{-30}^{-5} [f(y)]^2 \, dy$$

$$= \pi \int_{-30}^{-5} (900 - y^2) \, dy$$

$$= \pi \left[900y - \frac{1}{3} y^3 \right]_{-30}^{-5}$$

$$= \pi \left[\left(-4500 + \frac{125}{3} \right) - (-27000 + 9000) \right]$$

$$= \frac{40625}{3} \pi$$

Therefore, the tank contains $\frac{1}{3}(40625)\pi$ cubic feet of oil.

20. The region bounded by a pentagon having vertices at $(-4, 4), (-2, 0), (0, 8),$ $(2, 0),$ and $(4, 4)$ is revolved about the x-axis. Find the volume of the solid generated.

SOLUTION: A sketch of the pentagon and a plane section of the solid of revolution is shown in Fig. 8.2.20. Since the pentagon is symmetric with respect to the y-axis, the solid of revolution is symmetric with respect to the plane that is perpendicular to the x-axis at the origin. Thus one half the volume of the solid is found by calculating the volume of the solid generated by revolving G, the first quadrant part of the pentagon, about the x-axis. We divide the region G into two parts: G_1 is the part of the pentagon to the right of the line $x = 2$; G_2 is the part of the pentagon between the y-axis and the line $x = 2$.

In G_1 we take an element of area perpendicular to the x-axis of width $\Delta_i x$ units and revolve the element about the x-axis. This results in an element of volume that is a circular ring with units of thickness given by $\Delta_i h = \Delta_i x$. An equation of the line that contains the points $(0, 8)$ and $(4, 4)$ is $y = -x + 8$. If $f(x) = -x + 8$, then the outer radius of the ring is given by $R_i = f(\bar{x}_i)$. An equation of the line that contains the points $(2, 0)$ and $(4, 4)$ is $y = 2x - 4$. If $g(x) = 2x - 4$, then the inner radius of the ring is given by $r_i = g(\bar{x}_i)$. Furthermore, the region G_1 is bounded on the left by the line $x = 2$ and bounded on the right by the line $x = 4$. Therefore,

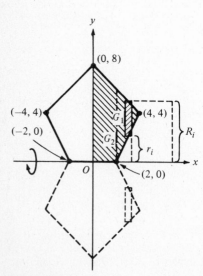

Figure 8.2.20

$$V_1 = \lim_{\|\Delta\| \to 0} \sum_{i=1}^{n} \pi(R_i^2 - r_i^2)\, \Delta_i h$$

$$= \lim_{\|\Delta\| \to 0} \sum_{i=1}^{n} \pi([f(\bar{x}_i)]^2 - [g(\bar{x}_i)]^2)\, \Delta_i x$$

$$= \pi \int_2^4 ([f(x)]^2 - [g(x)]^2)\, dx$$

$$= \pi \int_2^4 [(-x+8)^2 - (2x-4)^2]\, dx$$

$$= \pi \int_2^4 (-3x^2 + 48)\, dx$$

$$= \pi[-x^3 + 48x]_2^4$$
$$= 40\pi$$

In G_2 we take an element of area perpendicular to the x-axis of width $\Delta_i x$ units and revolve it about the x-axis. This results in an element of volume that is a circular disk with units of thickness given by $\Delta_i h = \Delta_i x$ and radius given by $R_i = f(\bar{x}_i)$. Furthermore, the region G_2 is bounded on the left by the line $x = 0$ and bounded on the right by the line $x = 2$. Therefore

$$V_2 = \lim_{\|\Delta\| \to 0} \sum_{i=1}^{n} \pi R_i^2\, \Delta_i h$$

$$= \lim_{\|\Delta\| \to 0} \sum_{i=1}^{n} \pi[f(\bar{x}_i)]^2\, \Delta_i x$$

$$= \pi \int_0^2 [f(x)]^2\, dx$$

$$= \pi \int_0^2 (-x+8)^2\, dx$$

$$= \pi \int_0^2 (x-8)^2\, dx$$

$$= \pi\left[\frac{1}{3}(x-8)^3\right]_0^2$$

$$= \frac{\pi}{3}[(-6)^3 - (-8)^3]$$

$$= \frac{296\pi}{3}$$

Because $V = 2(V_1 + V_2)$, we have

$$V = 2\left(40\pi + \frac{296\pi}{3}\right)$$

$$= \frac{832\pi}{3}$$

The volume of the solid of revolution is $\frac{832}{3}\pi$ cubic units.

8.3 VOLUME OF A SOLID OF REVOLUTION: CYLINDRICAL-SHELL METHOD

The formula given in the text for finding the volume of a solid of revolution applies only to the special case in which the y-axis is the axis of the revolution and the region is bounded by the curve $y = f(x)$ and the x-axis. When this formula does not apply, use the following more general methods which are based on the formula for the volume of a cylindrical shell.

$$V = 2\pi rh\Delta r$$

where r units is the mean of the inner radius and the outer radius of the shell, h units is the altitude of the shell, and Δr units is the thickness of the shell (the distance between the inner surface and the outer surface).

Let G be a region in the xy plane that is bounded above by the curve $y = f(x)$, bounded below by the curve $y = g(x)$, bounded on the left by the line $x = a$, and bounded on the right by the line $x = b$. To find the volume of the solid of revolution that results if G is revolved about a line L parallel to the y-axis (where L does not intersect the interior of G), take an element of area perpendicular to the x-axis of width $\Delta_i x$ units and revolve the element of area about the line L. This results in an element of volume that is a cylindrical shell whose thickness is given by $\Delta_i r = \Delta_i x$, whose mean radius is given by $r(\bar{x}_i)$, and whose altitude is given by $h(\bar{x}_i)$. The volume of the solid is given by

$$V = \lim_{\|\Delta\| \to 0} \sum_{i=1}^{n} 2\pi r(\bar{x}_i) \cdot h(\bar{x}_i)\, \Delta_i x$$

$$= 2\pi \int_{a}^{b} r(x) \cdot h(x)\, dx$$

Let H be a region in the xy plane that is bounded on the right by the curve $x = f(y)$, bounded on the left by the curve $x = g(y)$, bounded below by the line $y = a$, and bounded above by the line $y = b$. To find the volume of the solid of revolution that results if H is revolved about a line L parallel to the x-axis (where L does not intersect the interior of H), take an element of area perpendicular to the y-axis of width $\Delta_i y$ units and revolve the element about the line L. This results in an element of volume that is a cylindrical shell those thickness is given by $\Delta_i r = \Delta_i y$, whose mean radius is given by $r(\bar{y}_i)$, and whose altitude is given by $h(\bar{y}_i)$. The volume of the solid is given by

$$V = \lim_{\|\Delta\| \to 0} \sum_{i=1}^{n} 2\pi r(\bar{y}_i) \cdot h(\bar{y}_i)\, \Delta_i y$$

$$= 2\pi \int_{a}^{b} r(y) \cdot h(y)\, dy$$

Note that in this section the elements of area are taken parallel to the axis of revolution, and the resulting elements of volume are cylindrical shells. In Section 8.2 the elements of area are taken perpendicular to the axis of revolution, and the resulting elements of volume are circular rings. Sometimes the volume of a solid of revolution may be found by using either of the two methods.

Exercises 8.3

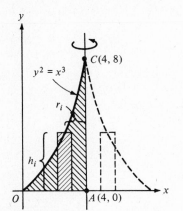

Figure 8.3.2

1-8. Solve Exercises 1-8 in Section 8.2 by taking the rectangular elements parallel to the axis of revolution.

2. OAC about the line AC

SOLUTION: Fig. 8.3.2 shows a sketch of the region OAC and a plane section of the solid of revolution. Because the line AC is the axis of revolution, we take an element of area perpendicular to the x-axis of width $\Delta_i x$ units and revolve the element about the line AC. This results in an element of volume that is a cylindrical shell with thickness given by $\Delta_i r = \Delta_i x$. Because the mean radius of the shell is the horizontal distance between the line $x = 4$ and a point on the curve, then $r_i = 4 - \bar{x}_i$. Because the altitude of the shell is the vertical distance between the x-axis and a point on the curve $y^2 = x^3$, we solve the equation for y, obtaining $y = \pm x^{3/2}$. If $f(x) = x^{3/2}$, then $h_i = f(\bar{x}_i)$. Furthermore, the region OAC is bounded on the left by the line $x = 0$ and bounded on the right by the line $x = 4$. Therefore,

$$V = \lim_{\|\Delta\| \to 0} \sum_{i=1}^{n} 2\pi r_i h_i \Delta_i r$$

$$= \lim_{\|\Delta\| \to 0} \sum_{i=1}^{n} 2\pi (4 - \bar{x}_i) f(\bar{x}_i) \Delta_i x$$

$$= 2\pi \int_0^4 (4 - x) \cdot f(x)\, dx$$

$$= 2\pi \int_0^4 (4 - x) x^{3/2}\, dx$$

$$= 2\pi \int_0^4 (4x^{3/2} - x^{5/2})\, dx$$

$$= 2\pi \left[\frac{8}{5} x^{5/2} - \frac{2}{7} x^{7/2} \right]_0^4$$

$$= \frac{1024\pi}{35}$$

Thus, the volume of the solid of revolution is $\frac{1024}{35}\pi$ cubic units.

8. OBC about the x-axis.

SOLUTION: Fig. 8.3.8 shows a sketch of the region OBC and a plane section of the solid of revolution. Because the x-axis is the axis of revolution, we take an element of area perpendicular to the y-axis of width $\Delta_i y$ units and revolve the element about the y-axis. This results in an element of volume that is a cylindrical shell with thickness given by $\Delta_i r = \Delta_i y$. Because the mean radius of the shell is the vertical distance between the x-axis and a point on the curve, we have $r_i = \bar{y}_i$. Because the altitude of the shell is the horizontal distance between the y-axis and a point on the curve $y^2 = x^3$, we solve the equation for x, obtaining $x = y^{2/3}$. If $f(y) = y^{2/3}$, we have $h_i = f(\bar{y}_i)$. Furthermore, the region OBC is bounded below by the line $y = 0$ and bounded above by the line $y = 8$. Therefore,

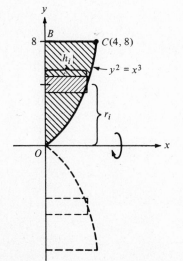

Figure 8.3.8

$$V = \lim_{\|\Delta\| \to 0} \sum_{i=1}^{n} 2\pi r_i h_i \Delta_i r$$

$$= \lim_{\|\Delta\| \to 0} \sum_{i=1}^{n} 2\pi \bar{y}_i f(\bar{y}_i) \Delta_i y$$

$$= 2\pi \int_0^8 y f(y) \, dy$$

$$= 2\pi \int_0^8 y^{5/3} \, dy$$

$$= 2\pi \left[\frac{3}{8} y^{8/3} \right]_0^8$$

$$= 192\pi$$

Thus, the volume of the solid of revolution is 192π cubic units.

14. R is the region bounded by the curve $y^2 = x$, the y-axis, and the line $y = 1$. R is revolved about the line $y = 2$. Find the volume of the solid of revolution if the rectangular elements are perpendicular to the axis of rotation.

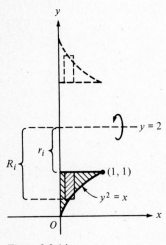

Figure 8.3.14

SOLUTION: Fig. 8.3.14 shows a sketch of the region R and a plane section of the solid of revolution. We take an element of area perpendicular to the x-axis of width $\Delta_i x$ units and revolve the element about the line $y = 2$. This results in an element of volume that is a circular ring with thickness given by $\Delta_i h = \Delta_i x$. Because the outer radius of the ring is the vertical distance between the line $y = 2$ and a point on the curve $y^2 = x$, we solve the equation of the curve for y, obtaining $y = \pm\sqrt{x}$. If we let $f(x) = \sqrt{x}$, then the outer radius of the ring is given by $R_i = 2 - f(\bar{x}_i)$. Because the inner radius of the ring is the vertical distance between the lines $y = 2$ and $y = 1$, we have $r_i = 1$. Furthermore, the region R is bounded on the left by the line $x = 0$ and bounded on the right by the line $x = 1$. Therefore,

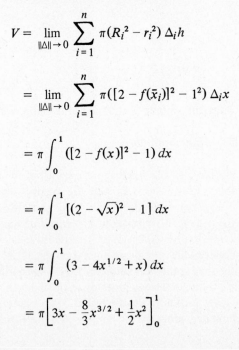

$$V = \lim_{\|\Delta\| \to 0} \sum_{i=1}^{n} \pi(R_i^2 - r_i^2) \Delta_i h$$

$$= \lim_{\|\Delta\| \to 0} \sum_{i=1}^{n} \pi([2 - f(\bar{x}_i)]^2 - 1^2) \Delta_i x$$

$$= \pi \int_0^1 ([2 - f(x)]^2 - 1) \, dx$$

$$= \pi \int_0^1 [(2 - \sqrt{x})^2 - 1] \, dx$$

$$= \pi \int_0^1 (3 - 4x^{1/2} + x) \, dx$$

$$= \pi \left[3x - \frac{8}{3} x^{3/2} + \frac{1}{2} x^2 \right]_0^1$$

$$= \frac{5\pi}{6}$$

Thus the volume of the solid of revolution is $\frac{5}{6}\pi$ cubic units.

18. Find the volume of the solid generated by revolving the region bounded by the curve $x^{2/3} + y^{2/3} = a^{2/3}$ about the y-axis.

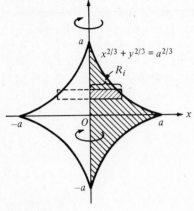

Figure 8.3.18

SOLUTION: The curve is symmetric with respect to both the x-axis and the y-axis. A sketch of the curve is shown in Fig. 8.3.18. Because the y-axis is the axis of revolution and this axis must not intersect the interior of the region, we use only that part of the region which is to the right of the y-axis as the region H, indicated by shading in the figure. We take an element of area perpendicular to the y-axis of width $\Delta_i y$ units and revolve the element about the y-axis. This results in an element of volume that is a circular disk with thickness given by $\Delta_i h = \Delta_i y$. Because the radius of the disk is the horizontal distance between the y-axis and a point on the curve $x^{2/3} + y^{2/3} = a^{2/3}$, we solve the equation for x, obtaining $x = f(y) = (a^{2/3} - y^{2/3})^{3/2}$. Thus, $R_i = f(\bar{y}_i)$. Furthermore, the region H is bounded below by the line $y = -a$ and bounded above by the line $y = a$. Therefore,

$$V = \lim_{\|\Delta\| \to 0} \sum_{i=1}^{n} \pi R_i^2 \, \Delta_i h$$

$$= \lim_{\|\Delta\| \to 0} \sum_{i=1}^{n} \pi [f(\bar{y}_i)]^2 \, \Delta_i y$$

$$= \pi \int_{-a}^{a} [f(y)]^2 \, dy$$

$$= \pi \int_{-a}^{a} [(a^{2/3} - y^{2/3})^{3/2}]^2 \, dy$$

$$= \pi \int_{-a}^{a} (a^{2/3} - y^{2/3})^3 \, dy$$

$$= \pi \int_{-a}^{a} (a^2 - 3a^{4/3}y^{2/3} + 3a^{2/3}y^{4/3} - y^2) \, dy$$

$$= \pi \left[a^2 y - \frac{9}{5}a^{4/3}y^{5/3} + \frac{9}{7}a^{2/3}y^{7/3} - \frac{1}{3}y^3 \right]_{-a}^{a}$$

$$= \pi \left[\left(a^3 - \frac{9}{5}a^{4/3}a^{5/3} + \frac{9}{7}a^{2/3}a^{7/3} - \frac{1}{3}a^3 \right) - \left(-a^3 + \frac{9}{5}a^{4/3}a^{5/3} - \frac{9}{7}a^{2/3}a^{7/3} + \frac{1}{3}a^3 \right) \right]$$

$$= \frac{32\pi a^3}{105}$$

Thus, the volume of the solid of revolution is $\frac{32}{105}\pi a^3$ cubic units.

20. Through a spherically shaped solid of radius 6 in, a hole of radius 2 in is bored, with the axis of the hole a diameter of the sphere. Find the volume of the part of the solid that remains.

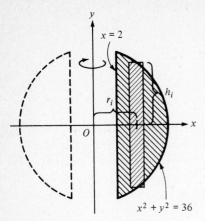

Figure 8.3.20

SOLUTION: Let the origin be at the center of the sphere with the y-axis the axis of the hole. Let S be the solid that remains. Fig. 8.3.20 shows a plane section of S, which is a solid of revolution if the region H, indicated by the shading in the figure, is revolved about the y-axis. Because the radius of the spherically shaped solid is 6 in, an equation of the circle that forms part of the boundary of H is $x^2 + y^2 = 36$. Because the radius of the hole is 2 in, H is bounded on the left by the line $x = 2$. We take an element of area perpendicular to the x-axis of width $\Delta_i x$ in and revolve the element about the y-axis. This results in an element of volume that is a cylindrical shell whose thickness is given by $\Delta_i r = \Delta_i x$ and whose mean radius is given by $r_i = \bar{x}_i$. Because the altitude of the shell is twice the vertical distance between the x-axis and the circle, we solve the equation of the circle for y, obtaining $y = \pm\sqrt{36 - x^2}$. If we let $f(x) = \sqrt{36 - x^2}$, then $h_i = 2f(\bar{x}_i)$. Therefore

$$
\begin{aligned}
V &= \lim_{\|\Delta\| \to 0} \sum_{i=1}^{n} 2\pi r_i h_i \Delta_i r \\
&= \lim_{\|\Delta\| \to 0} \sum_{i=1}^{n} 2\pi \bar{x}_i \cdot 2f(\bar{x}_i) \Delta_i x \\
&= 2\pi \int_{2}^{6} 2x\sqrt{36 - x^2}\, dx \\
&= -2\pi \int_{2}^{6} (36 - x^2)^{1/2}(-2x\, dx) \\
&= -2\pi \left[\frac{2}{3}(36 - x^2)^{3/2} \right]_{2}^{6} \\
&= \frac{512\pi\sqrt{2}}{3}
\end{aligned}
$$

Thus, the volume of the solid that remains is $\frac{512}{3}\pi\sqrt{2}$ cubic inches.

8.4 VOLUME OF A SOLID HAVING KNOWN PARALLEL PLANE SECTIONS

8.4.2 Definition

Let S be a solid such that S lies between planes drawn perpendicular to the x-axis at a and b. If the measure of the area of the plane section of S drawn perpendicular to the x-axis at x is given by $A(x)$, where A is continuous on $[a, b]$, then the measure of the volume of S is given by

$$
V = \lim_{\|\Delta\| \to 0} \sum_{i=1}^{n} A(\bar{x}_i)\, \Delta_i x = \int_{a}^{b} A(x)\, dx
$$

Exercises 8.4

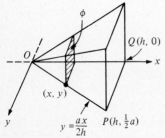

Figure 8.4.4

4. Find the volume of a right pyramid having a height of h units and a square base of side a units.

SOLUTION: Take the origin at the vertex of the pyramid, with the square base perpendicular to the x-axis situated so that two sides of the base are parallel to the y-axis. Fig. 8.4.4 shows a solid S, which is the part of the pyramid that lies above the xy plane and on the positive side of the y-axis, one fourth of the entire solid. We find the volume of S and multiply by 4 to obtain the volume of the required pyramid. Because the height of the given pyramid is h units and point Q is at the center of the base, then $Q = (h, 0)$. Because the line segment PQ is one half the side of the base of the given pyramid, then $|\overline{PQ}| = \frac{1}{2}a$. Thus, $P = (h, \frac{1}{2}a)$, and an equation of line OP is

$$y = \frac{ax}{2h} \tag{1}$$

Consider ϕ to be any plane section of S that is perpendicular to the x-axis. Let (x, y) be the point where ϕ intersects line OP. Because A, the area of ϕ, is given by $A = y^2$, by Eq. (1) we have

$$A(x) = \left(\frac{ax}{2h}\right)^2 \tag{2}$$

Furthermore, solid S lies between the planes drawn perpendicular to the x-axis at 0 and h. If V is the measure of the volume of the required pyramid, then $\frac{1}{4}V$ is the measure of the volume of solid S, and thus by Definition 8.4.2 and Eq. (2), we have

$$\frac{1}{4}V = \lim_{\|\Delta\| \to 0} \sum_{i=1}^{n} A(\bar{x}_i)\,\Delta_i x$$

$$\frac{1}{4}V = \int_0^h \left(\frac{ax}{2h}\right)^2 dx$$

$$V = \frac{a^2}{h^2} \int_0^h x^2\,dx$$

$$V = \frac{a^2}{3h^2} \cdot x^3 \Big]_0^h$$

$$= \frac{a^2 h}{3}$$

Thus, the volume of the pyramid is $\frac{1}{3}a^2 h$ cubic units.

6. The base of a solid is a circle with radius 4 in, and each plane section perpendicular to a fixed diameter of the base is an isosceles triangle having an altitude of 10 in and a chord of the circle as a base. Find the volume of the solid.

SOLUTION: Take the circle in the xy plane with center at the origin and the fixed diameter along the x-axis. Therefore, an equation of the circle is $x^2 + y^2 = 16$. Fig. 8.4.6 shows S, the part of the solid that is on the positive sides of the x and y axes. S represents one fourth of the entire solid. Let ϕ be any plane section of S that is perpendicular to the x-axis, and let (x, y) be the point where ϕ intersects the circle. Because ϕ is a triangle with altitude 10 in and base y in, the measure of the area of ϕ is given by $A = \frac{1}{2}y \cdot 10 = 5y$. Because $x^2 + y^2 = 16$, we have

$$A(x) = 5\sqrt{16 - x^2} \tag{1}$$

Furthermore, the solid S lies between the planes drawn perpendicular to the x-axis at 0 and 4. If V is the measure of the volume of the entire solid, then $\frac{1}{4}V$ is the measure of the volume of solid S. Thus, by Definition 8.4.2 and Eq. (1), we have

$$\frac{1}{4}V = \lim_{\|\Delta\| \to 0} \sum_{i=1}^{n} 5\sqrt{16 - \bar{x}_i^2}\,\Delta_i x$$

$$\frac{1}{4}V = \int_0^4 5\sqrt{16 - x^2}\,dx$$

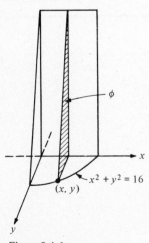

ϕ

$x^2 + y^2 = 16$

(x, y)

Figure 8.4.6

$$V = 20 \int_0^4 \sqrt{16 - x^2} \, dx \tag{2}$$

At this time we cannot use the fundamental theorem of the calculus to evaluate this integral, because we cannot find an antiderivative for the integral. (In Section 11.3 we learn a method for finding an antiderivative.) However, the graph of the function $f(x) = \sqrt{16 - x^2}$ when x is in $[0, 4]$ is the first quadrant arc of the circle $x^2 + y^2 = 16$. Because $\int_0^4 \sqrt{16 - x^2} \, dx$ gives the area of the region bounded by this arc and the x and y axes, and the area of this region is one-fourth the area of the circle $x^2 + y^2 = 16$, using the formula for the area of a circle, $A = \pi r^2$, we obtain

$$\int_0^4 \sqrt{16 - x^2} \, dx = \frac{1}{4} \pi 4^2 = 4\pi \tag{3}$$

Substituting from (3) into (2), we have $V = 80\pi$. The volume of the solid is 80π in^3.

8. Two right circular cylinders, each having a radius of r units, have axes that intersect at right angles. Find the volume of the solid common to the two cylinders.

SOLUTION: Take one cylinder perpendicular to the xy plane with center at the origin and the other cylinder having the y-axis as its axis. Fig. 8.4.8 illustrates S, the part of the solid common to the two cylinders that lies above the xy plane and on the positive sides of the x-axis and y-axis. The volume of S is one eighth of the volume of the entire solid common to the two cylinders. Let ϕ be any plane section of S perpendicular to the x-axis, and let (x, y) be the point at which ϕ intersects the circular part of the boundary of the base of S in the xy plane. Because ϕ is a square with side y, the measure of its area is given by $A = y^2$. Because $x^2 + y^2 = r^2$ is an equation of the circular boundary, we have

$$A(x) = r^2 - x^2 \tag{1}$$

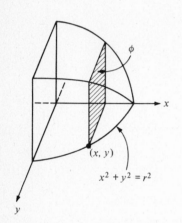

Figure 8.4.8

Furthermore, the solid S lies between the planes drawn perpendicular to the x-axis at 0 and r. If V is the measure of the volume of the solid common to the two cylinders, then $\frac{1}{8} V$ is the measure of the volume of solid S. Thus, by Definition 8.4.2 and Eq. (1), we have

$$\frac{1}{8} V = \lim_{\|\Delta\| \to 0} \sum_{i=1}^{n} (r^2 - \bar{x}_i^2) \, \Delta_i x$$

$$\frac{1}{8} V = \int_0^r (r^2 - x^2) \, dx$$

$$V = 8 \left[r^2 x - \frac{1}{3} x^3 \right]_0^r$$

$$= \frac{16}{3} r^3$$

The volume of the solid common to the two cylinders is $\frac{16}{3} r^3$ cubic units.

8.5 WORK   If a constant force of F units acting on a body causes a displacement of D units, where the force and the displacement are in the same direction, then the work done by the force is given by

$$W = F \cdot D \tag{1}$$

The following definition can sometimes be used to calculate the work done by a variable force. Its use is illustrated in Exercises 4 and 12.

8.5.1 Definition Let the function f be continuous on the closed interval $[a, b]$ and $f(x)$ be the number of units in the force acting on an object at the point x on the x-axis. Then if W units is the work done by the force as the object moves from a to b, W is given by

$$W = \lim_{\|\Delta\| \to 0} \sum_{i=1}^{n} f(\bar{x}_i)\, \Delta_i x = \int_a^b f(x)\, dx$$

When Definition 8.5.1 does not apply, we make a partition of some interval $[a, b]$ on the x-axis and use $\Delta_i F$, D_i, and $\Delta_i W$, the elements of force, displacement, and work, respectively, and Formula (1) to obtain

$$\Delta_i W = \Delta_i F \cdot D_i$$

from which we have

$$W = \lim_{\|\Delta\| \to 0} \sum_{i=1}^{n} D_i \Delta_i F \qquad (2)$$

If the sum in (2) can be expressed as a Riemann sum, then W can be found by evaluating the definite integral that corresponds to the limit of the Riemann sum. This is illustrated in Exercises 8 and 14.

Exercises 8.5

---

4. A spring has a natural length of 6 in. A 1200-lb force compresses it to $5\frac{1}{2}$ in. Find the work done in compressing it from 6 to $4\frac{1}{2}$ in.

SOLUTION: Place the spring along the x-axis with the origin at the point where the compression begins. Let

$x =$ the number of inches the spring is compressed
$f(x) =$ the number of pounds in the force acting on the spring when the spring is compressed x in

By Hooke's law, $f(x) = kx$. Because a 1200-lb force compresses the spring $\frac{1}{2}$ in, we have $f(\frac{1}{2}) = 1200$. Thus,

$$1200 = k \cdot \frac{1}{2}$$

$$k = 2400$$

Therefore,

$$f(x) = 2400x \qquad (1)$$

Let W inch-pounds be the work done in compressing the spring from 6 to $4\frac{1}{2}$ in. Then x is in $[0, \frac{3}{2}]$, and by Definition 8.5.1 and Eq. (1), we have

$$W = \lim_{\|\Delta\| \to 0} \sum_{i=1}^{n} 2400\bar{x}_i\, \Delta_i x$$

$$= \int_0^{3/2} 2400x\, dx$$

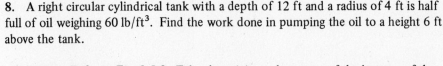

$$= 1200x^2 \big]_0^{3/2}$$

$$= 2700$$

The work done is 2700 inch-pounds.

8. A right circular cylindrical tank with a depth of 12 ft and a radius of 4 ft is half full of oil weighing 60 lb/ft³. Find the work done in pumping the oil to a height 6 ft above the tank.

SOLUTION: Refer to Fig. 8.5.8. Take the origin at the center of the bottom of the tank with positive direction upward. Definition 8.5.1 does not apply. Because the oil is 6 ft deep, we partition the closed interval $[0, 6]$. For the ith subinterval let $\Delta_i F$ be the element of force, D_i the element of displacement, and $\Delta_i W$ the element of work. We are given that work equals force times displacement. Thus,

$$\Delta_i W = \Delta_i F \cdot D_i \tag{1}$$

Furthermore, because the force in this case is the weight of the element of oil, and the oil weighs 60 lb/ft³, we have

$$\Delta_i F = 60 \, \Delta_i V \tag{2}$$

where $\Delta_i V$ is the element of volume.

Because the element of volume is a circular disk with radius 4 ft and thickness $\Delta_i x$ ft, we have

$$\Delta_i V = 16\pi \, \Delta_i x \tag{3}$$

Substituting from (3) into (2) we obtain

$$\Delta_i F = 960\pi \, \Delta_i x \tag{4}$$

Furthermore, because the oil is pumped to a height 6 ft above the top of the tank, the oil is pumped to a point 18 ft above the origin, and thus

$$D_i = 18 - \bar{x}_i \tag{5}$$

where $\bar{x}_i$ is a number in the ith subinterval. By substituting from Eqs. (4) and (5) into (1), we obtain

$$\Delta_i W = 960\pi (18 - \bar{x}_i) \, \Delta_i x$$

Therefore,

$$W = \lim_{\|\Delta\| \to 0} \sum_{i=1}^{n} 960\pi (18 - \bar{x}_i) \, \Delta_i x$$

$$= \int_0^6 960\pi (18 - x) \, dx$$

$$= 960\pi \left[18x - \frac{1}{2}x^2 \right]_0^6$$

$$= 86400\pi$$

The work done is $86,400\pi$ foot-pounds.

12. As a water tank is being raised, water spills out at a constant rate of 2 ft³ per foot of rise. If the tank originally contained 1000 ft³ of water, find the work done in raising the tank 20 ft.

SOLUTION: Let the x-axis be directed upward with the origin at the point where the

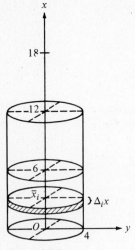

Figure 8.5.8

tank is originally. Let

$x =$ the number of feet the tank is raised

$f(x) =$ the number of pounds in the force acting on the tank when it is x ft above the origin

We are given that the number of ft^3 in the volume of water in the tank is $1000 - 2x$. Therefore, if the density of water is w pounds per ft^3, we have

$$f(x) = w(1000 - 2x) \tag{1}$$

Because the tank is raised 20 ft, then x is in $[0, 20]$. If W foot-pounds is the work done in raising the tank, by Definition 8.5.1 and Eq. (1), we have

$$W = \lim_{\|\Delta\| \to 0} \sum_{i=1}^{n} w(1000 - 2\bar{x}_i) \, \Delta_i x$$

$$= \int_0^{20} w(1000 - 2x) \, dx$$

$$= w[1000x - x^2]_0^{20}$$
$$= 19{,}600w$$

The work done is $19{,}600w$ foot-pounds.

14. A cylindrical tank 10 ft high and 5 ft in radius is standing on a platform 50 ft high. Find the depth of the water when one half of the work required to fill the tank from the ground level through a pipe in the bottom has been done.

SOLUTION: Refer to Fig. 8.5.14. We take the origin at the center of the bottom of the tank with the x-axis directed upward and partition the closed interval $[0, 10]$ on the x-axis. For the ith subinterval, let $\Delta_i F$ be the element of force, D_i the element of displacement, and $\Delta_i W$ the element of work. Then

$$\Delta_i W = \Delta_i F \cdot D_i \tag{1}$$

To find the element of force, we multiply the element of volume by w, the density of water. Because the element of volume is a circular disk with radius 5 ft and thickness $\Delta_i x$ ft, we have

$$\Delta_i F = 25w\pi \, \Delta_i x \tag{2}$$

Although water enters the tank through a pipe in the bottom, each element of volume must be raised to the level of the surface of the water in the tank. Thus, if $\bar{x}_i$ is a number in the ith subinterval, then

$$D_i = \bar{x}_i + 50 \tag{3}$$

Substituting from (2) and (3) into (1) we get

$$\Delta_i W = 25w\pi(\bar{x}_i + 50) \, \Delta_i x$$

Thus, the total work required to fill the tank is given by

$$W = \lim_{\|\Delta\| \to 0} \sum_{i=1}^{n} 25w\pi(\bar{x}_i + 50) \, \Delta_i x$$

$$= \pi \int_0^{10} 25w(x + 50) \, dx$$

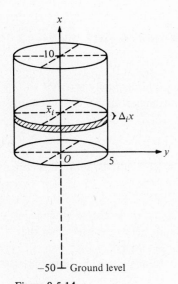

Figure 8.5.14

$$= 25w\pi \int_0^{10} (x + 50) \, dx \tag{4}$$

Let b ft be the depth of the water in the tank when one-half the work required to fill the tank has been completed. Then we partition the interval $[0, b]$, and as before, we derive the following integral

$$\frac{1}{2}W = 25w\pi \int_0^b (x + 50) \, dx$$

$$W = 50w\pi \int_0^b (x + 50) \, dx \tag{5}$$

Equating the right members of (4) and (5), we get

$$50w\pi \int_0^b (x + 50) \, dx = 25w\pi \int_0^{10} (x + 50) \, dx$$

$$2\left[\frac{1}{2}x^2 + 50x\right]_0^b = \left[\frac{1}{2}x^2 + 50x\right]_0^{10}$$

$$b^2 + 100b = 550$$

Solving for b, we obtain

$$b = -50 + \sqrt{3050} \approx 5.23$$

Thus, the depth of the water is approximately 5.23 ft when one-half of the work is done.

8.6 LIQUID PRESSURE

If a flat surface with area A square units is submerged horizontally at a depth of h units in a liquid with density w pounds per cubic unit, then the total force due to liquid pressure on the surface is F pounds, given by

$$F = whA \tag{1}$$

The following definition can sometimes be used to calculate the total force due to liquid pressure. Its use is illustrated in Exercises 4 and 14.

8.6.1 Definition

Suppose that a flat plate is submerged vertically in a liquid of weight w pounds per cubic unit. The length of the plate at a depth of x units below the surface of the liquid is $f(x)$ units, where f is continuous on the closed interval $[a, b]$ and $f(x) \geqslant 0$ on $[a, b]$. Then F, the number of pounds of force caused by liquid pressure on the plate, is given by

$$F = \lim_{\|\Delta\| \to 0} \sum_{i=1}^n w\bar{x}_i f(\bar{x}_i) \, \Delta_i x = \int_a^b wxf(x) \, dx$$

When Definition 8.6.1 does not apply, we make a partition of some interval $[a, b]$ on the x-axis and use $\Delta_i A$, h_i, and $\Delta_i F$, the elements of area, depth, and force, respectively, and Formula (1) to obtain

$$\Delta_i F = wh_i \Delta_i A$$

from which we have

$$F = \lim_{\|\Delta\| \to 0} \sum_{i=1}^n wh_i \Delta_i A \tag{2}$$

If the sum in (2) can be expressed as a Riemann sum, then F can be found by evaluating the definite integral that corresponds to the limit of this Riemann sum. This is illustrated in Exercises 10 and 12.

Exercises 8.6

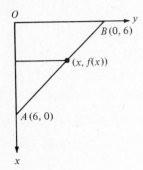

Figure 8.6.4

4. A plate in the shape of an isosceles right triangle is submerged vertically in a tank of water, with one leg lying in the surface. The legs are each 6 ft long. Find the force due to liquid pressure on one side of the plate.

SOLUTION: Fig. 8.6.4 illustrates the plate AOB. The y-axis is in the surface of the water. Because an equation of line AB is $y = -x + 6$, we let $f(x) = -x + 6$. The length of the plate at a depth of x units is $f(x)$ units with x in $[0, 6]$. Thus, we use Definition 8.6.1 to find the total force F due to liquid pressure on the plate.

$$
\begin{aligned}
F &= \lim_{\|\Delta\| \to 0} \sum_{i=1}^{n} w\bar{x}_i f(\bar{x}_i) \, \Delta_i x \\[2mm]
&= \int_a^b w x f(x) \, dx \\[2mm]
&= \int_0^6 w x(-x + 6) \, dx \\[2mm]
&= w\left[-\frac{1}{3}x^3 + 3x^2\right]_0^6 \\[2mm]
&= 36w
\end{aligned}
$$

The total force due to liquid pressure is $36w$ pounds, where w pounds per cubic ft is the density of water.

10. The face of a gate of a dam is vertical and in the shape of an isosceles trapezoid 3 ft wide at the top, 4 ft wide at the bottom, and 3 ft high. If the upper base is 20 ft below the surface of the water, find the total force due to liquid pressure on the gate.

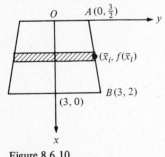

Figure 8.6.10

SOLUTION: Fig. 8.6.10 shows the face of the gate. The origin is at the center of the upper base of the trapezoid. Because the surface of the water is not at the origin, x units does not represent the depth of the water, and we cannot use Definition 8.6.1. However, we may partition the interval $[0, 3]$ on the x-axis, and let $\Delta_i A$, h_i, and $\Delta_i F$ be the elements of area, depth, and force, respectively, for the ith subinterval. Then we have

$$
\Delta_i F = w \, h_i \, \Delta_i A \tag{1}
$$

Because an equation of line AB is $y = \frac{1}{6}x + \frac{3}{2}$, we let $f(x) = \frac{1}{6}x + \frac{3}{2}$. Then the element of area is a rectangle with width $\Delta_i x$ ft and length $2f(\bar{x}_i)$ ft where $\bar{x}_i$ is some number in the ith subinterval. Thus, the element of area is given by

$$
\Delta_i A = 2f(\bar{x}_i) \, \Delta_i x = \left(\frac{1}{3}\bar{x}_i + 3\right)\Delta_i x \tag{2}
$$

Because the upper base is 20 ft below the surface of the water, the element of depth is given by

$$
h_i = \bar{x}_i + 20 \tag{3}
$$

Substituting from (2) and (3) into (1), we obtain

$$\Delta_i F = w\left(\frac{1}{3}\bar{x}_i + 3\right)(\bar{x}_i + 20)\,\Delta_i x$$

Therefore,

$$F = \lim_{\|\Delta\| \to 0} \sum_{i=1}^{n} w\left(\frac{1}{3}\bar{x}_i + 3\right)(\bar{x}_i + 20)\,\Delta_i x$$

$$= \int_0^3 w\left(\frac{1}{3}x + 3\right)(x + 20)\,dx$$

$$= w\int_0^3 \left(\frac{1}{3}x^2 + \frac{29}{3}x + 60\right)dx$$

$$= w\left[\frac{1}{9}x^3 + \frac{29}{6}x^2 + 60x\right]_0^3$$

$$= 226.5w$$

The total force is $226.5w$ pounds.

12. The face of a dam adjacent to the water is inclined at an angle of $30°$ from the vertical. The shape of the face is an isosceles trapezoid 120 ft wide at the top, 80 ft wide at the bottom, and with a slat height of 40 ft. If the dam is full of water, find the total force due to liquid pressure on the face.

SOLUTION: Fig. 8.6.12(a) illustrates the face of the dam. The origin is at the center of the top of the dam, and the x-axis runs down the center of the face of the dam. Because the face of the dame is not vertical, we cannot use Definition 8.6.1. However, we may partition the interval $[0, 40]$ on the x-axis. As in Exercise 10, we have

$$\Delta_i F = wh_i\,\Delta_i A \tag{1}$$

Because an equation of line AB is $y = -\frac{1}{2}x + 60$, we let $f(x) = -\frac{1}{2}x + 60$. The element of area is a rectangle with width $\Delta_i x$ ft and length $2 \cdot f(\bar{x}_i)$ ft. Thus,

$$\Delta_i A = 2 \cdot f(\bar{x}_i)\,\Delta_i x = (-\bar{x}_i + 120)\,\Delta_i x \tag{2}$$

Because the x-axis is not vertical, the element of depth h_i ft is not given by $\bar{x}_i$. However, as shown in Fig. 8.6.12(b), we have

$$\frac{h_i}{\bar{x}_i} = \cos 30°$$

$$h_i = \frac{1}{2}\sqrt{3}\,\bar{x}_i \tag{3}$$

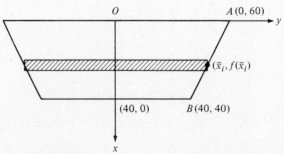

Figure 8.6.12(a)   Figure 8.6.12(b)

Substituting from (2) and (3) into (1), we obtain

$$\Delta_i F = \frac{1}{2} w \sqrt{3} \, \bar{x}_i (-\bar{x}_i + 120) \, \Delta_i x$$

Therefore,

$$F = \lim_{\|\Delta\| \to 0} \sum_{i=1}^{n} \frac{1}{2} w \sqrt{3} \, \bar{x}_i (-\bar{x}_i + 120) \, \Delta_i x$$

$$= \int_0^{40} \frac{1}{2} w \sqrt{3} \, x (-x + 120) \, dx$$

$$= \frac{1}{2} w \sqrt{3} \int_0^{40} (-x^2 + 120x) \, dx$$

$$= \frac{1}{2} w \sqrt{3} \left[-\frac{1}{3} x^3 + 60 x^2 \right]_0^{40}$$

$$= \frac{112000 \sqrt{3} \, w}{3}$$

$$\approx 64{,}663 w$$

Thus, the total force is approximately $64{,}663 w$ pounds.

14. If the end of a water tank is in the shape of a rectangle and the tank is full, show that the measure of the force due to liquid pressure on the end is the product of the measure of the area of the end and the measure of the force at the geometrical center.

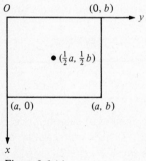

Figure 8.6.14

SOLUTION: Fig. 8.6.14 illustrates the end of the tank. We use Definition 8.6.1 to find the total force due to liquid pressure. Because the origin is at the surface of the water and the tank is a ft deep, then x units is the distance below the surface of the water for x in the interval $[0, a]$. Because the rectangle is b units long, we let $f(x) = b$. Hence, the total force is given by

$$F = \lim_{\|\Delta\| \to 0} \sum_{i=1}^{n} w \bar{x}_i b \, \Delta_i x$$

$$= \int_0^a w x b \, dx$$

$$= \frac{1}{2} w b x^2 \Big]_0^a$$

$$= \frac{1}{2} w a^2 b \qquad (1)$$

Furthermore, the center of the rectangle is at the point $(\frac{1}{2}a, \frac{1}{2}b)$. Thus, $\frac{1}{2}aw$ pounds represents the force per square unit of area at this point. Because the area of the rectangle is ab square units and $(\frac{1}{2}aw)(ab) = \frac{1}{2}wa^2 b$, by comparison with (1), we conclude that the total force due to liquid pressure on the end of the tank is the product of the measure of the area of the end and the measure of the force at the geometrical center.

8.7 CENTER OF MASS OF A ROD

If a system of n particles is located on the x-axis at the points $x_1, x_2, \ldots, x_n$ and the mass of the ith particle is given by m_i, then the center of mass for the system is at the point $\bar{x}$ on the x-axis, where

$$\bar{x} = \frac{\displaystyle\sum_{i=1}^{n} m_i x_i}{\displaystyle\sum_{i=1}^{n} m_i} \tag{1}$$

8.7.1 Definition

A rod of length L ft has its left endpoint at the origin. If the number of slugs per foot in the linear density at a point x ft from the origin is $\rho(x)$, where ρ is continuous on $[0, L]$, then the total *mass* of the rod is M slugs, where

$$M = \lim_{\|\Delta\| \to 0} \sum_{i=1}^{n} \rho(\xi_i)\,\Delta_i x = \int_0^L \rho(x)\,dx$$

8.7.2 Definition

A rod of length L ft has its left endpoint at the origin and the number of slugs per foot in the linear density at a point x ft from the origin is $\rho(x)$, where ρ is continuous on $[0, L]$. The *moment of mass* of the rod with respect to the origin is M_0 slug-ft, where

$$M_0 = \lim_{\|\Delta\| \to 0} \sum_{i=1}^{n} \xi_i \rho(\xi_i)\,\Delta_i x = \int_0^L x\rho(x)\,dx$$

The center of mass of the rod described in Definitions 8.7.1 and 8.7.2 is at the point $\bar{x}$, where

$$\bar{x} = \frac{M_0}{M}$$

Because $M_0 = M\bar{x}$, the integral given in Definition 8.7.2 gives $M \cdot \bar{x}$.

Exercises 8.7

4. A system of particles is located on the x-axis. The number of slugs in the mass of each particle and the coordinate of its position are given. Distance is measured in feet. Find the center of mass of the system: $m_1 = 5$ at -7; $m_2 = 3$ at -2; $m_3 = 5$ at 0; $m_4 = 1$ at 2; $m_5 = 8$ at 10.

SOLUTION: We use Eq. (1) with m_i as given for $i = 1, 2, 3, 4, 5$ and $x_1 = -7$, $x_2 = -2$, $x_3 = 0$, $x_4 = 2$, $x_5 = 10$. Thus,

$$\bar{x} = \frac{m_1 x_1 + m_2 x_2 + m_3 x_3 + m_4 x_4 + m_5 x_5}{m_1 + m_2 + m_3 + m_4 + m_5}$$

$$= \frac{5(-7) + 3(-2) + (5 \cdot 0) + (1 \cdot 2) + (8 \cdot 10)}{5 + 3 + 5 + 1 + 8}$$

$$= \frac{41}{22}$$

Thus the center of mass is $\frac{41}{22}$ ft to the right of the origin.

8. A rod is 10 ft long, and the measure of the linear density at a point is a linear function of the measure of the distance from the center of the rod. The linear density at each end of the rod is 5 slugs/ft, and at the center the linear density is $3\frac{1}{2}$ slugs/ft. Find the total mass of the rod and the center of mass.

SOLUTION: Position the rod so that it lies along the x-axis with left endpoint at the origin. Let ρ be the linear density function. We must find $\rho(x)$. Because the rod is 10 ft long, its center is at the point where $x = 5$, and thus $|x - 5|$ is the measure of the distance from the point x to the center of the rod, with x in $[0, 10]$. Because the linear density is a linear function of $|x - 5|$, there are constants a and b such that

$$\rho(x) = a + b\,|x - 5| \tag{1}$$

Because the linear density is $\frac{7}{2}$ slugs/ft at the center of the rod, we are given that $\rho(5) = \frac{7}{2}$. From (1) we get $\rho(5) = a$. Thus $a = \frac{7}{2}$. Because the linear density is 5 slugs/ft at the end of the rod, we are given that $\rho(0) = 5$. From (1) we have $\rho(0) = a + 5b$, and because $a = \frac{7}{2}$, then $5 = \frac{7}{2} + 5b$. Hence, $b = \frac{3}{10}$. Therefore, from Eq. (1) with $a = \frac{7}{2}$ and $b = \frac{3}{10}$, we have

$$\rho(x) = \frac{7}{2} + \frac{3}{10}\,|x - 5| \tag{2}$$

To find the total mass of the rod we use Definition 8.7.1 and Eq. (2). Thus,

$$M = \lim_{\|\Delta\| \to 0} \sum_{i=1}^{n} \left[\frac{7}{2} + \frac{3}{10}\,|\bar{x}_i - 5| \right] \Delta_i x$$

$$= \int_{0}^{10} \left[\frac{7}{2} + \frac{3}{10}\,|x - 5| \right] dx \tag{3}$$

Because $|x - 5| = -(x - 5)$ if $0 \leqslant x \leqslant 5$ and $|x - 5| = x - 5$ if $5 \leqslant x \leqslant 10$, we separate the integral in (3) into two integrals. Thus,

$$M = \int_{0}^{5} \left[\frac{7}{2} - \frac{3}{10}(x - 5) \right] dx + \int_{5}^{10} \left[\frac{7}{2} + \frac{3}{10}(x - 5) \right] dx$$

$$= \int_{0}^{5} \left(-\frac{3}{10}x + 5 \right) dx + \int_{5}^{10} \left(\frac{3}{10}x + 2 \right) dx$$

$$= \left[-\frac{3}{20}x^2 + 5x \right]_{0}^{5} + \left[\frac{3}{20}x^2 + 2x \right]_{5}^{10}$$

$$= \frac{85}{4} + \frac{85}{4}$$

$$= \frac{85}{2} \tag{4}$$

Thus the total mass of the rod is $\frac{85}{2}$ slugs.

To find $\bar{x}$, the center of mass, we use Definition 8.7.2 and Eq. (2). Thus

$$M \cdot \bar{x} = \lim_{\|\Delta\| \to 0} \sum_{i=1}^{n} \left[\frac{7}{2} + \frac{3}{10}\,|\bar{x}_i - 5| \right] \bar{x}_i\,\Delta_i x$$

$$= \int_{0}^{10} \left[\frac{7}{2} + \frac{3}{10}\,|x - 5| \right] x\,dx$$

$$= \int_0^5 \left(-\frac{3}{10}x^2 + 5x \right) dx + \int_5^{10} \left(\frac{3}{10}x^2 + 2x \right) dx$$

$$= \left[-\frac{1}{10}x^3 + \frac{5}{2}x^2 \right]_0^5 + \left[\frac{1}{10}x^3 + x^2 \right]_5^{10}$$

$$= \frac{425}{2} \tag{5}$$

Therefore, substituting from (4) into (5), we get

$$\frac{85}{2}\bar{x} = \frac{425}{2}$$

$$\bar{x} = 5$$

Thus, the center of mass is at the point where $x = 5$, the center of the rod.

12. The total mass of a rod of length L ft is M slugs, and the measure of the linear density at a point x ft from the left end is proportional to the measure of the distance of the point from the right end. Show that the linear density at a point on the rod x ft from the left end is $2M(L - x)/L^2$ slugs/ft.

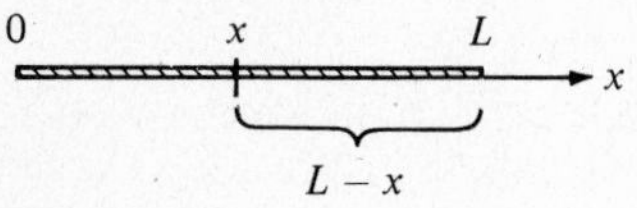

Figure 8.7.12

SOLUTION: Position the rod with its left endpoint at the origin and right endpoint at the point where $x = L$, as illustrated in Fig. 8.7.12. Let ρ be the linear density function. Because $(L - x)$ ft is the distance to the right endpoint from any point that is x ft from the origin, we are given that

$$\rho(x) = k(L - x) \tag{1}$$

We must find the constant k. By Definition 8.7.1 and Eq. (1), we obtain

$$M = \lim_{\|\Delta\| \to 0} \sum_{i=1}^{n} k(L - \bar{x}_i)\,\Delta_i x$$

$$= \int_0^L k(L - x)\,dx$$

$$= k\left[Lx - \frac{1}{2}x^2 \right]_0^L$$

$$= k\left(L^2 - \frac{1}{2}L^2 \right)$$

$$= \frac{1}{2}kL^2$$

Solving for k, we obtain

$$k = \frac{2M}{L^2}$$

Substituting this value of k into (1) we get

$$\rho(x) = \frac{2M}{L^2}(L - x)$$

which is what we want to prove.

8.8 CENTER OF MASS OF A PLANE REGION

The formulas given in the text for finding the centroid of a region in the xy plane apply only to the case in which the region is bounded below by the x-axis. When these formulas do not apply, we use the following more general method.

Let R be a region in the xy plane that is bounded above by the curve $y = f(x)$, bounded below by the curve $y = g(x)$, bounded on the left by the line $x = a$, and bounded on the right by the line $x = b$. To find $(\bar{x}, \bar{y})$, the centroid of R, first use the method of Section 8.1 to find A, the measure of the area of R. Then, because the centroid of the vertical element of area is at the point $(\bar{x}_i, \frac{1}{2}[f(\bar{x}_i) + g(\bar{x}_i)])$, we have

$$A \cdot \bar{x} = \lim_{\|\Delta\| \to 0} \sum_{i=1}^{n} [f(\bar{x}_i) - g(\bar{x}_i)]\bar{x}_i \, \Delta_i x$$

$$= \int_{a}^{b} [f(x) - g(x)]x \, dx$$

$$A \cdot \bar{y} = \lim_{\|\Delta\| \to 0} \sum_{i=1}^{n} \frac{1}{2}[f(\bar{x}_i) - g(\bar{x}_i)][f(\bar{x}_i) + g(\bar{x}_i)] \, \Delta_i x$$

$$= \frac{1}{2} \int_{a}^{b} [f(x) - g(x)][f(x) + g(x)] \, dx$$

Let R be a region in the xy plane that is bounded on the right by the curve $x = f(y)$, bounded on the left by the curve $x = g(y)$, bounded below by the line $y = a$, and bounded above by the line $y = b$. To find $(\bar{x}, \bar{y})$, the centroid of R, first use the method of Section 8.1 to find A, the measure of the area of R. Then, because the centroid of the horizontal element of area is at the point $(\frac{1}{2}[f(\bar{y}_i) + g(\bar{y}_i)], \bar{y}_i)$, we have

$$A \cdot \bar{x} = \lim_{\|\Delta\| \to 0} \sum_{i=1}^{n} \frac{1}{2}[f(\bar{y}_i) - g(\bar{y}_i)][f(\bar{y}_i) + g(\bar{y}_i)] \, \Delta_i y$$

$$= \frac{1}{2} \int_{a}^{b} [f(y) - g(y)][f(y) + g(y)] \, dy$$

$$A \cdot \bar{y} = \lim_{\|\Delta\| \to 0} \sum_{i=1}^{n} [f(\bar{y}_i) - g(\bar{y}_i)]\bar{y}_i \, \Delta_i y$$

$$= \int_{a}^{b} [f(y) - g(y)]y \, dy$$

Sometimes the following theorem can be used to find the centroid.

8.8.1 Theorem If the plane region R has the line L as an axis of symmetry, the centroid of R lies on L.

If R is a vertical plane region that is completely submerged in a liquid with density w, then if F lb is the force due to liquid pressure on R,

$$F = w\bar{h}a \tag{15}$$

where $\bar{h}$ units is the depth of the centroid of R and A square units is the area of R.

Exercises 8.8

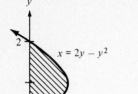

Figure 8.8.4

In Exercises 4-11, find the centroid of the region with the indicated boundaries.

4. The parabola $x = 2y - y^2$ and the y-axis.

SOLUTION: Fig. 8.8.4 shows a sketch of the region R. The region is bounded on the right by the curve $x = f(y) = 2y - y^2$, bounded on the left by the line $x = g(y) = 0$, bounded below by the line $y = 0$, and bounded above by the line $y = 2$. Therefore, the area of R is given by

$$A = \lim_{\|\Delta\| \to 0} \sum_{i=1}^{n} [f(\bar{y}_i) - g(\bar{y}_i)] \, \Delta_i y$$

$$= \int_0^2 [f(y) - g(y)] \, dy$$

$$= \int_0^2 (2y - y^2) \, dy$$

$$= y^2 - \frac{1}{3} y^3 \Big]_0^2$$

$$= \frac{4}{3}$$

If $(\bar{x}, \bar{y})$ is the centroid of region R, then

$$A \cdot \bar{x} = \lim_{\|\Delta\| \to 0} \sum_{i=1}^{n} \frac{1}{2} [f(\bar{y}_i) - g(\bar{y}_i)][f(\bar{y}_i) + g(\bar{y}_i)] \, \Delta_i y$$

$$= \frac{1}{2} \int_0^2 [f(y) - g(y)][f(y) + g(y)] \, dy$$

$$= \frac{1}{2} \int_0^2 (2y - y^2)^2 \, dy$$

$$= \frac{1}{2} \int_0^2 (4y^2 - 4y^3 + y^4) \, dy$$

$$= \frac{1}{2} \left[\frac{4}{3} y^3 - y^4 + \frac{1}{5} y^5 \right]_0^2$$

$$= \frac{8}{15}$$

Because $A = \frac{4}{3}$, we have

$$\frac{4}{3} \bar{x} = \frac{8}{15}$$

$$\bar{x} = \frac{2}{5}$$

Furthermore,

$$A \cdot \bar{y} = \lim_{\|\Delta\| \to 0} \sum_{i=1}^{n} [f(\bar{y}_i) - g(\bar{y}_i)] \, \bar{y}_i \, \Delta_i y$$

$$= \int_0^2 [f(y) - g(y)] \, y \, dy$$

$$= \int_0^2 (2y^2 - y^3) \, dy$$

$$= \frac{2}{3} y^3 - \frac{1}{4} y^4 \Big]_0^2$$

$$= \frac{4}{3}$$

Because $A = \frac{4}{3}$, we have

$$\frac{4}{3}\bar{y} = \frac{4}{3}$$

$$\bar{y} = 1$$

Therefore, the centroid of region R is $(\frac{2}{5}, 1)$.

8. The lines $y = 2x + 1$, $x + y = 7$, and $x = 8$.

SOLUTION: Fig. 8.8.8 shows a sketch of the region R. Because R is bounded above by the line $y = 2x + 1$, we take $f(x) = 2x + 1$. Because R is bounded below by the line $x + y = 7$, or, equivalently, by the line $y = -x + 7$, we take $g(x) = -x + 7$. R is bounded on the left by the line $x = 2$ and bounded on the right by the line $x = 8$. Therefore, the area of R is given by

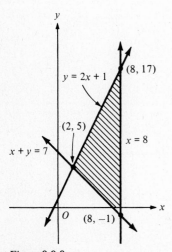

Figure 8.8.8

$$A = \lim_{\|\Delta\| \to 0} \sum_{i=1}^{n} [f(\bar{x}_i) - g(\bar{x}_i)] \, \Delta_i x$$

$$= \int_2^8 [f(x) - g(x)] \, dx$$

$$= \int_2^8 [(2x + 1) - (-x + 7)] \, dx$$

$$= \int_2^8 (3x - 6) \, dx$$

$$= \frac{3}{2} x^2 - 6x \Big]_2^8$$

$$= 54$$

and

$$A \cdot \bar{x} = \lim_{\|\Delta\| \to 0} \sum_{i=1}^{n} [f(\bar{x}_i) - g(\bar{x}_i)] \, \bar{x}_i \, \Delta_i x$$

$$= \int_2^8 [f(x) - g(x)] x \, dx$$

$$= \int_2^8 (3x^2 - 6x) \, dx$$

$$= x^3 - 3x^2 \big]_2^8$$
$$= 324$$

Therefore,

$$54\bar{x} = 324$$
$$\bar{x} = 6$$

Furthermore,

$$A \cdot \bar{y} = \lim_{\|\Delta\| \to 0} \sum_{i=1}^n \frac{1}{2} [f(\bar{x}_i) - g(\bar{x}_i)][f(\bar{x}_i) + g(\bar{x}_i)] \, \Delta_i x$$

$$= \frac{1}{2} \int_2^8 [f(x) - g(x)][f(x) + g(x)] \, dx$$

$$= \frac{1}{2} \int_2^8 (3x - 6)(x + 8) \, dx$$

$$= \frac{3}{2} \int_2^8 (x^2 + 6x - 16) \, dx$$

$$= \frac{3}{2} \left[\frac{1}{3} x^3 + 3x^2 - 16x \right]_2^8$$

$$= 378$$

Therefore

$$54\bar{y} = 378$$
$$\bar{y} = 7$$

The centroid of R is $(6, 7)$.

12. Prove that the distance from the centroid of a triangle to any side of the triangle is equal to one-third the length of the altitude to that side.

SOLUTION: Let PQ be any side of the triangle, and position the triangle so that side PQ lies on the x-axis and vertex R is on the positive y-axis as shown in Fig. 8.8.12. Let $P = (a, 0)$, $Q = (b, 0)$, and $R = (0, c)$. Because the length of the base of triangle PQR is $(b - a)$ units and the length of the altitude is c units, then A square units is the area of the triangle, with

$$A = \frac{1}{2}(b - a)c \tag{1}$$

We find $\bar{y}$, the y coordinate of the centroid of triangle PQR. The intercept form of an equation of line QR is

$$\frac{x}{b} + \frac{y}{c} = 1$$

Solving for x, we obtain

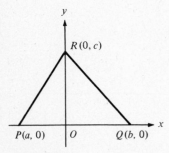

Figure 8.8.12

$$x = f(y) = -\frac{b}{c}y + b \tag{2}$$

In a similar manner we find that an equation for line PR is

$$x = g(y) = -\frac{a}{c}y + a \tag{3}$$

Furthermore, triangle PQR is bounded below by the line $y = 0$ and bounded above by the line $y = c$. Therefore,

$$A \cdot \bar{y} = \lim_{\|\Delta\| \to 0} \sum_{i=1}^{n} [f(\bar{y}_i) - g(\bar{y}_i)]\, \bar{y}_i\, \Delta_i y$$

$$= \int_0^c [f(y) - g(y)]\, y\, dy$$

$$= \int_0^c \left[\left(-\frac{b}{c}y + b\right) - \left(-\frac{a}{c}y + a\right) \right] y\, dy$$

$$= \int_0^c \left[\frac{a-b}{c}y^2 - (a-b)y \right] dy$$

$$= \frac{a-b}{c} \cdot \frac{y^3}{3} - (a-b) \cdot \frac{y^2}{2} \Big]_0^c$$

$$= \frac{1}{3}(a-b)c^2 - \frac{1}{2}(a-b)c^2$$

$$= \frac{1}{6}(b-a)c^2 \tag{4}$$

Substituting from Eq. (1) into (4), we obtain

$$\frac{1}{2}(b-a)c\,\bar{y} = \frac{1}{6}(b-a)c^2$$

$$\bar{y} = \frac{1}{3}c \tag{5}$$

Because $\bar{y}$ units is the distance from the centroid to side PQ and c units is the length of the altitude to side PQ, Eq. (5) proves the desired result.

18. The ends of a trough are equilateral triangles having sides with lengths of 2 ft. If the water in the trough is 1 ft deep, find the force due to liquid pressure on one end. Use formula (15) of this section.

SOLUTION: By Formula (15), the force F is given by

$$F = w\bar{x}A \tag{1}$$

where w pounds per square foot is the density of water, $\bar{x}$ ft is the depth of the centroid of the vertical region that is immersed in the water, and A square feet is the area of that region. Fig. 8.8.18 shows one end of the trough with the water level in line POQ and the centroid of triangle PQR at point C. We are given that $|\overline{OR}| = 1$. In Exercise 12 we proved that the distance from the centroid of a triangle to any side of the triangle is one-third the length of the altitude to that side. We conclude that $\bar{x} = \frac{1}{3}$. Next, we find the area of triangle PQR. Because the end of the trough is an equilateral triangle, angle PRQ has degree measure 60, and thus angle ORQ has

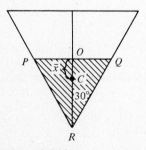

Figure 8.8.18

degree measure 30. Hence, $|\overline{OQ}|/|\overline{OR}| = \tan 30° = \frac{1}{3}\sqrt{3}$. Because, $|\overline{OR}| = 1$, we have $|\overline{OQ}| = \frac{1}{3}\sqrt{3}$. Thus, the number of square feet in the area of triangle PQR is $|\overline{OQ}| \cdot |\overline{OR}| = \frac{1}{3}\sqrt{3}$. Substituting these values for $\bar{x}$ and the area of triangle PQR into (1), we obtain

$$F = w\left(\frac{1}{3}\right)\left(\frac{1}{3}\sqrt{3}\right)$$

We conclude that the force due to liquid pressure is $\frac{1}{9}\sqrt{3}\,w$ foot-pounds.

20. Find the center of mass of the lamina bounded by the parabola $2y^2 = 18 - 3x$ and the y-axis if the area density at any point (x, y) is x slugs/ft^2.

SOLUTION: If ρ is the density function, we are given that $\rho(x) = x$. Fig. 8.8.20 shows a sketch of the region R. First, we find the mass of the lamina. Because ρ is a function of x, we take elements of area that are perpendicular to the x-axis with width $\Delta_i x$ ft. Solving the given equation for y, we obtain $y = \pm\sqrt{9 - \frac{3}{2}x}$. Thus, we let $f(x) = \sqrt{9 - \frac{3}{2}x}$, and the measure of the element of area is given by $2f(\bar{x}_i)\,\Delta_i x$. Because the element of mass is the product of the element of area and the element of density the element of mass is given by

$$\Delta_i M = 2\rho(\bar{x}_i)f(\bar{x}_i)\,\Delta_i x \tag{1}$$

Therefore,

$$M = \lim_{\|\Delta\| \to 0} \sum_{i=1}^{n} 2\rho(\bar{x}_i)f(\bar{x}_i)\,\Delta_i x$$

Because x is in the interval $[0, 6]$, we have

$$M = \int_0^6 2\rho(x)f(x)\,dx$$

$$= \int_0^6 2x\sqrt{9 - \frac{3}{2}x}\,dx \tag{2}$$

Let $u = \sqrt{9 - \frac{3}{2}x}$. Then $x = \frac{2}{3}(9 - u^2)$; $dx = -\frac{4}{3}u\,du$; $u = 3$ when $x = 0$; and $u = 0$ when $x = 6$. Thus, from (2) we obtain

$$M = \int_3^0 2\left[\frac{2}{3}(9 - u^2)\right]u\left(-\frac{4}{3}u\,du\right)$$

$$= \frac{16}{9}\int_0^3 (9u^2 - u^4)\,du$$

$$= \frac{16}{9}\left[3u^3 - \frac{1}{5}u^5\right]_0^3$$

$$= \frac{288}{5} \tag{3}$$

Multiplying on both sides of (1) by $\bar{x}_i$, the abscissa of the centroid of the element of mass, we obtain

$$\Delta_i M \cdot \bar{x} = 2\rho(\bar{x}_i)f(\bar{x}_i)\bar{x}_i\,\Delta_i x$$

Therefore,

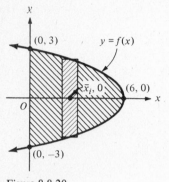

Figure 8.8.20

$$M \cdot \bar{x} = \lim_{\|\Delta\| \to 0} \sum_{i=1}^{n} 2\rho(\bar{x}_i) f(\bar{x}_i) \bar{x}_i \, \Delta_i x$$

$$= \int_0^6 2x^2 f(x) \, dx$$

$$= \int_0^6 2x^2 \sqrt{9 - \frac{3}{2}x} \, dx \qquad (4)$$

With the same choice of u as above, from (4) we obtain

$$M \cdot \bar{x} = \int_3^0 2\left[\frac{2}{3}(9 - u^2)\right]^2 u \left(-\frac{4}{3} u \, du\right)$$

$$= \frac{32}{27} \int_0^3 (81u^2 - 18u^4 + u^6) \, du$$

$$= \frac{32}{27} \left[27u^3 - \frac{18}{5}u^5 + \frac{1}{7}u^7\right]_0^3$$

$$= \frac{6912}{35} \qquad (5)$$

Substituting from (3) into (5), we get

$$\frac{288}{5}\bar{x} = \frac{6912}{35}$$

$$\bar{x} = \frac{24}{7}$$

Because $\bar{y}_i = 0$ for each element of area, $\Delta_i M \, \bar{y}_i = 0$, and thus $\bar{y} = 0$. We conclude that the center of mass of the given lamina is $\left(\frac{24}{7}, 0\right)$.

22.   Use the theorem of Pappus to find the volume of the torus (doughnut-shaped) generated by revolving a circle with a radius of r units about a line in its plane at a distance of b units from its center where $b > r$.

SOLUTION: By the theorem of Pappus, if a region R is revolved about a line L in the plane of R that does not cut the region R, then the measure of the volume of the resulting solid of revolution is given by

$$V = 2\pi \bar{r} A$$

where $\bar{r}$ units is the distance between the line L and the centroid of the region R, and A is the measure of the area of R. We have $\bar{r} = b$ and $A = \pi r^2$. Thus

$$V = 2\pi b(\pi r^2)$$
$$2\pi^2 b r^2$$

The volume of the torus is $2\pi^2 b r^2$ cubic units.

26. Let R be the region bounded by the semicircle $y = \sqrt{r^2 - x^2}$ and the x-axis. Use the theorem of Pappus to find the volume of the solid of revolution generated by revolving R about the line $x - y = r$. (*Hint:* Use the result of Exercise 24 in Section 5.3.)

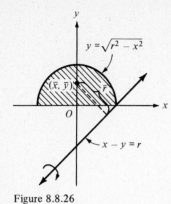

Figure 8.8.26

SOLUTION: Fig. 8.8.26 shows a sketch of the region R and the line $x - y = r$. As in Exercise 22, the volume of the solid of revolution is given by

$$V = 2\pi \bar{r} A \qquad (1)$$

Because the radius of the circle that forms part of the boundary of the region R is r, the area of R is given by

$$A = \frac{1}{2}\pi r^2 \qquad (2)$$

Let $(\bar{x}, \bar{y})$ be the centroid of region R. By symmetry and Theorem 8.8.1, $\bar{x} = 0$. Let $f(x) = \sqrt{r^2 - x^2}$. We have

$$A\bar{y} = \lim_{\|\Delta\| \to 0} \sum_{i=1}^{n} \frac{1}{2}[f(\bar{x}_i)]^2 \, \Delta_i x$$

$$= \int_{-r}^{r} \frac{1}{2}[f(x)]^2 \, dx$$

$$= \frac{1}{2} \int_{-r}^{r} (r^2 - x^2) \, dx$$

$$= \frac{1}{2} \left[r^2 x - \frac{1}{3}x^3 \right]_{-r}^{r}$$

$$= \frac{2}{3}r^3 \qquad (3)$$

Substituting from (2) into (3), we have

$$\frac{1}{2}\pi r^2 \bar{y} = \frac{2}{3}r^3$$

$$\bar{y} = \frac{4r}{3\pi}$$

By Exercise 24 in Section 5.3, the distance from the point (x_1, y_1) to the line $Ax + By + C = 0$ is

$$\frac{|Ax_1 + By_1 + C|}{\sqrt{A^2 + B^2}}$$

We use this formula for the given line $x - y - r = 0$ and substitute the coordinates of the centroid $(\bar{x}, \bar{y})$. Thus

$$\bar{r} = \frac{|\bar{x} - \bar{y} - r|}{\sqrt{2}}$$

$$= \frac{\left| 0 - \dfrac{4r}{3\pi} - r \right|}{\sqrt{2}}$$

$$= \frac{4r + 3\pi r}{3\sqrt{2}\,\pi} \qquad (4)$$

Substituting from (4) and (2) into (1), we obtain

$$V = 2\pi \left(\frac{4r + 3\pi r}{3\sqrt{2}\,\pi} \right)\left(\frac{1}{2}\pi r^2 \right)$$

$$= \frac{(4 + 3\pi)\pi r^3}{3\sqrt{2}} \tag{5}$$

Thus the volume of the solid of revolution is V cubic units, where V is the number given in Eq. (5).

8.9 CENTER OF MASS OF A SOLID OF REVOLUTION

The formulas given in the text for finding the centroid of S, a solid of revolution, apply only to special cases. When these formulas do not apply, we use the following more general method for finding $(\bar{x}, \bar{y}, \bar{z})$, the centroid of S.

1. Use the methods of Sections 8.2 and 8.3 to find V, the measure of the volume of S.

2. Let $\displaystyle\sum_{i=1}^{n} V_i$ be the Riemann sum used in step 1 to find V, and let the centroid (geometric center) of the element of volume used be at the point $(\bar{x}_i, \bar{y}_i, 0)$. Then

$$V \cdot \bar{x} = \lim_{\|\Delta\| \to 0} \sum_{i=1}^{n} V_i \bar{x}_i$$

$$V \cdot \bar{y} = \lim_{\|\Delta\| \to 0} \sum_{i=1}^{n} V_i \bar{y}_i$$

$$V \cdot \bar{z} = 0$$

3. Express each of the above sums as a Riemann sum, find the corresponding definite integral, and evaluate the integral.

If either $\bar{x}_i$ or $\bar{y}_i$ is a constant, then it is not necessary to calculate the corresponding definite integral. If $\bar{x}_i$ is a constant, then $\bar{x} = \bar{x}_i$. And if $\bar{y}_i$ is a constant, then $\bar{y} = \bar{y}_i$.

Exercises 8.9

In Exercises 1-16, find the centroid of the solid of revolution generated by revolving the given region about the indicated line.

4. The region bounded by $x + 2y = 2$, the x-axis, and the y-axis, about the x-axis. Take the rectangular elements parallel to the axis of revolution.

SOLUTION: Fig. 8.9.4 shows the region R. We take an element of area perpendicular to the y-axis of width $\Delta_i y$ units and revolve the element about the x-axis. This results in an element of volume that is a cylindrical shell with the number of units in the thickness given by $\Delta_i r = \Delta_i y$ and the number of units in the mean radius given by $r_i = \bar{y}_i$. Solving the equation of the line for x gives $x = f(y) = -2y + 2$, and thus the measure of the altitude of the shell is given by $h_i = f(\bar{y}_i)$. Therefore,

$$V = \lim_{\|\Delta\| \to 0} \sum_{i=1}^{n} 2\pi r_i h_i \, \Delta_i r$$

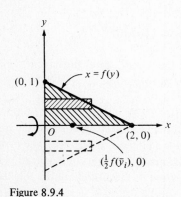

Figure 8.9.4

$$= \lim_{\|\Delta\| \to 0} \sum_{i=1}^{n} 2\pi \bar{y}_i f(\bar{y}_i) \, \Delta_i y \tag{1}$$

$$V = 2\pi \int_0^1 y\, f(y)\, dy$$

$$= 2\pi \int_0^1 y(-2y + 2)\, dy$$

$$= 4\pi \int_0^1 (-y^2 + y)\, dy$$

$$= 4\pi \left[-\frac{1}{3}y^3 + \frac{1}{2}y^2 \right]_0^1$$

$$= \frac{2}{3}\pi \tag{2}$$

Because the centroid of the cylindrical shell is $(\frac{1}{2}f(\bar{y}_i), 0, 0)$, from Eq. (1) we have

$$V \cdot \bar{x} = \lim_{\|\Delta\| \to 0} \sum_{i=1}^n \pi \bar{y}_i [f(\bar{y}_i)]^2\, \Delta_i y$$

$$= \pi \int_0^1 y[f(y)]^2\, dy$$

$$= \pi \int_0^1 y(-2y + 2)^2\, dy$$

$$= 4\pi \int_0^1 (y^3 - 2y^2 + y)\, dy$$

$$= 4\pi \left[\frac{1}{4}y^4 - \frac{2}{3}y^3 + \frac{1}{2}y^2 \right]_0^1$$

$$= \frac{1}{3}\pi \tag{3}$$

Substituting from (2) into (3), we get

$$\frac{2}{3}\pi\bar{x} = \frac{1}{3}\pi$$

$$\bar{x} = \frac{1}{2}$$

Therefore, the centroid of the solid of revolution is $(\frac{1}{2}, 0, 0)$.

8. The region bounded by $y = x^3$, $x = 2$, and the x-axis, about the line $x = 2$. Take the rectangular elements parallel to the axis of revolution.

SOLUTION: Figure 8.9.8 shows a sketch of the region. We take an element of area perpendicular to the x-axis of width $\Delta_i x$ units and revolve the element about the line $x = 2$. This results in an element of volume that is a cylindrical shell with the number of units of thickness given by $\Delta_i r = \Delta_i x$ and the number of units in the mean radius given by $r_i = 2 - \bar{x}_i$. If $f(x) = x^3$ then the measure of the altitude of the shell is given by $h_i = f(\bar{x}_i)$. Therefore,

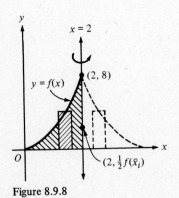

Figure 8.9.8

$$V = \lim_{\|\Delta\| \to 0} \sum_{i=1}^n 2\pi r_i h_i\, \Delta_i r$$

$$= \lim_{\|\Delta\| \to 0} \sum_{i=1}^{n} 2\pi(2 - \bar{x}_i)f(\bar{x}_i)\, \Delta_i x \qquad \text{(1)}$$

$$= 2\pi \int_{0}^{2} (2 - x)f(x)\, dx$$

$$= 2\pi \int_{0}^{2} (2x^3 - x^4)\, dx$$

$$= 2\pi \left[\frac{1}{2}x^4 - \frac{1}{5}x^5 \right]_{0}^{2}$$

$$= \frac{16}{5}\pi \qquad \text{(2)}$$

Because the centroid of the cylindrical shell is $(2, \frac{1}{2}f(\bar{x}_i), 0)$, from Eq. (1) we obtain

$$V \cdot \bar{y} = \lim_{\|\Delta\| \to 0} \sum_{i=1}^{n} \pi(2 - \bar{x}_i)[f(\bar{x}_i)]^2\, \Delta_i x$$

$$= \pi \int_{0}^{2} (2 - x)[f(x)]^2\, dx$$

$$= \pi \int_{0}^{2} (2x^6 - x^7)\, dx$$

$$= \pi \left[\frac{2}{7}x^7 - \frac{1}{8}x^8 \right]_{0}^{2}$$

$$= \frac{32}{7}\pi \qquad \text{(3)}$$

Substituting from (2) into (3), we obtain

$$\frac{16}{5}\pi\bar{y} = \frac{32}{7}\pi$$

$$\bar{y} = \frac{10}{7}$$

We conclude that the centroid of the solid of revolution is $(2, \frac{10}{7}, 0)$.

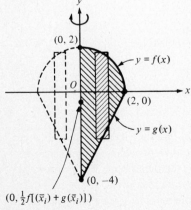

(0, 2)

$y = f(x)$

O

(2, 0)

$y = g(x)$

(0, −4)

$(0, \frac{1}{2}f[(\bar{x}_i) + g(\bar{x}_i)])$

Figure 8.9.12

12. The region bounded by the portion of the circle $x^2 + y^2 + 4$ in the first quadrant, the portion of the line $2x - y = 4$ in the fourth quadrant, and the y-axis, about the y-axis.

SOLUTION: Fig. 8.9.12 shows a sketch of the region. We take an element of area perpendicular to the x-axis of width $\Delta_i x$ units and revolve the element about the y-axis. This results in an element of volume that is a cylindrical shell. The measure of the thickness of the shell is given by $\Delta_i r = \Delta_i x$ and the measure of the mean radius of the shell is given by $r_i = \bar{x}_i$. Solving the equation of the circle for y, we have $y = \pm\sqrt{4 - x^2}$, and solving the equation of the line for y we have $y = 2x - 4$. Let $f(x) = \sqrt{4 - x^2}$ and $g(x) = 2x - 4$. Then the measure of the altitude of the shell is given by $h_i = f(\bar{x}_i) - g(\bar{x}_i)$. Therefore,

$$V = \lim_{\|\Delta\| \to 0} \sum_{i=1}^{n} 2\pi \bar{x}_i [f(\bar{x}_i) - g(\bar{x}_i)] \, \Delta_i x \tag{1}$$

$$= 2\pi \int_0^2 x [\sqrt{4 - x^2} - (2x - 4)] \, dx$$

$$= -\pi \int_0^2 (4 - x^2)^{1/2}(-2x \, dx) - 4\pi \int_0^2 (x^2 - 2x) \, dx$$

$$= -\frac{2}{3}\pi (4 - x^2)^{3/2} \Big]_0^2 - 4\pi \left[\frac{1}{3}x^3 - x^2 \right]_0^2$$

$$= \frac{16}{3}\pi + \frac{16}{3}\pi$$

$$= \frac{32}{3}\pi \tag{2}$$

Because the centroid of the shell is at the point $(0, \frac{1}{2}[f(\bar{x}_i) + g(\bar{x}_i)])$, from (1) we obtain

$$V \cdot \bar{y} = \lim_{\|\Delta\| \to 0} \sum_{i=1}^{n} \pi \bar{x}_i [f(\bar{x}_i) - g(\bar{x}_i)][f(\bar{x}_i) + g(\bar{x}_i)] \, \Delta_i x$$

$$= \pi \int_0^2 x ([f(x)]^2 - [g(x)]^2) \, dx$$

$$= \pi \int_0^2 (-5x^3 + 16x^2 - 12x) \, dx$$

$$= \pi \left[-\frac{5}{4}x^4 + \frac{16}{3}x^3 - 6x^2 \right]_0^2$$

$$= -\frac{4}{3}\pi$$

Substituting from (2) into (3), we get

$$\frac{32}{3}\pi \bar{y} = -\frac{4}{3}\pi$$

$$\bar{y} = -\frac{1}{8}$$

The centroid of the solid is at the point $(0, -\frac{1}{8}, 0)$.

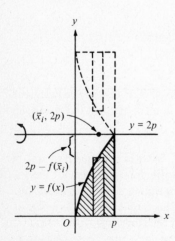

Figure 8.9.16

16. The region bounded by $y = \sqrt{4px}$, the x-axis, and the line $x = p$, about the line $y = 2p$.

SOLUTION: Fig. 8.9.16 shows a sketch of the region. We take an element of area perpendicular to the x-axis of width $\Delta_i x$ units and revolve the element about the line $y = 2p$. This results in an element of volume that is a circular ring. The measure of the thickness of the ring is given by $\Delta_i h = \Delta_i x$, and the measure of the outer radius is given by $R_i = 2p$. We let $f(x) = \sqrt{4px}$. Then the measure of the inner radius of the ring is given by $r_i = 2p - f(\bar{x}_i)$. Thus,

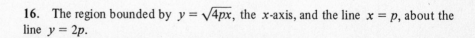

$$V = \lim_{\|\Delta\| \to 0} \sum_{i=1}^{n} \pi(R_i^2 - r_i^2)\, \Delta_i h$$

$$= \lim_{\|\Delta\| \to 0} \sum_{i=1}^{n} \pi([2p]^2 - [2p - f(\bar{x}_i)]^2)\, \Delta_i x \tag{1}$$

$$= \pi \int_0^p ([2p]^2 - [2p - f(x)]^2)\, dx$$

$$= \pi \int_0^p (4pf(x) - [f(x)]^2)\, dx$$

$$= \pi \int_0^p [4p\sqrt{4px} - 4px]\, dx$$

$$= 4p\pi \int_0^p (2\sqrt{p}\, x^{1/2} - x)\, dx$$

$$= 4p\pi \left[\frac{4}{3}\sqrt{p}\, x^{3/2} - \frac{1}{2}x^2 \right]_0^p$$

$$= \frac{10}{3}\pi p^3 \tag{2}$$

Because the centroid of the ring is at the point $(\bar{x}_i, 2p)$, from (1) we obtain

$$V \cdot \bar{x} = \lim_{\|\Delta\| \to 0} \sum_{i=1}^{n} \pi([2p]^2 - [2p - f(\bar{x}_i)]^2\, \bar{x}_i\, \Delta_i x$$

$$= \int_0^p \pi[4p^2 - (2p - \sqrt{4px})^2]x\, dx$$

$$= \pi \int_0^p (-4px^2 + 8p^{3/2}x^{3/2})\, dx$$

$$= \pi \left[-\frac{4}{3}px^3 + \frac{16}{5}p^{3/2}x^{5/2} \right]_0^p$$

$$= \frac{28}{15}\pi p^4 \tag{3}$$

Substituting from (2) into (3), we get

$$\frac{10}{3}\pi p^3 \bar{x} = \frac{28}{15}\pi p^4$$

$$\bar{x} = \frac{14}{25}p$$

We conclude that the centroid of the solid of revolution is at the point $(\tfrac{14}{25}p, 2p, 0)$.

20. Suppose that a cylindrical hole with radius of r units is bored through a solid wooden hemisphere of radius $2r$ units, so that the axis of the cylinder is the same as the axis of the hemisphere. Find the centroid of the solid remaining.

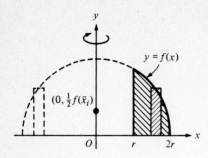

Figure 8.9.20

SOLUTION: The solid remaining is a solid of revolution if the region R, indicated by shading in Fig. 8.9.20, is revolved about the y-axis. R is bounded above by the circle $x^2 + y^2 = 4r^2$, bounded below by the x-axis, bounded on the left by the line $x = r$, and bounded on the right by the line $x = 2r$. We take an element of area perpendicular to the x-axis of width $\Delta_i x$ units and revolve the element about the y-axis. This results in an element of volume that is a cylindrical shell. The measure of the thickness of the shell is given by $\Delta_i r = \Delta_i x$ and the measure of the mean radius of the shell is given by $r_i = \bar{x}_i$. Solving the equation of the circle for y, we obtain $y = \pm\sqrt{4r^2 - x^2}$. Then let $f(x) = \sqrt{4r^2 - x^2}$, and the measure of the altitude of the shell is given by $h_i = f(\bar{x}_i)$. Therefore,

$$V = \lim_{\|\Delta\| \to 0} \sum_{i=1}^{n} 2\pi \bar{x}_i f(\bar{x}_i) \, \Delta_i x \tag{1}$$

$$= \int_{r}^{2r} 2\pi x \sqrt{4r^2 - x^2} \, dx$$

$$= -\pi \int_{r}^{2r} (4r^2 - x^2)^{1/2} (-2x \, dx)$$

$$= -\frac{2}{3}\pi (4r^2 - x^2)^{3/2} \Big]_{r}^{2r}$$

$$= 2\pi\sqrt{3}\, r^3 \tag{2}$$

Because the centroid of the shell is at the point $(0, \frac{1}{2}f(x_i))$, from (1) we obtain

$$V \cdot \bar{y} = \sum_{i=1}^{n} \pi \bar{x}_i [f(\bar{x}_i)]^2 \, \Delta_i x$$

$$= \int_{r}^{2r} \pi x \left(\sqrt{4r^2 - x^2}\right)^2 \, dx$$

$$= \pi \int_{r}^{2r} (4r^2 x - x^3) \, dx$$

$$= \pi \left[2r^2 x^2 - \frac{1}{4}x^4\right]_{r}^{2r}$$

$$= \frac{9}{4}\pi r^4 \tag{3}$$

Substituting from (2) into (3), we get

$$2\pi\sqrt{3}\, r^3 \bar{y} = \frac{9}{4}\pi r^4$$

$$\bar{y} = \frac{3}{8}\sqrt{3}\, r$$

We conclude that the centroid of the solid remaining is $\frac{3}{8}\sqrt{3}\, r$ units above the center of the base.

8.10 LENGTH OF ARC OF A PLANE CURVE

8.10.3 Theorem   If the function f and its derivative f' are continuous on the closed interval $[a, b]$, then the length of arc of the curve $y = f(x)$ from the point $(a, f(a))$ to the point $(b, f(b))$ is given by

$$L = \int_a^b \sqrt{1 + [f'(x)]^2} \, dx$$

8.10.4 Theorem If the function F and its derivative F' are continuous on the closed interval $[c, d]$, then the length of arc of the curve $x = F(y)$ from the point $(F(c), c)$ to the point $(F(d), d)$ is given by

$$L = \int_c^d \sqrt{1 + [F'(y)]^2} \, dy$$

Exercises 8.10

---

4. Use Theorem 8.10.3 to find the length of the arc of the curve $y^3 = 8x^2$ from the point $(1, 2)$ to the point $(27, 18)$.

SOLUTION: Solving the given equation for y, we obtain $y = 2x^{2/3}$. We let $f(x) = 2x^{2/3}$. Then

$$f'(x) = \frac{4}{3} x^{-1/3}$$

$$[f'(x)]^2 = \frac{16}{9} x^{-2/3}$$

$$1 + [f'(x)]^2 = 1 + \frac{16}{9} x^{-2/3}$$

$$1 + [f'(x)]^2 = \frac{9x^{2/3} + 16}{9x^{2/3}}$$

$$\sqrt{1 + [f'(x)]^2} = \frac{\sqrt{9x^{2/3} + 16}}{3x^{1/3}} \quad \text{(because } x > 0)$$

Therefore, by Theorem 8.10.3, we have

$$L = \frac{1}{3} \int_1^{27} \sqrt{9x^{2/3} + 16} \cdot x^{-1/3} \, dx \tag{1}$$

Let $u = 9x^{2/3} + 16$. Then $du = 6x^{-1/3} \, dx$; when $x = 1, u = 25$; when $x = 27$, $u = 97$. Therefore, from (1) we get

$$L = \frac{1}{18} \int_{25}^{97} u^{1/2} \, du$$

$$= \frac{1}{27} u^{3/2} \Big]_{25}^{97}$$

$$= \frac{1}{27} [97^{3/2} - 25^{3/2}]$$

$$\approx 31$$

Thus, the arc is approximately 31 units long.

8. Find the length of the curve $6xy = y^4 + 3$ from the point where $y = 1$ to the point where $y = 2$.

SOLUTION: Solving the given equation for x, we obtain $x = \frac{1}{6}y^3 + \frac{1}{2}y^{-1}$. We let $F(y) = \frac{1}{6}y^3 + \frac{1}{2}y^{-1}$. Then

$$F'(y) = \frac{1}{2}y^2 - \frac{1}{2}y^{-2}$$

$$= \frac{y^4 - 1}{2y^2}$$

$$1 + [F'(y)]^2 = 1 + \left(\frac{y^4 - 1}{2y^2}\right)^2$$

$$= \frac{y^8 + 2y^4 + 1}{4y^4}$$

$$= \left(\frac{y^4 + 1}{2y^2}\right)^2$$

Thus,

$$\sqrt{1 + [F'(y)]^2} = \frac{y^4 + 1}{2y^2}$$

Therefore, by Theorem 8.10.4, we have

$$L = \int_1^2 \frac{y^4 + 1}{2y^2}\, dy$$

$$= \frac{1}{2}\left[\frac{1}{3}y^3 - y^{-1}\right]_1^2$$

$$= \frac{17}{12}$$

Thus the curve is $\frac{17}{12}$ units long.

10. Find the length of the curve $9y^2 = 4(1 + x^2)^3$ in the first quadrant from $x = 0$ to $x = 2\sqrt{2}$.

SOLUTION: Solving the given equation for y, we obtain $y = \pm\frac{2}{3}(1 + x^2)^{3/2}$. Because the curve is in the first quadrant, we take the positive square root and let $f(x) = \frac{2}{3}(1 + x^2)^{3/2}$. Then

$$f'(x) = 2x(1 + x^2)^{1/2}$$
$$1 + [f'(x)]^2 = 1 + 4x^2(1 + x^2)$$
$$= 4x^4 + 4x^2 + 1$$
$$= (2x^2 + 1)^2$$

Thus,

$$\sqrt{1 + [f'(x)]^2} = 2x^2 + 1$$

Therefore, by Theorem 8.10.3, we have

$$L = \int_0^{2\sqrt{2}} (2x^2 + 1)\, dx$$

$$= \frac{2}{3}x^3 + x \Big]_0^{2\sqrt{2}}$$

$$= \frac{38}{3}\sqrt{2}$$

The arc is $\frac{38}{3}\sqrt{2}$ units long.

Review Exercises

4. A container has the same shape and dimensions as a solid of revolution formed by revolving about the y-axis the region in the first quadrant bounded by the parabola $x^2 = 4py$, the y-axis, and the line $y = p$. If the container is full of water, find the work done in pumping all the water up to a point $3p$ ft above the top of the container.

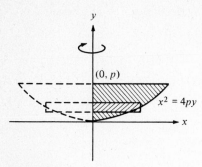

Figure 8.4R

SOLUTION: Fig. 8.4R shows a sketch of the container. We partition the interval $[0, p]$ on the y-axis. The element of volume is a circular disk with the measure of the altitude given by $\Delta_i h = \Delta_i y$. Solving the equation $x^2 = 4py$ for x, we obtain $x = \pm\sqrt{4py}$. We let $f(y) = \sqrt{4py}$. Then the measure of the radius of the disk is given by $R_i = f(\bar{y}_i)$. Thus

$$\Delta_i V = \pi R_i^2\, \Delta_i y = \pi[f(\bar{y}_i)]^2\, \Delta_i y = 4\pi p\, \bar{y}_i\, \Delta_i y$$

And multiplying $\Delta_i V$ by w, we have the element of force.

$$\Delta_i F = 4\pi wp\, \bar{y}_i\, \Delta_i y \tag{1}$$

Because the water is pumped to a point $3p$ ft above the top of the container, this point is $(0, 4p)$, and thus the element of displacement is

$$D_i = 4p - \bar{y}_i \tag{2}$$

Therefore, from (1) and (2) we have

$$W = \lim_{\|\Delta\| \to 0} \sum_{i=1}^{n} D_i \cdot \Delta_i F$$

$$= \lim_{\|\Delta\| \to 0} \sum_{i=1}^{n} (4p - \bar{y}_i)(4\pi wp\, \bar{y}_i\, \Delta_i y)$$

$$= 4\pi wp \int_0^p (4p - y)y\, dy$$

$$= 4\pi wp \left[2py^2 - \frac{1}{3}y^3 \right]_0^p$$

$$= \frac{20}{3}\pi wp^4$$

The total work is $\frac{20}{3}\pi wp^4$ foot-pounds.

8. Find the area of the region bounded by the curves $y = |x| + |x - 1|$ and $y = x + 1$.

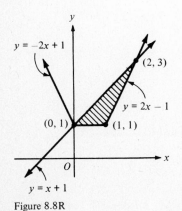

Figure 8.8R

SOLUTION:

If $x < 0$, then $|x| + |x - 1| = -x - (x - 1) = -2x + 1$.
If $0 \leqslant x \leqslant 1$, then $|x| + |x - 1| = x - (x - 1) = 1$.
If $x > 1$, then $|x| + |x - 1| = x + (x - 1) = 2x - 1$.

Therefore, the equation $y = |x| + |x - 1|$ defines the function F, where

$$F(x) = \begin{cases} -2x + 1 & \text{if } x < 0 \\ 1 & \text{if } 0 \leqslant x \leqslant 1 \\ 2x - 1 & \text{if } x > 0 \end{cases}$$

Let G be the function defined by $G(x) = x + 1$. In Fig. 8.8R we show the graphs of $y = F(x)$, $y = G(x)$, and the region R, which is bounded by the graphs. Because R is bounded on the right by the line $y = 2x - 1$, we solve the equation for x, obtaining $x = \frac{1}{2}(y + 1)$, and let $f(y) = \frac{1}{2}(y + 1)$. Because R is bounded on the left by the line $y = x + 1$, we solve the equation for x, obtaining $x = y - 1$, and let $g(y) = y - 1$. Furthermore, R is bounded below by the line $y = 1$ and bounded above by the line $y = 3$. Therefore

$$
\begin{aligned}
A &= \lim_{\|\Delta\| \to 0} \sum_{i=1}^{n} [f(\bar{y}_i) - g(\bar{y}_i)]\, \Delta_i y \\[2mm]
&= \int_1^3 [f(y) - g(y)]\, dy \\[2mm]
&= \int_1^3 \left[\frac{1}{2}(y + 1) - (y - 1) \right] dy \\[2mm]
&= \int_1^3 \left(-\frac{1}{2}y + \frac{3}{2} \right) dy \\[2mm]
&= -\frac{1}{4}y^2 + \frac{3}{2}y \Big]_1^3 \\[2mm]
&= 1
\end{aligned}
$$

The area of the region is 1 square unit. Note that the region is a triangle with base 1 and altitude 2. Using the formula $A = \frac{1}{2}bh$, we obtain $A = \frac{1}{2} \cdot 1 \cdot 2 = 1$, which agrees with the result found by integration.

14. The length of a rod is 8 in and the linear density of the rod at a point x in from the left end is $2\sqrt{x + 1}$ slugs/in. Find the total mass of the rod and the center of mass.

SOLUTION: Let ρ be the linear density function. Then we are given that $\rho(x) = 2\sqrt{x + 1}$ for x in $[0, 8]$. Thus

$$
\begin{aligned}
M &= \lim_{\|\Delta\| \to 0} \sum_{i=1}^{n} \rho(\bar{x}_i)\, \Delta_i x \\[2mm]
&= \int_0^8 \rho(x)\, dx \\[2mm]
&= \int_0^8 2\sqrt{x + 1}\, dx \\[2mm]
&= \frac{4}{3}(x + 1)^{3/2} \Big]_0^8 \\[2mm]
&= \frac{4}{3}[9^{3/2} - 1^{3/2}] \\[2mm]
&= \frac{104}{3}
\end{aligned}
$$

(1)

The total mass is $\frac{104}{3}$ slugs. Furthermore,

$$M \cdot \bar{x} = \lim_{\|\Delta\| \to 0} \sum_{i=1}^{n} \rho(\bar{x}_i)\bar{x}_i \, \Delta_i x$$

$$= \int_0^8 \rho(x)x \, dx$$

$$= \int_0^8 2x\sqrt{x+1} \, dx \tag{2}$$

Let $u = \sqrt{x+1}$. Then $x = u^2 - 1$; $dx = 2u \, du$; when $x = 0, u = 1$; and when $x = 8, u = 3$. Thus, from (2) we have

$$M \cdot \bar{x} = \int_1^3 2(u^2 - 1)u(2u \, du)$$

$$= 4 \int_1^3 (u^4 - u^2) \, du$$

$$= 4\left[\frac{1}{5}u^5 - \frac{1}{3}u^3\right]_1^3$$

$$= \frac{2384}{15} \tag{3}$$

Substituting from (1) into (3), we get

$$\frac{104}{3} \cdot \bar{x} = \frac{2384}{15}$$

$$\bar{x} = \frac{298}{65} \approx 4.58$$

The center of mass is approximately 4.58 in from the left end of the rod.

16. Find the centroid of the region bounded by the loop of the curve $y^2 = x^2 - x^3$.

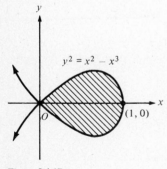

$y^2 = x^2 - x^3$

$(1, 0)$

Figure 8.16R

SOLUTION: Solving the given equation for y, we obtain $y = \pm x\sqrt{1-x}$. Thus, $x \leqslant 1$ and $y = 0$ when $x = 1$ or $x = 0$. A sketch of the graph is shown in Fig. 8.16R. Let $f(x) = x\sqrt{1-x}$ and $g(x) = -x\sqrt{1-x}$. Then the loop R is bounded above by the curve $y = f(x)$, bounded below by the curve $y = g(x)$, bounded on the left by the line $x = 0$, and bounded on the right by the line $x = 1$. Therefore,

$$A = \lim_{\|\Delta\| \to 0} \sum_{i=1}^{n} [f(\bar{x}_i) - g(\bar{x}_i)] \, \Delta_i x$$

$$= \int_0^1 [f(x) - g(x)] \, dx$$

$$= \int_0^1 2x\sqrt{1-x} \, dx \tag{1}$$

Let $u = \sqrt{1-x}$. Then $x = 1 - u^2$; $dx = -2u \, du$; when $x = 0, u = 1$; and when $x = 1, u = 0$. From (1) we have

$$A = \int_1^0 2(1-u^2)u(-2u\,du)$$

$$= 4\int_0^1 (u^2 - u^4)\,du$$

$$= 4\left[\frac{1}{3}u^3 - \frac{1}{5}u^5\right]_0^1$$

$$= \frac{8}{15} \tag{2}$$

Furthermore, from (1) we obtain

$$A \cdot \bar{x} = \int_0^1 2x^2\sqrt{1-x}\,dx$$

With the same choice for u as before, we have

$$A \cdot \bar{x} = \int_1^0 2(1-u^2)^2 u(-2u\,du)$$

$$= 4\int_0^1 (u^2 - 2u^4 + u^6)\,du$$

$$= 4\left[\frac{1}{3}u^3 - \frac{2}{5}u^5 + \frac{1}{7}u^7\right]_0^1$$

$$= \frac{32}{105} \tag{3}$$

Substituting from (2) into (3), we get

$$\frac{8}{15} \cdot \bar{x} = \frac{32}{105}$$

$$\bar{x} = \frac{4}{7}$$

Because the region is symmetric with respect to the x-axis, then $\bar{y} = 0$. Thus, the centroid is at the point $(\frac{4}{7}, 0)$.

20. Use integration to find the volume of a segment of a sphere if the sphere has a radius of r units and the altitude of the segment is h units.

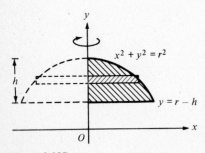

Figure 8.20R

SOLUTION: Let S be the segment of the sphere. Then S is a solid of revolution if the region G, shown in Fig. 8.20R, is revolved about the y-axis. G is bounded on the right by the circle $x^2 + y^2 = r^2$, bounded on the left by the y-axis, bounded below by the line $y = r - h$, and bounded above by the line $y = r$. We take a horizontal element of area of width $\Delta_i y$ units and revolve the element about the y-axis. This results in an element of volume that is a circular disk with the measure of the thickness given by $\Delta_i h = \Delta_i y$. Solving the equation of the circle for x, we obtain $x = \pm\sqrt{r^2 - y^2}$. We let $f(y) = \sqrt{r^2 - y^2}$, and thus the measure of the radius of the disk is given by $R_i = f(\bar{y}_i)$. Therefore,

$$V = \lim_{\|\Delta\| \to 0} \sum_{i=1}^{n} \pi R_i^2\,\Delta_i h$$

$$= \lim_{\|\Delta\| \to 0} \sum_{i=1}^{n} \pi(r^2 - \bar{y}_i^2)\, \Delta_i y$$

$$= \pi \int_{r-h}^{r} (r^2 - y^2)\, dy$$

$$= \pi[r^2 y]_{r-h}^{r} - \frac{1}{3}\pi[y^3]_{r-h}^{r}$$

$$= \pi[r^3 - r^2(r-h)] - \frac{1}{3}\pi[r^3 - (r-h)^3]$$

$$= \pi r^2 h - \frac{1}{3}\pi(3r^2 h - 3rh^2 + h^3)$$

$$= \pi r^2 h - \pi r^2 h + \pi r h^2 - \frac{1}{3}\pi h^3$$

$$= \frac{1}{3}\pi h^2 (3r - h)$$

The volume of the segment is $\frac{1}{3}\pi h^2 (3r - h)$ cubic units.

26. The work necessary to stretch a spring from 9 in to 10 in is $\frac{3}{2}$ times the work necessary to stretch it from 8 in to 9 in. What is the natural length of the spring?

SOLUTION: Let a in be the natural length of the spring and place the spring along the x-axis so that its left end is at the origin and its right end is at the point where $x = a$ when the spring is in its natural state. When the spring is stretched until its right end is at x, then $x - a$ is the number of inches that it has been stretched, and Hooke's law states that the force on the spring is $F(x)$ pounds where

$$F(x) = k(x - a)$$

Let W_1 be the measure of the work required to stretch the spring from 9 in to 10 in. We have

$$W_1 = \lim_{\|\Delta\| \to 0} \sum_{i=1}^{n} F(\bar{x}_i)\, \Delta_i x$$

$$= \int_{9}^{10} F(x)\, dx$$

$$= k \int_{9}^{10} (x - a)\, dx$$

$$= k \frac{1}{2}(x - a)^2 \Big]_{9}^{10}$$

$$= \frac{1}{2}k[(10 - a)^2 - (9 - a)^2]$$

$$= \frac{1}{2}k(19 - 2a) \tag{1}$$

Let W_2 be the measure of the work required to stretch the spring from 8 in to 9 in. Then

$$W_2 = k \int_8^9 (x - a)\, dx$$

$$= k \left[\frac{1}{2}(x - a)^2 \right]_8^9$$

$$= \frac{1}{2} k[(9 - a)^2 - (8 - a)^2]$$

$$= \frac{1}{2} k(17 - 2a) \tag{2}$$

Because we are given that $W_1 = \frac{3}{2} W_2$, by substituting from (1) and (2) into this equation we obtain

$$\frac{1}{2} k(19 - 2a) = \frac{3}{2} \left[\frac{1}{2} k(17 - 2a) \right]$$

$$2(19 - 2a) = 3(17 - 2a)$$

$$a = \frac{13}{2}$$

Thus, the natural length of the spring is $\frac{13}{2}$ inches.

28. A semicircular plate with a radius of 3 ft is submerged vertically in a tank of water, with its diameter lying in the surface. Use Formula (15) of Section 8.8 to find the force due to liquid pressure on one side of the plate.

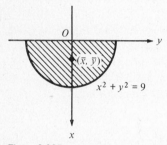

Figure 8.28R

SOLUTION: Fig. 8.28R shows the plate. Formula (15) of Section 8.8 states that the force due to liquid pressure on one side of the plate is given by

$$F = w\bar{x}A \tag{1}$$

where $\bar{x}$ ft is the depth of the centroid of the plate and A square feet is the area of the plate. In Exercise 26 of Exercises 8.8 we found that the centroid of any semicircular region of radius r is on the axis of the region $4r/3\pi$ units from the center of the circle. Because $r = 3$ for the plate, we conclude that

$$\bar{x} = \frac{4}{\pi} \tag{2}$$

Furthermore, the area of the plate is given by the formula $A = \frac{1}{2}\pi r^2$, with $r = 3$. Hence

$$A = \frac{9}{2}\pi \tag{3}$$

Substituting from (2) and (3) into (1), we obtain

$$F = w\left(\frac{4}{\pi}\right)\left(\frac{9}{2}\pi\right)$$

$$= 18w$$

We conclude that the force on the plate is $18w$ pounds.

32. The region bounded by the parabola $x^2 = 4y$, the x-axis, and the line $x = 4$ is revolved about the y-axis. Find the centroid of the solid of revolution formed.

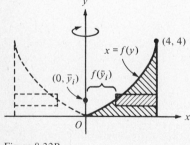

Figure 8.32R

SOLUTION: A sketch of the region H and a plane section of the solid of revolution if H is revolved about the y-axis is shown in Fig. 8.32R. The element of volume is a circular ring. The measure of the thickness of the ring is given by $\Delta_i h = \Delta_i y$. The measure of the outer radius of the ring is given by $R_i = 4$. Solving the equation $x^2 = 4y$ for x, we get $x = \pm 2\sqrt{y}$. Let $f(y) = 2\sqrt{y}$. Then the measure of the inner radius of the ring is given by $r_i = f(\bar{y}_i)$. Therefore,

$$V = \lim_{\|\Delta\| \to 0} \sum_{i=1}^{n} \pi (R_i^2 - r_i^2)\, \Delta_i h$$

$$= \lim_{\|\Delta\| \to 0} \sum_{i=1}^{n} \pi (16 - [f(\bar{y}_i)]^2)\, \Delta_i y$$

$$= \pi \int_0^4 (16 - [f(y)]^2)\, dy$$

$$= \pi \int_0^4 (16 - 4y)\, dy \tag{1}$$

$$= \pi [16y - 2y^2]_0^4$$
$$= 32\pi \tag{2}$$

Because the centroid of the element of volume is at the point $(0, \bar{y}_i, 0)$, from Eq. (1), we have

$$V \cdot \bar{y} = \pi \int_0^4 (16 - 4y)y\, dy \tag{3}$$

$$= 4\pi \int_0^4 (4y - y^2)\, dy$$

$$= 4\pi \left[2y^2 - \frac{1}{3}y^3 \right]_0^4$$

$$= \frac{128}{3}\pi \tag{4}$$

Substituting from (2) into (4), we get

$$32\pi\bar{y} = \frac{128}{3}\pi$$

$$\bar{y} = \frac{4}{3}$$

We conclude that the centroid is at the point $(0, \frac{4}{3}, 0)$.

34. Find the center of mass of the solid of revolution of Exercise 32 if the measure of the volume density of the solid at any point is equal to the measure of the distance of the point from the xz plane.

SOLUTION: Because the distance of a point from the xz plane is y, we are given that the density function is defined by $\rho(y) = y$. Refer to Fig. 8.32R and the solution to Exercise 32. Because the Riemann sum for the mass M of the solid in this Exercise can be found by multiplying the Riemann sum for the volume V of the solid in Exercise 32 by $\rho(\bar{y}_i)$, we conclude that the integrand of the definite integral for M is y times the integrand of the definite integral for V. Therefore from Eq. (1) in Exercise 32, we obtain

$$M = \pi \int_0^4 (16 - 4y)y \, dy \tag{1}$$

Because the integral for M is the same as that in Eq. (3) of Exercise 32, we conclude that

$$M = \frac{128}{3}\pi \tag{2}$$

As in Exercise 32, we use Eq. (1) to obtain

$$M \cdot \bar{y} = \pi \int_0^4 (16 - 4y)y^2 \, dy$$

$$= 4\pi \int_0^4 (4y^2 - y^3) \, dy$$

$$= 4\pi \left[\frac{4}{3}y^3 - \frac{1}{4}y^4 \right]_0^4$$

$$= \frac{256}{3}\pi \tag{3}$$

Substituting from (2) into (3), we obtain

$$\frac{128}{3}\pi \cdot \bar{y} = \frac{256}{3}\pi$$

$$\bar{y} = 2$$

The center of mass of the solid is $(0, 2, 0)$.

Appendix

CHAPTER TESTS

CHAPTER TESTS

After taking each test with the time limit as indicated, turn to the page on which the solutions are given and correct your paper. It may be most beneficial to you if you try to simulate the conditions under which an actual test might be given for your class. Thus, do not refer to the examples in the text or to your notes, unless you will be allowed to do this in class.

TEST FOR CHAPTER 1   **(45 minutes)**   **Solutions on page 304.**

1. Find the solution set of the following inequality.

$$|2x - 1| < x$$

2. Find an equation of the line that is the perpendicular bisector of the segment with endpoints $(-1, 3)$ and $(5, -1)$.

3. Find the center and radius of the circle that is the graph of the following equation.

$$4x^2 + 4y^2 - 16x + 24y + 47 = 0$$

4. Determine k so that the triangle ABC is an isosceles triangle with its vertex at point C if $A = (2, 1)$, $B = (5, 6)$ and $C = (4, k)$.

5. Describe the symmetry of the graph of the following equation and draw a sketch of the graph. Label the intercept points.

$$y^2 = x + 4$$

6. Draw a sketch of the graph of the function f and find the domain and range of f.

$$f = \left\{ (x, y) | y = \frac{x^2 - x - 2}{x - 2} \right\}$$

7. Find the set of all replacements for x that result in a real value for the following expression and illustrate the set on the number line.

$$\sqrt{\frac{x}{2x + 1}}$$

8. Let $f(x) = \sqrt{2x - 1}$ and $g(x) = 3x^2 + 1$. Find the expression represented by each of the following and simplify your answer.

(a) $f[g(x)]$ (b) $g(x + 2) - g(x)$

TEST FOR CHAPTER 2   (50 minutes)   Solutions on page 305.

1. For each of the following use any limit theorems to find the limit.

(a) $\displaystyle \lim_{x \to 3} \sqrt{\frac{x^2 - 9}{x^2 - 3x}}$

(b) $\displaystyle \lim_{x \to 0} \frac{2 - \sqrt{4 + x}}{x}$

(c) $\displaystyle \lim_{x \to 0^-} \frac{x^2 - 2x}{|x|}$

(d) $\displaystyle \lim_{x \to 2^+} \frac{x - 1}{2x - x^2}$

2. For the function f defined below find $\displaystyle \lim_{x \to -1^+} f(x)$, $\displaystyle \lim_{x \to -1^-} f(x)$, and $\displaystyle \lim_{x \to -1} f(x)$ if it exists. Draw a sketch of the graph of the function f.

$$f(x) = \begin{cases} x + 1 & \text{if } x \leqslant -1 \\ -x + 1 & \text{if } x > -1 \end{cases}$$

3. Determine k so that the function f is continuous at 2.

$$f(x) = \begin{cases} kx^2 & \text{if } x < 2 \\ x + k & \text{if } x \geqslant 2 \end{cases}$$

4. Find each number at which the function f is discontinuous and use limits to show whether the discontinuity is removable or essential.

$$f(x) = \frac{x^2 - 1}{x^2 - x}$$

5. (a) Write a statement using inequalities that defines what is meant by the following:

$$\lim_{x \to b^-} f(x) = +\infty$$

(b) If f is a function of x, write a statement about the limit of f that implies that f is continuous at the number c.

6. Let $f(x) = \frac{1}{2}x^2$, $a = 4$, $L = 8$. For any $\epsilon > 0$ find $\delta > 0$ so that $|f(x) - L| < \epsilon$ whenever $0 < |x - a| < \delta$. Assume $\delta \leqslant 1$.

TEST FOR CHAPTER 3   (75 minutes)   Solutions on page 306.

1. Use the differentiation formulas to find $f'(x)$ and factor your answer completely.

   (a) $f(x) = \dfrac{x^2}{2x+1}$

   (b) $f(x) = (x+1)^2 (4x-1)^3$
   (c) $f(x) = \sqrt{2x+1}\,(x-3)^3$

2. Use implicit differentiation to find $D_x y$.

   $2y = x^2 y^3 + 1$

3. Find an equation of the line that is tangent to the given curve at the point $(2, -1)$.

   $y = (4x^{-2} - x)^3$

4. A particle is moving along a vertical line, and after t seconds its directed distance from the starting point is s feet with $s = t^3 - 3t^2$ for $t \geqslant 0$ and positive direction upward.

   (a) Find the position of the particle when its instantaneous velocity is 24 ft/sec.
   (b) Find the instantaneous acceleration of the particle at the moment when the particle reverses its direction.

5. Answer true or false

   (a) If $f'(a)$ exists, then f is continuous at a.
   (b) If $f'(a)$ does not exist, then f is not continuous at a.
   (c) If f is continuous at a, then $f'(a)$ exists.
   (d) If f is not continuous at a, then $f'(a)$ does not exist.

6. Let $f(x) = \sqrt{x^2 + 5}$ and find

   $\displaystyle \lim_{x \to 2} \frac{f(x) - f(2)}{x - 2}$

7. Write the equation that defines $f'(x)$ and use the definition to find the derivative of the function given by

   $f(x) = \dfrac{3}{x^2}$

8. Boyle's Law states that $PV = C$, where P lb/in^2 is the pressure of a gas, V in^3 is the volume of the gas, and C is a constant. Find the instantaneous rate of change of P per unit change in V when the pressure is 20 lb/in^2 and the volume is 15 in^3.

9. An automobile is approaching an intersection at the rate of 20 meters per second. When the auto is 80 meters from the intersection, a truck moving at 15 meters per second crosses the intersection on a road that is at right angles to the one on which the auto is traveling. Find the rate at which the distance between the auto and the truck is changing 2 seconds after the truck has left the intersection.

TEST FOR CHAPTER 4   (75 minutes)   Solutions on page 307.

1. Find the indicated limit.

   $\displaystyle \lim_{x \to -\infty} \frac{x}{4x + \sqrt{x^2 + 1}}$

2. Use limits to find the horizontal and vertical asymptotes of the graph of the equation and make a sketch of the graph, using only the asymptotes and the intercepts.

$$y = \frac{x+1}{(x-1)^2}$$

3. Determine whether the function f is continuous or discontinuous on each of the indicated intervals.

$$f(x) = \sqrt{\frac{x^2 - 4}{x}}$$

$(-\infty, -2), [-2, 0], (-2, 0), [-2, 0), (0, 2), [2, +\infty)$

4. Find the critical numbers of the function f.

$$f(x) = x^{5/3} + x^{2/3}$$

5. Find the absolute extrema of the function f on the indicated interval.

$$f(x) = x^4 - 6x^2, \ -3 \leqslant x \leqslant 1$$

6. Consider the mean-value theorem for the function f and the interval $[x_1, x_2]$.

(a) What is the hypothesis of the theorem?
(b) Write the conclusion of the theorem.

7. If f is a function such that $x - f(x) \leqslant x f(x) \leqslant x + f(x)$, use the squeeze theorem to show $\lim\limits_{x \to +\infty} f(x) = 1$.

8. A manufacturer can make a profit of \$10 on each item if not more than 20 items are produced each day. But for each item over 20 that is produced, the profit per item on the entire lot is decreased by \$0.25. Find the greatest possible total profit per day.

9. Find the radius and altitude of the right circular cylinder with maximum volume that can be inscribed in a right circular cone with radius 6 in and altitude 8 in.

TEST FOR CHAPTER 5   **(65 minutes)**   Solutions on page 310.

1. Let $f(x) = 3x^4 + 4x^3$

(a) Find each interval on which f is increasing and each interval on which f is decreasing.
(b) Find each relative extremum of f.

2. Let $f(x) = 6x^2 - x^3$

(a) Find where the graph of the curve $y = f(x)$ is concave upward and where it is concave downward.
(b) Find the coordinates of each point of inflection of the graph of $y = f(x)$ and find the slope of the line tangent to the curve at each point of inflection.

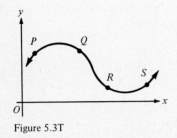

Figure 5.3T

3. Consider only the points P, Q, R, S on the graph of $y = f(x)$ which is shown in Fig. 5.3T. At which of these points are the given conditions satisfied?

(a) $f'(x) > 0$ and $f''(x) > 0$
(b) $f'(x) < 0$ and $f''(x) < 0$
(c) $f'(x) > 0$ and $f''(x) < 0$
(d) $f'(x) < 0$ and $f''(x) > 0$

4. What are the conditions in the second-derivative test that must be satisfied to conclude that the function f has a relative maximum value at the number x_0?

5. Let $f(x) = 3x^{2/3} - x$.

   (a) Find the coordinates of each relative extremum of f.
   (b) Find the slope of the line tangent to the curve $y = f(x)$ at each relative extremum, if possible.
   (c) Find each point of inflection of the graph of $y = f(x)$.
   (d) Draw a sketch of the graph of the curve $y = f(x)$.

6. The total cost of producing x items is $C(x)$ dollars, where $C(x) = \frac{1}{8}x^2 + 10x + 200$.

   (a) Find the marginal cost when $x = 30$.
   (b) How many items should be produced to make the average cost per item the least?

7. The demand equation for a certain commodity is $10p + x = 500$, where x is the number of items demanded each day when the price is p cents per item.

   (a) Find the greatest possible daily revenue.
   (b) If it costs a monopolist 10 cents to produce each item, what price should he set to make the greatest profit?

8. If x units is the breadth and y units is the depth of a rectangular beam, then the number of units in its strength is given by $S = kxy^2$, where k is a constant. Find the dimensions of the strongest rectangular beam that can be cut from a cylindrical log with diameter 6 inches.

9. If the material used for the top of a 12 ounce beer can costs twice as much per square unit as that used for the sides and bottom, what is the ratio of height to base radius that will result in the least total cost of material?

TEST FOR CHAPTER 6   (55 minutes)   Solutions on page 312.

1. Let $y = 3x^2$ and calculate the exact values for Δy and dy when $x = 5.0$ and $\Delta x = 0.02$.

2. The diameter of a sphere is measured and found to be 10.0 inches with a possible error of ± 0.05 inches. The measured value is then used to calculate V, the number of cubic inches in the volume of the sphere. Use differentials to find a two-place decimal approximation for the possible error in V.

3. Let $y = \sqrt{5x + 4}$ and $x^2 t + t = 3x + 1$. Use differentials to find the value of $D_t y$ when $x = 1$.

4. Find the most general antiderivative

$$\int x\sqrt{x + 1}\, dx$$

5. Find the particular solution in the form $y = f(x)$ for the differential equation $dy = 2xy^2\, dx$ if $y = 1$ when $x = 1$.

6. Find an equation of the curve $y = f(x)$ if the line $y = x + 2$ is tangent to the curve at the point $(1, 3)$ and if $d^2y/dx^2 = 6x$ at every point on the curve.

7. When rolling on a level surface the velocity of a certain ball will decrease at the rate of 4 ft/sec². Suppose that the ball is given an initial velocity of 30 ft/sec.

   (a) How far will the ball roll during the first 5 seconds?
   (b) How far will the ball roll before coming to rest?

8. If the marginal cost function is given by $60(3x + 8)^{-1/3}$ and the fixed cost is zero, find the total cost function.

TEST FOR CHAPTER 7   (60 minutes)   Solutions on page 314.

1. Use the properties of sigma to find this sum.

$$\sum_{i=1}^{20} (\sqrt{3i + 4} - \sqrt{3i + 1})$$

2. Given:

$$\sum_{i=1}^{n} i = \frac{n(n + 1)}{2} \qquad \sum_{i=1}^{n} i^2 = \frac{n(n + 1)(2n + 1)}{6}$$

   (a) Use the method of rectangles to calculate the area of the region that is bounded by the curve $y = x^2 + x$, the coordinate axes, and the line $x = 2$.
   (b) Use the limit of a Riemann sum to calculate this definite integral. Do *not* use the fundamental theorem.

$$\int_{-1}^{4} x^2 \, dx$$

3. Calculate the Riemann sum for the function f on the interval $[1, 2]$ if $f(x) = 2x + 1$ and Δ is as indicated in Table 3.

Table 3

i	x_i	$\bar{x}_i$
0	1.0	
1	1.3	1.1
2	1.8	1.5
3	2.0	1.9

4. Given:

$$\int_{1}^{2} f(x) \, dx = \frac{7}{3} \quad \text{and} \quad \int_{1}^{4} f(x) \, dx = 21$$

   Assume that f is continuous and use the properties of the definite integral to calculate this integral.

$$\int_{2}^{4} [3f(x) + 5] \, dx$$

5. Use the fundamental theorem and the mean-value theorem for integrals to find two numbers at which the average value of the function f over the interval $[0, 3]$ occurs.

$$f(x) = 4x - x^2$$

6. Use the fundamental theorem to calculate the following definite integral.

$$\int_0^3 \frac{x\,dx}{\sqrt{x+1}}$$

TEST FOR CHAPTER 8   (90 minutes)   Solutions on page 316.

1. Let R be the region bounded by the curve $y^3 = 8x$, the line $y = 2$, and the y-axis.

   (a) Set up a definite integral of the form $\int_a^b f(x)\,dx$ that gives the measure of the area of R.
   (b) Set up a definite integral of the form $\int_c^d g(y)\,dy$ that gives the measure of the area of R.

2. Let R be the region bounded by the curve $y = 3x^2$, the line $x = 2$, and the x-axis.

   (a) Use the circular-ring method to set up a definite integral that gives the measure of the volume of the solid of revolution if R is revolved about the y-axis.
   (b) Use the cylindrical-shell method to set up a definite integral that gives the measure of the volume of the solid of revolution if R is revolved about the vertical line $x = 3$.
   (c) Set up a definite integral of the form $\int_a^b f(x)\,dx$ that gives the measure of the volume of the solid whose base is the region R if each plane section perpendicular to the x-axis is an equilateral triangle.

3. A hemispherical tank with radius 8 ft is filled with water to a depth of 6 ft. Take the origin at the center of the top of the tank with the positive direction of the x-axis downward. Set up a definite integral that gives the number of foot-pounds in the total work required to empty the tank by pumping the water to the top of the tank.

4. A dam with vertical face has a gate that is in the shape of an isosceles triangle 3 ft wide at the top and 2 ft high, and the upper edge of the gate is 10 ft below the surface of the water. Find the total force on the gate due to water pressure.

5. The linear density of a 2-ft rod is $\sqrt{4x+1}$ slugs/ft, where x is the number of feet from the end of the rod. Find the total mass of the rod.

6. Find the centroid of the region in the first quadrant that is bounded by the curve $y = x^3$ and the line $y = x$.

7. Find the centroid of the solid of revolution if the region bounded by the curve $y = x^2$ and the line $y = 4$ is revolved about the y-axis.

8. Find the length of arc of the curve $6xy = x^4 + 3$ from the point $(1, \frac{2}{3})$ to the point $(3, \frac{14}{3})$.

SOLUTIONS FOR CHAPTER TESTS

SOLUTIONS FOR TEST 1

1. $2x - 1 < x$ and $2x - 1 > -x$

   $x < 1$ and $\quad x > \dfrac{1}{3}$

   Solution set is $(\frac{1}{3}, 1)$.

2. $m = \dfrac{-1 - 3}{5 - (-1)} = \dfrac{-4}{6} = -\dfrac{2}{3}$

   Thus the slope of the perpendicular line is $\frac{3}{2}$.

   $$(\bar{x}, \bar{y}) = (2, 1)$$

   $$y - 1 = \frac{3}{2}(x - 2)$$

   $$2y - 2 = 3x - 6$$
   $$3x - 2y - 4 = 0$$

3.
   $$4x^2 - 16x + 4y^2 + 24y = -47$$
   $$4(x^2 - 4x + 4) + 4(y^2 + 6y + 9) = -47 + 16 + 36$$
   $$4(x - 2)^2 + 4(y + 3)^2 = 5$$

   $$(x - 2)^2 + (y + 3)^2 = \frac{5}{4}$$

   Center $= (2, -3)$; radius $= \dfrac{1}{2}\sqrt{5}$

4. $\sqrt{(4 - 2)^2 + (k - 1)^2} = \sqrt{(4 - 5)^2 + (k - 6)^2}$
   $$4 + k^2 - 2k + 1 = 1 + k^2 - 12k + 36$$
   $$10k = 32$$

   $$k = \frac{16}{5}$$

5. Symmetric with respect to the x-axis. Graph is Fig. 1.5T.

6. $y = \dfrac{(x - 2)(x + 1)}{x - 2} = x + 1 \quad$ if $x \neq 2$

   domain $= \{x \mid x \neq 2\}$
   range $= \{y \mid y \neq 3\}$

   Graph is Fig. 1.6T.

7. $\dfrac{x}{2x + 1} \geq 0$

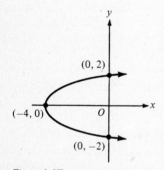

Figure 1.5T

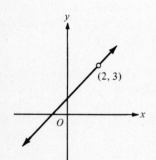

Figure 1.6T

	$-\frac{1}{2}$	0	
x	$-$	$-$	$+$
$2x + 1$	$-$	$+$	$+$
$\dfrac{x}{2x + 1}$	$+$	$-$	$+$

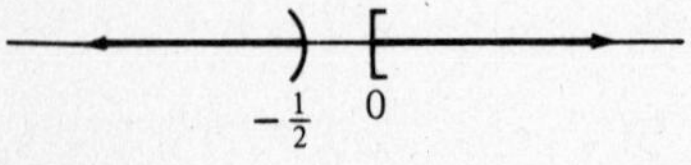

Thus, the required set is $(-\infty, -\frac{1}{2}) \cup [0, +\infty)$ and is illustrated in Fig. 1.7T.

Figure 1.7T

8. (a) $f[g(x)] = f[3x^2 + 1] = \sqrt{2(3x^2 + 1) - 1} = \sqrt{6x^2 + 1}$

   (b) $g(x + 2) - g(x) = [3(x + 2)^2 + 1] - [3x^2 + 1]$
   $$= 3x^2 + 12x + 12 + 1 - 3x^2 - 1$$
   $$= 12x + 12$$

SOLUTIONS FOR TEST 2   1. (a) $\displaystyle\lim_{x \to 3} \sqrt{\frac{x^2 - 9}{x^2 - 3x}} = \lim_{x \to 3} \sqrt{\frac{(x + 3)(x - 3)}{x(x - 3)}}$

   $$= \lim_{x \to 3} \sqrt{\frac{x + 3}{x}}$$

   $$= \sqrt{2}$$

   (b) $\displaystyle\lim_{x \to 0} \frac{2 - \sqrt{4 + x}}{x} \cdot \frac{2 + \sqrt{4 + x}}{2 + \sqrt{4 + x}} = \lim_{x \to 0} \frac{4 - (4 + x)}{x(2 + \sqrt{4 + x})} = \lim_{x \to 0} \frac{-1}{2 + \sqrt{4 + x}}$$

   $$= -\frac{1}{4}$$

   (c) $\displaystyle\lim_{x \to 0^-} \frac{x^2 - 2x}{|x|} = \lim_{x \to 0^-} \frac{x(x - 2)}{-x} = \lim_{x \to 0^-} (-x + 2) = 2$

   (d) $\displaystyle\lim_{x \to 2^+} \frac{x - 1}{2x - x^2} = \lim_{x \to 2^+} \frac{x - 1}{x(2 - x)} = -\infty$

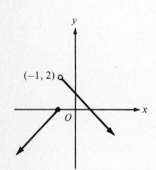

Figure 2.2T

2. $\displaystyle\lim_{x \to -1^+} f(x) = \lim_{x \to -1^+} (-x + 1) = 2$

   $\displaystyle\lim_{x \to -1^-} f(x) = \lim_{x \to -1^-} (x + 1) = 0$

   $\displaystyle\lim_{x \to -1} f(x)$ does not exist.

   Graph is Fig. 2.2T.

3. $\displaystyle\lim_{x \to 2^-} f(x) = \lim_{x \to 2^-} (kx^2) = 4k$

   $\displaystyle\lim_{x \to 2^+} f(x) = \lim_{x \to 2^+} (x + k) = 2 + k$

   $$4k = 2 + k$$

   $$k = \frac{2}{3}$$

4. Discontinuous at $x = 0$ and $x = 1$.

   $$\lim_{x \to 0} \frac{x^2 - 1}{x^2 - x} = \pm\infty$$

   Discontinuity at $x = 0$ is essential.

   $$\lim_{x \to 1} \frac{x^2 - 1}{x^2 - x} = \lim_{x \to 1} \frac{(x + 1)(x - 1)}{x(x - 1)} = 2$$

   Discontinuity at $x = 1$ removable.

5. (a) For any $N > 0$ there is some $\delta > 0$ such that $f(x) > N$ whenever
   $b - \delta < x < b$.

   (b) $\displaystyle\lim_{x \to c} f(x) = f(c)$

6. $\left|\dfrac{1}{2}x^2 - 8\right| < \epsilon$

if and only if

$|x + 4||x - 4| < 2\epsilon$

If $|x - 4| < \delta$ and $\delta \leqslant 1$, then $7 < |x + 4| < 9$. Thus,

$|x + 4||x - 4| < 9\delta$

We want $9\delta \leqslant 2\epsilon$, or $\delta \leqslant \frac{2}{9}\epsilon$. Therefore, take $\delta = \min(1, \frac{2}{9}\epsilon)$.

SOLUTIONS FOR TEST 3

1. (a) $f'(x) = \dfrac{(2x + 1)(2x) - x^2 \cdot 2}{(2x + 1)^2} = \dfrac{2x^2 + 2x}{(2x + 1)^2} = \dfrac{2x(x + 1)}{(2x + 1)^2}$

(b) $f'(x) = (x + 1)^2 (3)(4x - 1)^2 (4) + (4x - 1)^3 (2)(x + 1)$
$= 2(x + 1)(4x - 1)^2 [6(x + 1) + (4x - 1)]$
$= 10(x + 1)(4x - 1)^2 (2x + 1)$

(c) $f'(x) = (2x + 1)^{1/2} (3)(x - 3)^2 + (x - 3)^3 \left(\dfrac{1}{2}\right)(2x + 1)^{-1/2} (2)$

$= (2x + 1)^{-1/2}(x - 3)^2 [3(2x + 1) + (x - 3)]$

$= \dfrac{7x(x - 3)^2}{(2x + 1)^{1/2}}$

2. $\qquad 2D_x y = x^2 (3y^2) D_x y + y^3 (2x)$
$(2 - 3x^2 y^2) D_x y = 2xy^3$

$D_x y = \dfrac{2xy^3}{2 - 3x^2 y^2}$

3. $\qquad D_x y = 3(4x^{-2} - x)^2 (-8x^{-3} - 1)$
$m(2) = 3(4 \cdot 2^{-2} - 2)^2 (-8 \cdot 2^{-3} - 1) = 3(-2) = -6$
$y + 1 = -6(x - 2)$
$6x + y - 11 = 0$

4. (a) $v = D_t s = 3t^2 - 6t$
$3t^2 - 6t = 24$
$t^2 - 2t - 8 = 0$
$(t + 2)(t - 4) = 0$
$t = 4$

$s(4) = 4^3 - 3 \cdot 4^2 = 16$

Position is 16 feet above the origin.

(b) $a = D_t v = 6t - 6$

Let $v = 0$. Then

$3t^2 - 6t = 0$
$t(t - 2) = 0$
$t = 2$

$a(2) = 6 \cdot 2 - 6 = 6$

The acceleration is 6 ft/sec².

5. (a) true
(b) false
(c) false
(d) true

6. $\lim\limits_{x \to 2} \dfrac{\sqrt{x^2 + 5} - 3}{x - 2} \cdot \dfrac{\sqrt{x^2 + 5} + 3}{\sqrt{x^2 + 5} + 3} = \lim\limits_{x \to 2} \dfrac{(x + 2)(x - 2)}{(x - 2)(\sqrt{x^2 + 5} + 3)} = \dfrac{2}{3}.$

7. $f'(x) = \lim\limits_{\Delta x \to 0} \dfrac{f(x + \Delta x) - f(x)}{\Delta x}$

$= \lim\limits_{\Delta x \to 0} \dfrac{\dfrac{3}{(x + \Delta x)^2} - \dfrac{3}{x^2}}{\Delta x}$

$= \lim\limits_{\Delta x \to 0} \dfrac{3[x^2 - x^2 - 2x\,\Delta x - (\Delta x)^2]}{\Delta x\,(x + \Delta x)^2 x^2}$

$= \lim\limits_{\Delta x \to 0} \dfrac{3(-2x - \Delta x)}{(x + \Delta x)^2 x^2}$

$= -\dfrac{6}{x^3}$

8. $PV = C$

$PD_V V + V D_V P = 0$

$D_V P = -\dfrac{P}{V} = -\dfrac{20}{15} = -\dfrac{4}{3}$

The rate of change is $-\frac{4}{3}$.

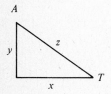

Figure 3.9T

9. See Fig. 3.9T.

Given: $D_t x = 15$; $D_t y = -20$. Find: $D_t z$ when $y = 40$ and $x = 30$.

$$x^2 + y^2 = z^2$$
$$2x\,D_t x + 2y\,D_t y = 2z\,D_t z$$

$$D_t z = \frac{x\,D_t x + y\,D_t y}{z} = \frac{15x - 20y}{z}$$

When $y = 40$ and $x = 30$, then $z = 50$. Thus,

$$D_t z = \frac{15(30) - 20(40)}{50} = -7$$

Therefore the distance is decreasing at the rate of 7 mps.

SOLUTIONS FOR TEST 4

1. $\lim\limits_{x \to -\infty} \dfrac{x}{4x + \sqrt{x^2 + 1}} \cdot \dfrac{\dfrac{1}{x}}{\dfrac{1}{x}} = \lim\limits_{x \to -\infty} \dfrac{1}{(4x)\left(\dfrac{1}{x}\right) + \sqrt{x^2 + 1}\left(-\dfrac{1}{\sqrt{x^2}}\right)}$

$= \lim\limits_{x \to -\infty} \dfrac{1}{4 - \sqrt{1 + \dfrac{1}{x^2}}}$

$= \dfrac{1}{3}$

2. $\lim\limits_{x \to \pm\infty} \dfrac{x + 1}{(x - 1)^2} = \lim\limits_{x \to \pm\infty} \dfrac{\dfrac{1}{x} + \dfrac{1}{x^2}}{\left(1 - \dfrac{1}{x}\right)^2} = 0$

Thus, the x-axis is a horizontal asymptote.

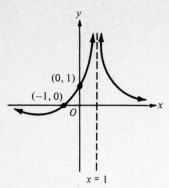

(0, 1)

(−1, 0)

O

$x = 1$

Figure 4.2T

$$\lim_{x \to 1^+} \frac{x+1}{(x-1)^2} = +\infty \quad \text{and} \quad \lim_{x \to 1^-} \frac{x+1}{(x-1)^2} = +\infty$$

Thus, the line $x = 1$ is a vertical asymptote. See Fig. 4.2T for the graph.

3. See Table 3. Thus, $(x^2 - 4)/x \geqslant 0$ if $-2 \leqslant x < 0$ or $x \geqslant 2$. Then f is continuous on $(-2, 0)$, $[-2, 0)$, $[2, +\infty)$, and f is discontinuous on $(-\infty, -2)$, $[-2, 0]$, $(0, 2)$.

Table 3

	$-\infty$	-2	0	2	$+\infty$
$x + 2$		$-$	$+$	$+$	$+$
$x - 2$		$-$	$-$	$-$	$+$
x		$-$	$-$	$+$	$+$
$\dfrac{x^2 - 4}{x}$		$-$	$+$	$-$	$+$

4. $f'(x) = \dfrac{5}{3}x^{2/3} + \dfrac{2}{3}x^{-1/3}$

$\quad\quad = \dfrac{1}{3}x^{-1/3}(5x + 2)$

$\quad\quad = \dfrac{5x + 2}{3x^{1/3}}$

Critical numbers are 0 and $-\frac{2}{5}$.

5. $f'(x) = 4x^3 - 12x = 4x(x^2 - 3)$

Critical numbers in $[-3, 1]$ are 0 and $-\sqrt{3}$.

x	-3	$-\sqrt{3}$	0	1
$f(x)$	27	-9	0	-5

Thus, the absolute maximum value is 27, and the absolute minimum value is -9.

6. (a) f is continuous on $[x_1, x_2]$, and f is differentiable on (x_1, x_2).
   (b) There is a number $\bar{x}$ between x_1 and x_2 such that

$$f'(\bar{x}) = \frac{f(x_2) - f(x_1)}{x_2 - x_1}$$

7. If $x - f(x) \leqslant x f(x)$, then

$\quad\quad x \leqslant x f(x) + f(x)$
$\quad\quad x \leqslant (x + 1)f(x)$

$\quad\quad \dfrac{x}{x + 1} \leqslant f(x) \quad$ if $x + 1 > 0$

If $x f(x) \leqslant x + f(x)$, then

$\quad\quad x f(x) - f(x) \leqslant x$
$\quad\quad (x - 1)f(x) \leqslant x$

$\quad\quad f(x) \leqslant \dfrac{x}{x - 1} \quad$ if $x - 1 > 0$

Thus,

$$\frac{x}{x+1} \leqslant f(x) \leqslant \frac{x}{x-1} \quad \text{if } x > 1$$

Because,

$$\lim_{x \to +\infty} \frac{x}{x+1} = \lim_{x \to +\infty} \frac{1}{1 + \dfrac{1}{x}} = 1$$

$$\lim_{x \to +\infty} \frac{x}{x-1} = \lim_{x \to +\infty} \frac{1}{1 - \dfrac{1}{x}} = 1$$

Then, by the squeeze theorem,

$$\lim_{x \to +\infty} f(x) = 1$$

8. Let x = number of items and P = number of dollars total profit on x items. Then

$$P(x) = \begin{cases} 10x & \text{if } 0 \leqslant x \leqslant 20 \\ [10 - 0.25(x-20)]x & \text{if } x > 20 \end{cases}$$

$$P(x) = \begin{cases} 10x & \text{if } 0 \leqslant x \leqslant 20 \\ 15x - \dfrac{1}{4}x^2 & \text{if } x > 20 \end{cases}$$

$$P'(x) = \begin{cases} 10 & \text{if } 0 < x < 20 \\ 15 - \dfrac{1}{2}x & \text{if } x > 20 \end{cases}$$

Critical numbers are 20 and 30.

$$P(20) = 200 \quad \text{and} \quad P(30) = 15(30) - \frac{1}{4}(30)^2 = 225$$

Thus, the greatest profit is $225 per day when 30 items are produced.

9. See Fig. 4.9T. We have

$$V = \pi r^2 h$$

By similar triangles

$$\frac{h}{6-r} = \frac{8}{6}$$

$$h = \frac{4}{3}(6-r)$$

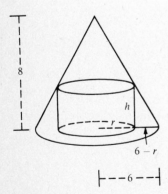

Figure 4.9T

Thus,

$$V = \pi r^2 \left(\frac{4}{3}\right)(6-r)$$

$$= \frac{4}{3}\pi(6r^2 - r^3)$$

$$D_r V = \frac{4}{3}\pi(12r - 3r^2)$$

If $D_r V = 0$, then $r = 0$ or $r = 4$. If $r = 4$, then $h = \frac{8}{3}$. Thus, the radius is 4 in and the altitude is $\frac{8}{3}$ in.

SOLUTIONS FOR TEST 5   1. $f'(x) = 12x^3 + 12x^2 = 12x^2(x+1)$

$$
f'(x) \quad
\begin{array}{|c|c|c|}
\hline
- & + & + \\
\hline
\end{array}
\quad
\begin{array}{ll}
-1 & 0
\end{array}
$$

(a) f is increasing on $(-1, 0)$ and on $(0, +\infty)$, and f is decreasing on $(-\infty, -1)$.
(b) $f(-1) = -1$. Thus, -1 is a relative minimum value of f, and f does not have a relative maximum value.

2. $f'(x) = 12x - 3x^2$
$f''(x) = 12 - 6x$

$$
f''(x) \quad
\begin{array}{|c|c|}
\hline
+ & - \\
\hline
\end{array}
\quad 2
$$

(a) The graph is concave upward in $(-\infty, 2)$ and concave downward in $(2, +\infty)$.
(b) $f(2) = 16$. Thus, $(2, 16)$ is a point of inflection. $f'(2) = 12$. Thus, the slope of the tangent line is 12.

3.   (a) S (b) Q (c) P (d) R

4. $f'(x_0) = 0$ and $f''(x_0) < 0$

5. $f'(x) = 2x^{-1/3} - 1 = x^{-1/3}(2 - x^{1/3}) = \dfrac{2 - x^{1/3}}{x^{1/3}}$

$f''(x) = -\dfrac{2}{3}x^{-4/3} = \dfrac{-2}{3x^{4/3}}$

(a) $f'(x) = 0$ if $x^{1/3} = 2$, or, equivalently, if $x = 8$.

$$
f'(x) \quad
\begin{array}{|c|c|c|}
\hline
- & + & - \\
\hline
\end{array}
\quad
\begin{array}{ll}
0 & 8
\end{array}
$$

$f(0) = 0$ and $f(8) = 4$

Thus, $(0, 0)$ is a relative minimum point and $(8, 4)$ is a relative maximum point.
(b) $f'(0) = \infty$ and $f'(8) = 0$

Thus, slope is not defined at $(0, 0)$, and slope is zero at $(8, 4)$.
(c) Because $f''(x) < 0$ for all $x \neq 0$, there is no point of inflection. The graph is concave downward in $(-\infty, 0)$ and in $(0, +\infty)$.
(d) The graph is shown in Fig. 5.5T.

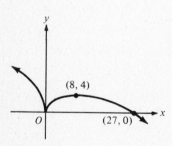

Figure 5.5T

6. $C'(x) = \dfrac{1}{4}x + 10$

(a) $C'(30) = \dfrac{1}{4}(30) + 10 = 17.5$

The marginal cost is $17.50.
(b) Let

$$
Q(x) = \frac{C(x)}{x} = \frac{1}{8}x + 10 + 200x^{-1}
$$

$$Q'(x) = \frac{1}{8} - 200x^{-2}$$

If $Q'(x) = 0$, then

$$\frac{1}{8} = \frac{200}{x^2}$$

$$x^2 = 1600$$
$$x = 40$$

Thus, 40 items should be produced.

7. (a) $p = 50 - \frac{1}{10}x$

Let

$$R(x) = xp = x\left(50 - \frac{1}{10}x\right) = 50x - \frac{1}{10}x^2$$

$$R'(x) = 50 - \frac{1}{5}x$$

If $R'(x) = 0$, then $x = 250$.

$$R(250) = 250(50 - 25) = 6250$$

The greatest possible daily revenue is \$62.50.

(b) Let

$$S(x) = R(x) - C(x) = 50x - \frac{1}{10}x^2 - 10x$$

$$S'(x) = 40 - \frac{1}{5}x$$

If $S'(x) = 0$, then $x = 200$. If $x = 200$, then $p = 30$. Thus, the price should be 30 cents.

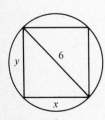

Figure 5.8T

8. See Fig. 5.8T. Because $x^2 + y^2 = 36$, we have

$$S = kx(36 - x^2) = k(36x - x^3)$$
$$D_x S = k(36 - 3x^2)$$

If $D_x S = 0$, then $x = \sqrt{12}$ and $y = \sqrt{24}$. Thus, the dimensions are $2\sqrt{3}$ inches by $2\sqrt{6}$ inches.

9. Let k dollars be the cost per square unit, and let C dollars be the total cost of material. Then we multiply the area of the top by $2k$ and the area of the sides and bottom by k.

$$C = 2k(\pi r^2) + k(\pi r^2) + k(2\pi r)$$
$$= k\pi(3r^2 + 2rh)$$

Thus,

$$D_r C = k\pi(6r + 2r\,D_r h + 2h) \tag{1}$$

Because the volume is constant,

$$V = \pi r^2 h$$
$$D_r V = \pi(r^2\,D_r h + 2rh) = 0$$

$$D_r h = -\frac{2h}{r} \tag{2}$$

Substituting from (2) into (1), we get

$$D_r C = k\pi(6r - 4h + 2h)$$
$$= 2k\pi(3r - h)$$

If $D_r C = 0$, then $h/r = 3$. Thus, the required ratio is 3.

SOLUTIONS FOR TEST 6

1. $\Delta y = 3(x + \Delta x)^2 - 3x^2 = 3(5.02)^2 - 3(5^2) = 0.6012$
   $dy = (6x)\Delta x = 30(0.02) = 0.6$

2. Let x inches be the diameter. Then $r = \frac{1}{2}x$. Thus,

$$V = \frac{4}{3}\pi r^3 = \frac{4}{3}\pi\left(\frac{1}{2}x\right)^3 = \frac{1}{6}\pi x^3$$

$$dV = \frac{1}{2}\pi x^2\, dx$$

$$= \frac{1}{2}\pi(100)(0.05)$$

$$= 7.85$$

The error is ±7.85 cubic inches.

3. $$D_t y = (D_x y)(D_t x) = \frac{dy}{dx} \cdot \frac{dx}{dt}$$

$$\frac{dy}{dx} = \frac{1}{2}(5x + 4)^{-1/2}(5)$$

$$\frac{dy}{dx}\bigg]_{x=1} = \frac{5}{2}(9^{-1/2}) = \frac{5}{6}$$

$$x^2\, dt + 2xt\, dx + dt = 3\, dx$$
$$(2xt - 3)dx = -(x^2 + 1)\, dt$$

$$\frac{dx}{dt} = -\frac{x^2 + 1}{2xt - 3}$$

When $x = 1$, $1^2(t) + t = 3(1) + 1$, or $t = 2$. Thus,

$$\frac{dx}{dt}\bigg]_{\substack{x=1 \\ t=2}} = -\frac{1^2 + 1}{2(1)(2) - 3} = -2$$

Thus,

$$D_t y = \frac{5}{6}(-2) = -\frac{5}{3}$$

4. Let $u = \sqrt{x + 1}$. Then $x = u^2 - 1$ and $dx = 2u\, du$.

$$\int x\sqrt{x + 1}\, dx = \int (u^2 - 1)u(2u\, du)$$

$$= 2\int (u^4 - u^2)\, du$$

$$= 2\left(\frac{1}{5}u^5 - \frac{1}{3}u^3\right) + C$$

$$= \frac{2}{5}(x + 1)^{5/2} - \frac{2}{3}(x + 1)^{3/2} + C$$

5. $y^{-2}\,dy = 2x\,dx$

$$\int y^{-2}\,dy = \int 2x\,dx$$

$$-y^{-1} = x^2 + C$$
$$-1 = 1 + C$$
$$C = -2$$
$$-y^{-1} = x^2 - 2$$
$$y^{-1} = -x^2 + 2$$

$$y = \frac{-1}{x^2 - 2}$$

6. Let $dy/dx = y'$. Then

$$\frac{dy'}{dx} = 6x$$

$$\int dy' = \int 6x\,dx$$

$$y' = 3x^2 + C$$

Because $y' = 1$ when $x = 1$, we have $C = -2$. Thus,

$$\frac{dy}{dx} = y' = 3x^2 - 2$$

$$\int dy = \int (3x^2 - 2)\,dx$$

$$y = x^3 - 2x + K$$

Because $y = 3$ when $x = 1$, we have $K = 4$. Thus,

$$y = x^3 - 2x + 4$$

7. (a) $a = \dfrac{dv}{dt} = -4$

$$\int dv = -4\int dt$$

$$v = -4t + v_0$$

Because $v = 30$ when $t = 0$, we have $v_0 = 30$. Thus,

$$\frac{ds}{dt} = v = -4t + 30$$

$$\int ds = \int (-4t + 30)\,dt$$

$$s = -2t^2 + 30t + s_0$$

We take $s_0 = 0$ and find s when $t = 5$. Thus,

$$s = -2(5^2) + 30(5) = 100$$

The ball will roll 100 ft.

(b) If $v = 0$, then $t = \frac{15}{2}$. If $t = \frac{15}{2}$, then

$$s = -2\left(\frac{15}{2}\right)^2 + 30\left(\frac{15}{2}\right)$$

$$= \frac{225}{2}$$

$$= 112.5$$

The ball rolls a distance of 112.5 feet before coming to rest.

8. $C'(x) = 60(3x + 8)^{-1/3}$

$$C(x) = 60 \int (3x + 8)^{-1/3} \, dx$$

$$= 20 \int (3x + 8)^{-1/3}(3dx)$$

$$= 20\left(\frac{3}{2}\right)(3x + 8)^{2/3} + C$$

Because $C(0) = 0$, we have

$$0 = 30(8^{2/3}) + C$$
$$C = -120$$

Thus,

$$C(x) = 30(3x + 8)^{2/3} - 120$$

SOLUTIONS FOR TEST 7
1. $F(i) = \sqrt{3i + 4}$
$F(i - 1) = \sqrt{3i + 1}$

Thus, the given sum is $F(20) - F(0) = \sqrt{64} - \sqrt{4} = 6$.

2. (a) $\Delta x = \dfrac{b - a}{n} = \dfrac{2}{n}$

$$x_i = a + i\,\Delta x = \frac{2i}{n}$$

$$A = \lim_{n \to +\infty} \sum_{i=1}^{n} \left[\left(\frac{2i}{n}\right)^2 + \left(\frac{2i}{n}\right)\right]\frac{2}{n}$$

$$= \lim_{n \to +\infty} \left[\frac{8}{n^3} \sum_{i=1}^{n} i^2 + \frac{4}{n^2} \sum_{i=1}^{n} i\right]$$

$$= \lim_{n \to +\infty} \left[\frac{8}{n^3} \cdot \frac{n(n+1)(2n+1)}{6} + \frac{4}{n^2} \cdot \frac{n(n+1)}{2}\right]$$

$$= \lim_{n \to +\infty} \left[\frac{4}{3}\left(1 + \frac{1}{n}\right)\left(2 + \frac{1}{n}\right) + 2\left(1 + \frac{1}{n}\right)\right]$$

$$= \frac{8}{3} + 2$$

$$= \frac{14}{3}$$

(b) $\Delta x = \dfrac{b-a}{n} = \dfrac{4-(-1)}{n} = \dfrac{5}{n}$

$x_i = a + i\,\Delta x = -1 + \dfrac{5i}{n}$

$$\int_{-1}^{4} x^2\,dx = \lim_{n\to+\infty} \sum_{i=1}^{n} \left(-1 + \frac{5i}{n}\right)^2 \frac{5}{n}$$

$$= \lim_{n\to+\infty}\left[\frac{5}{n}\sum_{i=1}^{n}1 - \frac{50}{n^2}\sum_{i=1}^{n}i + \frac{125}{n^3}\sum_{i=1}^{n}i^2\right]$$

$$= \lim_{n\to+\infty}\left[\frac{5}{n}\cdot n - \frac{50}{n^2}\cdot\frac{n(n+1)}{2} + \frac{125}{n^3}\cdot\frac{n(n+1)(2n+1)}{6}\right]$$

$$= \lim_{n\to+\infty}\left[5 - 25\left(1+\frac{1}{n}\right) + \frac{125}{6}\left(1+\frac{1}{n}\right)\left(2+\frac{1}{n}\right)\right]$$

$$= 5 - 25 + \frac{125}{6}\cdot 2$$

$$= \frac{65}{3}$$

3.

i	$\Delta_i x$	$f(\bar{x}_i)$	$f(\bar{x}_i)\cdot\Delta_i x$
1	0.3	3.2	0.96
2	0.5	4.0	2.00
3	0.2	4.8	.96

sum $= 3.92$

Riemann sum is 3.92.

4. $\displaystyle\int_{2}^{4}[3f(x)+5]\,dx = 3\int_{2}^{4}f(x)\,dx + 5\int_{2}^{4}dx$

$$= 3\left[\int_{1}^{4}f(x)\,dx - \int_{1}^{2}f(x)\,dx\right] + 5\int_{2}^{4}dx$$

$$= 3\left[21 - \frac{7}{3}\right] + 5[4-2]$$

$$= 66$$

5. A.V. $= \dfrac{1}{3}\displaystyle\int_{0}^{3}(4x - x^2)\,dx$

$$= \frac{1}{3}\left[2x^2 - \frac{1}{3}x^3\right]_{0}^{3}$$

$$= \frac{1}{3}[18 - 9]$$

$$= 3$$

Let $f(\bar{x}) = 3$. Thus,

$$4\bar{x} - \bar{x}^2 = 3$$
$$\bar{x}^2 - 4\bar{x} + 3 = 0$$
$$(\bar{x} - 3)(\bar{x} - 1) = 0$$

Hence, $\bar{x} = 3$ or $\bar{x} = 1$.

6. Let $u = \sqrt{x + 1}$. Then $x = u^2 - 1$ and $dx = 2u\,du$. When $x = 0$, $u = 1$; when $x = 3$, $u = 2$. Thus,

$$\int_0^3 \frac{x\,dx}{\sqrt{x + 1}} = \int_1^2 \frac{(u^2 - 1)(2u\,du)}{u}$$

$$= 2 \int_1^2 (u^2 - 1)\,du$$

$$= 2\left[\frac{1}{3}u^3 - u\right]_1^2$$

$$= 2\left[\left(\frac{8}{3} - 2\right) - \left(\frac{1}{3} - 1\right)\right]$$

$$= \frac{8}{3}$$

SOLUTIONS FOR TEST 8

1. The graph is shown in Fig. 8.1T.

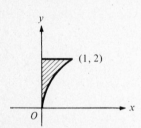

Figure 8.1T

(a) $\int_0^1 (2 - 2x^{1/3})\,dx$

(b) $\int_0^2 \left(\frac{1}{8}y^3\right)dy$

2. (a) See Fig. 8.2T(a).

$$\Delta h = \Delta y$$
$$R = 2$$
$$r = x = \left(\frac{1}{3}y\right)^{1/2}$$
$$V = \pi \int_0^{12} \left(4 - \frac{1}{3}y\right)dy$$

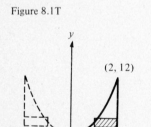

Figure 8.2.T(a)

(b) See Fig. 8.2T(b).

$$\Delta r = \Delta x$$
$$h = y = 3x^2$$
$$\bar{r} = 3 - x$$
$$V = 2\pi \int_0^2 (3 - x)(3x^2)\,dx$$

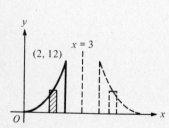

Figure 8.2T(b)

Figure 8.2T(c)

(c) See Fig. 8.2T(c).

$$h = b \sin 60° = b\left(\frac{1}{2}\sqrt{3}\right)$$

$$A = \frac{1}{2}bh = \frac{1}{4}\sqrt{3}\,b^2$$

$$b = y = 3x^2$$

$$A = \frac{1}{4}\sqrt{3}\,(3x^2)^2 = \frac{9}{4}\sqrt{3}\,x^4$$

$$V = \int_0^2 \frac{9}{4}\sqrt{3}\,x^4\,dx$$

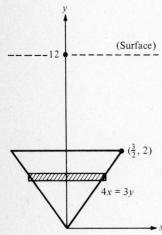

Figure 8.3T

3. See Fig. 8.3T.

$$\Delta V = \pi r^2\,\Delta h = \pi y^2\,\Delta x = \pi(64 - x^2)\,\Delta x$$
$$D = x$$

$$W = w\pi \int_2^8 (64 - x^2)x\,dx$$

4. See Fig. 8.4T.

$$\Delta A = 2x\,\Delta y = \frac{3}{2}y\,\Delta y$$

$$h = 12 - y$$

$$F = \frac{3}{2}w \int_0^2 (12 - y)y\,dy$$

$$= \frac{3}{2}w\left[6y^2 - \frac{1}{3}y^3\right]_0^2$$

$$= \frac{3}{2}w\left[24 - \frac{8}{3}\right]$$

$$= 32w$$

Work is $32w$ foot-pounds.

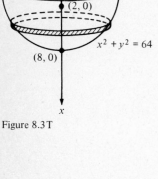

Figure 8.4T

5. $M = \int_0^2 \sqrt{4x + 1}\,dx$

$$= \frac{1}{4}\int_0^2 (4x + 1)^{1/2}\,(4dx)$$

$$= \frac{1}{6}(4x + 1)^{3/2}\Big]_0^2$$

$$= \frac{13}{3}$$

Total mass is $\frac{13}{3}$ slugs.

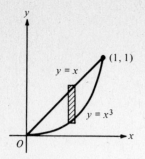

Figure 8.6T

6. See Fig. 8.6T.

$$A = \int_0^1 (x - x^3)\, dx$$

$$= \frac{1}{2}x^2 - \frac{1}{4}x^4 \Big]_0^1$$

$$= \frac{1}{4}$$

$$A \cdot \bar{x} = \int_0^1 (x - x^3)x\, dx$$

$$= \frac{1}{3}x^3 - \frac{1}{5}x^5 \Big]_0^1$$

$$= \frac{2}{15}$$

$$A \cdot \bar{y} = \frac{1}{2}\int_0^1 (x - x^3)(x + x^3)\, dx$$

$$= \frac{1}{2}\left[\frac{1}{3}x^3 - \frac{1}{7}x^7\right]_0^1$$

$$= \frac{2}{21}$$

Thus, $\bar{x} = \frac{2}{15} \div \frac{1}{4} = \frac{8}{15}$ and $\bar{y} = \frac{2}{21} \div \frac{1}{4} = \frac{8}{21}$. Centroid is $\left(\frac{8}{15}, \frac{8}{21}\right)$.

7. See Fig. 8.7T.

$$V = \pi \int_0^4 y\, dy$$

$$= \frac{1}{2}\pi y^2 \Big]_0^4$$

$$= 8\pi$$

$$V \cdot \bar{y} = \pi \int_0^4 y^2\, dy$$

$$= \frac{1}{3}\pi y^3 \Big]_0^4$$

$$= \frac{64}{3}\pi$$

Thus,

$$\bar{y} = \frac{64}{3}\pi \div 8\pi = \frac{8}{3}$$

The centroid is $\left(0, \frac{8}{3}, 0\right)$.

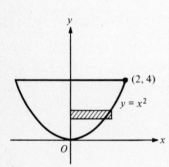

Figure 8.7T

8. $y = \frac{1}{6}(x^3 + 3x^{-1})$

$$\frac{dy}{dx} = \frac{1}{6}(3x^2 - 3x^{-2}) = \frac{x^4 - 1}{2x^2}$$

$$1 + \left(\frac{dy}{dx}\right)^2 = 1 + \frac{x^8 - 2x^4 + 1}{4x^4} = \frac{x^8 + 2x^4 + 1}{4x^4} = \left(\frac{x^4 + 1}{2x^2}\right)^2$$

$$L = \int_1^3 \sqrt{\left(\frac{x^4 + 1}{2x^2}\right)^2}\, dx$$

$$= \frac{1}{2}\int_1^3 (x^2 + x^{-2})\, dx$$

$$= \frac{1}{2}\left[\frac{1}{3}x^3 - x^{-1}\right]_1^3$$

$$= \frac{14}{3}$$

Printer and Binder: The Murray Printing Company

79 9 8 7 6 5